Student Solutions Manual and Study Guide

for

SERWAY AND JEWETT'S

PHYSICS

FOR SCIENTISTS AND ENGINEERS

VOLUME ONE
EIGHTH EDITION

John R. Gordon
Emeritus, James Madison University

Ralph V. McGrew
Broome Community College

Raymond A. Serway
Emeritus, James Madison University

BROOKS/COLE
CENGAGE Learning™

Australia • Brazil • Canada • Mexico • Singapore • Spain • United Kingdom • United States

ISBN-13: 978-1-4390-4854-2
ISBN-10: 1-4390-4854-1

Brooks/Cole
20 Channel Center Street
Boston, MA 02210
USA

Cengage Learning products are represented in Canada by Nelson Education, Ltd.

For product information and technology assistance, contact us at **Cengage Learning Customer & Sales Support, 1-800-354-9706**

For permission to use material from this text or product, submit all requests online at **www.cengage.com/permissions**
Further permissions questions can be emailed to **permissionrequest@cengage.com**

For your course and learning solutions, visit **www.cengage.com**

Purchase any of our products at your local college store or at our preferred online store **www.ichapters.com**

Printed in the United States of America
1 2 3 4 5 6 7 8 9 10 13 12 11 10 09

PREFACE

This *Student Solutions Manual and Study Guide* has been written to accompany the textbook **Physics for Scientists and Engineers,** Eighth Edition, by Raymond A. Serway and John W. Jewett, Jr. The purpose of this *Student Solutions Manual and Study Guide* is to provide students with a convenient review of the basic concepts and applications presented in the textbook, together with solutions to selected end-of-chapter problems from the textbook. This is not an attempt to rewrite the textbook in a condensed fashion. Rather, emphasis is placed upon clarifying typical troublesome points and providing further practice in methods of problem solving.

Every textbook chapter has in this book a matching chapter, which is divided into several parts. Very often, reference is made to specific equations or figures in the textbook. Each feature of this Study Guide has been included to ensure that it serves as a useful supplement to the textbook. Most chapters contain the following components:

- **Equations and Concepts:** This represents a review of the chapter, with emphasis on highlighting important concepts, and describing important equations and formalisms.

- **Suggestions, Skills, and Strategies:** This offers hints and strategies for solving typical problems that the student will often encounter in the course. In some sections, suggestions are made concerning mathematical skills that are necessary in the analysis of problems.

- **Review Checklist:** This is a list of topics and techniques the student should master after reading the chapter and working the assigned problems.

- **Answers to Selected Questions:** Suggested answers are provided for approximately 15 percent of the objective and conceptual questions.

- **Solutions to Selected End-of-Chapter Problems:** Solutions are shown for approximately 20 percent of the problems from the text, chosen to illustrate the important concepts of the chapter. The solutions follow the *Conceptualize—Categorize—Analyze—Finalize* strategy presented in the text.

A note concerning significant figures: When the statement of a problem gives data to three significant figures, we state the answer to three significant figures. The last digit is uncertain; it can for example depend on the precision of the values assumed for physical constants and properties. When a calculation involves several steps, we carry out intermediate steps to many digits, but we write down only three. We "round off" only at the end of any chain of calculations, never anywhere in the middle.

We sincerely hope that this *Student Solutions Manual and Study Guide* will be useful to you in reviewing the material presented in the text and in improving your ability to solve problems and score well on exams. We welcome any comments or suggestions which could help improve the content of this study guide in future editions, and we wish you success in your study.

John R. Gordon
Harrisonburg, Virginia

Ralph V. McGrew
Binghamton, New York

Raymond A. Serway
Leesburg, Virginia

Acknowledgments

We are glad to acknowledge that John Jewett and Hal Falk suggested significant improvements in this manual. We are grateful to Charu Khanna and the staff at MPS Limited for assembling and typing this manual and preparing diagrams and page layouts. Susan English of Durham Technical Community College checked the manual for accuracy and suggested many improvements. We thank Brandi Kirksey (Associate Developmental Editor), Mary Finch (Publisher), and Cathy Brooks (Senior Content Project Manager) of Cengage Learning, who coordinated this project and provided resources for it. Finally, we express our appreciation to our families for their inspiration, patience, and encouragement.

Suggestions for Study

We have seen a lot of successful physics students. The question, "How should I study this subject?" has no single answer, but we offer some suggestions that may be useful to you.

1. Work to understand the basic concepts and principles before attempting to solve assigned problems. Carefully read the textbook before attending your lecture on that material. Jot down points that are not clear to you, take careful notes in class, and ask questions. Reduce memorization of material to a minimum. Memorizing sections of a text or derivations would not necessarily mean you understand the material.

2. After reading a chapter, you should be able to define any new quantities that were introduced and discuss the first principles that were used to derive fundamental equations. A review is provided in each chapter of the Study Guide for this purpose, and the marginal notes in the textbook (or the index) will help you locate these topics. You should be able to correctly associate with each *physical quantity* the *symbol* used to represent that quantity (including vector notation if appropriate) and the SI *unit* in which the quantity is specified. Furthermore, you should be able to express each important formula or equation in a concise and accurate prose statement.

3. Try to solve plenty of the problems at the end of the chapter. The worked examples in the text will serve as a basis for your study. This Study Guide contains detailed solutions to about fifteen of the problems at the end of each chapter. You will be able to check the accuracy of your calculations for any odd-numbered problems, since the answers to these are given at the back of the text.

4. Besides what you might expect to learn about physics concepts, a very valuable skill you can take away from your physics course is the ability to solve complicated problems. The way physicists approach complex situations and break them down into manageable pieces is widely useful. At the end of Chapter 2, the textbook develops a general problem-solving strategy that guides you through the steps. To help you remember the steps of the strategy, they are called *Conceptualize, Categorize, Analyze,* and *Finalize.*

General Problem-Solving Strategy

Conceptualize

-   The first thing to do when approaching a problem is to *think about* and *understand* the situation. Read the problem several times until you are confident you understand what is being asked. Study carefully any diagrams, graphs, tables, or photographs that accompany the problem. Imagine a movie, running in your mind, of what happens in the problem.

-   If a diagram is not provided, you should almost always make a quick drawing of the situation. Indicate any known values, perhaps in a table or directly on your sketch.

- Now focus on what algebraic or numerical information is given in the problem. In the problem statement, look for key phrases such as "starts from rest" ($v_i = 0$), "stops" ($v_f = 0$), or "falls freely" ($a_y = -g = -9.80$ m/s^2). Key words can help simplify the problem.

- Next, focus on the expected result of solving the problem. Precisely what is the question asking? Will the final result be numerical or algebraic? If it is numerical, what units will it have? If it is algebraic, what symbols will appear in the expression?

- Incorporate information from your own experiences and common sense. What should a reasonable answer look like? What should its order of magnitude be? You wouldn't expect to calculate the speed of an automobile to be 5×10^6 m/s.

Categorize

- After you have a really good idea of what the problem is about, you need to *simplify* the problem. Remove the details that are not important to the solution. For example, you can often model a moving object as a particle. Key words should tell you whether you can ignore air resistance or friction between a sliding object and a surface.

- Once the problem is simplified, it is important to *categorize* the problem. How does it fit into a framework of ideas that you construct to understand the world? Is it a simple *plug-in problem*, such that numbers can be simply substituted into a definition? If so, the problem is likely to be finished when this substitution is done. If not, you face what we can call an *analysis problem*—the situation must be analyzed more deeply to reach a solution.

- If it is an analysis problem, it needs to be categorized further. Have you seen this type of problem before? Does it fall into the growing list of types of problems that you have solved previously? Being able to classify a problem can make it much easier to lay out a plan to solve it. For example, if your simplification shows that the problem can be treated as a particle moving under constant acceleration and you have already solved such a problem (such as the examples in Section 2.6), the solution to the new problem follows a similar pattern. From the textbook you can make an explicit list of the analysis models.

Analyze

- Now, you need to analyze the problem and strive for a mathematical solution. Because you have categorized the problem and identified an analysis model, you can select relevant equations that apply to the situation in the problem. For example, if your categorization shows that the problem involves a particle moving under constant acceleration, Equations 2.13 to 2.17 are relevant.

- Use algebra (and calculus, if necessary) to solve symbolically for the unknown variable in terms of what is given. Substitute in the appropriate numbers, calculate the result, and round it to the proper number of significant figures.

Finalize

- This final step is the most important part. Examine your numerical answer. Does it have the correct units? Does it meet your expectations from your conceptualization of the problem? What about the algebraic form of the result—before you substituted numerical values? Does it make sense? Try looking at the variables in it to see whether the answer would change in a physically meaningful way if they were drastically increased or decreased or even became zero. Looking at limiting cases to see whether they yield expected values is a very useful way to make sure that you are obtaining reasonable results.

- Think about how this problem compares with others you have done. How was it similar? In what critical ways did it differ? Why was this problem assigned? You should have learned something by doing it. Can you figure out what? Can you use your solution to expand, strengthen, or otherwise improve your framework of ideas? If it is a new category of problem, be sure you understand it so that you can use it as a model for solving future problems in the same category.

When solving complex problems, you may need to identify a series of subproblems and apply the problem-solving strategy to each. For very simple problems, you probably don't need this whole strategy. But when you are looking at a problem and you don't know what to do next, remember the steps in the strategy and use them as a guide.

Work on problems in this Study Guide yourself and compare your solutions to ours. Your solution does not have to look just like the one presented here. A problem can sometimes be solved in different ways, starting from different principles. If you wonder about the validity of an alternative approach, ask your instructor.

5. We suggest that you use this Study Guide to review the material covered in the text and as a guide in preparing for exams. You can use the sections Review Checklist, Equations and Concepts, and Suggestions, Skills, and Strategies to focus in on points that require further study. The main purpose of this Study Guide is to improve the efficiency and effectiveness of your study hours and your overall understanding of physical concepts. However, it should not be regarded as a substitute for your textbook or for individual study and practice in problem solving.

TABLE OF CONTENTS

1

Physics and Measurement

EQUATIONS AND CONCEPTS

The **density** of any substance is defined as the ratio of mass to volume. The SI units of density are kg/m^3. *Density is an example of a derived quantity.*

$$\rho \equiv \frac{m}{V} \qquad (1.1)$$

SUGGESTIONS, SKILLS, AND STRATEGIES

A general strategy for problem solving will be described in Chapter 2.

Appendix B of your textbook includes a review of mathematical techniques including:

- Scientific notation: using powers of ten to express large and small numerical values.

- Basic algebraic operations: factoring, handling fractions, and solving quadratic equations.

- Fundamentals of plane and solid geometry: graphing functions, calculating areas and volumes, and recognizing equations and graphs of standard figures (e.g. straight line, circle, ellipse, parabola, and hyperbola).

- Basic trigonometry: definition and properties of functions (e.g. sine, cosine, and tangent), the Pythagorean Theorem, and basic trigonometry identities.

REVIEW CHECKLIST

You should be able to:

- Describe the standards which define the SI units for the fundamental quantities length (meter, m), mass (kilogram, kg), and time (second, s). Identify and properly use prefixes and mathematical notations such as the following: $\propto$ (is proportional to), $<$ (is less than), $\approx$ (is approximately equal to), Δ (change in value), etc. (Section 1.1)

- Convert units from one measurement system to another (or convert units within a system). Perform a dimensional analysis of an equation containing physical quantities whose individual units are known. (Sections 1.3 and 1.4)

- Carry out order-of-magnitude calculations or estimates. (Section 1.5)

- Express calculated values with the correct number of significant figures. (Section 1.6)

ANSWER TO AN OBJECTIVE QUESTION

3.   Answer each question yes or no. Must two quantities have the same dimensions (a) if you are adding them? (b) If you are multiplying them? (c) If you are subtracting them? (d) If you are dividing them? (e) If you are equating them?

Answer.   (a) Yes. Three apples plus two jokes has no definable answer. (b) No. One acre times one foot is one acre-foot, a quantity of floodwater. (c) Yes. Three dollars minus six seconds has no definable answer. (d) No. The gauge of a rich sausage can be 12 kg divided by 4 m, giving 3 kg/m. (e) Yes, as in the examples given for parts (b) and (d). Thus we have (a) yes (b) no (c) yes (d) no (e) yes

ANSWER TO A CONCEPTUAL QUESTION

1.   Suppose the three fundamental standards of the metric system were length, *density*, and time rather than length, *mass,* and time. The standard of density in this system is to be defined as that of water. What considerations about water would you need to address to make sure the standard of density is as accurate as possible?

Answer.   There are the environmental details related to the water: a standard temperature would have to be defined, as well as a standard pressure. Another consideration is the quality of the water, in terms of defining an upper limit of impurities. A difficulty with this scheme is that density cannot be measured directly with a single measurement, as can length, mass, and time. As a combination of two measurements (mass and volume, which itself involves three measurements!), a density value has higher uncertainty than a single measurement.

SOLUTIONS TO SELECTED END-OF-CHAPTER PROBLEMS

9.   Which of the following equations are dimensionally correct?

     (a) $v_f = v_i + ax$ (b) $y = (2 \text{ m})\cos(kx)$ where $k = 2 \text{ m}^{-1}$

Solution

Conceptualize: It is good to check an unfamiliar equation for dimensional correctness to see whether it can possibly be true.

Categorize: We evaluate the dimensions as a combination of length, time, and mass for each term in each equation.

Analyze:

(a)   Write out dimensions for each quantity in the equation $v_f = v_i + ax$

     The variables v_f and v_i are expressed in unts of m/s, so $[v_f] = [v_i] = \text{LT}^{-1}$

The variable a is expressed in units of m/s^2 and $\qquad\qquad$ $[a] = LT^{-2}$

The variable x is expressed in meters. Therefore $\qquad\qquad$ $[ax] = L^2T^{-2}$

Consider the right-hand member (RHM) of equation (a): $\qquad$ $[RHM] = LT^{-1} + L^2T^{-2}$

Quantities to be added must have the same dimensions.
Therefore, equation (a) **is not** dimensionally correct. $\qquad\qquad\qquad\qquad$ ∎

(b) Write out dimensions for each quantity in the equation $\qquad$ $y = (2 \text{ m}) \cos (kx)$

For y, $\qquad\qquad\qquad\qquad\qquad\qquad\qquad\qquad\qquad\qquad$ $[y] = L$

for 2 m, $\qquad\qquad\qquad\qquad\qquad\qquad\qquad\qquad\qquad\qquad$ $[2 \text{ m}] = L$

and for (kx), $\qquad\qquad\qquad\qquad\qquad\qquad\qquad$ $[kx] = \left[\left(2 \text{ m}^{-1}\right)x\right] = L^{-1}L$

Therefore we can think of the quantity kx as an angle in radians, and we can take its cosine. The cosine itself will be a pure number with no dimensions. For the left-hand member (LHM) and the right-hand member (RHM) of the equation we have

$$[LHM] = [y] = L \qquad\qquad [RHM] = [2 \text{ m}][\cos (kx)] = L$$

These are the same, so equation (b) **is** dimensionally correct. $\qquad\qquad\qquad\qquad$ ∎

Finalize: We will meet an expression like $y = (2 \text{ m})\cos(kx)$, where $k = 2$ m^{-1}, as the wave function of a wave.

---

13. A rectangular building lot has a width of 75.0 ft and a length of 125 ft. Determine the area of this lot in square meters.

Solution

Conceptualize: We must calculate the area and convert units. Since a meter is about 3 feet, we should expect the area to be about $A \approx (25 \text{ m})(40 \text{ m}) = 1\,000$ m^2.

Categorize: We will use the geometrical fact that for a rectangle Area = Length × Width; and the conversion 1 m = 3.281 ft.

Analyze: $A = \ell \times w = (75.0 \text{ ft})\left(\dfrac{1 \text{ m}}{3.281 \text{ ft}}\right)(125 \text{ ft})\left(\dfrac{1 \text{ m}}{3.281 \text{ ft}}\right) = 871$ m^2 $\qquad$ ∎

Finalize: Our calculated result agrees reasonably well with our initial estimate and has the proper units of m^2. Unit conversion is a common technique that is applied to many problems. Note that one square meter is more than ten square feet.

---

15. A solid piece of lead has a mass of 23.94 g and a volume of 2.10 cm³. From these data, calculate the density of lead in SI units (kilograms per cubic meter).

Solution

Conceptualize: From Table 14.1, the density of lead is 1.13×10^4 kg/m³, so we should expect our calculated value to be close to this value. The density of water is 1.00×10^3 kg/m³, so we see that lead is about 11 times denser than water, which agrees with our experience that lead sinks.

Categorize: Density is defined as $\rho = m/V$. We must convert to SI units in the calculation.

Analyze:
$$\rho = \left(\frac{23.94 \text{ g}}{2.10 \text{ cm}^3}\right)\left(\frac{1 \text{ kg}}{1\,000 \text{ g}}\right)\left(\frac{100 \text{ cm}}{1 \text{ m}}\right)^3$$

$$= \left(\frac{23.94 \text{ g}}{2.10 \text{ cm}^3}\right)\left(\frac{1 \text{ kg}}{1\,000 \text{ g}}\right)\left(\frac{1\,000\,000 \text{ cm}^3}{1 \text{ m}^3}\right) = 1.14 \times 10^4 \text{ kg/m}^3 \qquad \blacksquare$$

Finalize: Observe how we set up the unit conversion fractions to divide out the units of grams and cubic centimeters, and to make the answer come out in kilograms per cubic meter. At one step in the calculation, we note that **one million** cubic centimeters make one cubic meter. Our result is indeed close to the expected value. Since the last reported significant digit is not certain, the difference from the tabulated values is possibly due to measurement uncertainty and does not indicate a discrepancy.

———————————————

21. One cubic meter (1.00 m³) of aluminum has a mass of 2.70×10^3 kg, and the same volume of iron has a mass of 7.86×10^3 kg. Find the radius of a solid aluminum sphere that will balance a solid iron sphere of radius 2.00 cm on an equal-arm balance.

Solution

Conceptualize: The aluminum sphere must be larger in volume to compensate for its lower density. Its density is roughly one-third as large, so we might guess that the radius is three times larger, or 6 cm.

Categorize: We require equal masses: $\quad m_{Al} = m_{Fe} \quad$ or $\quad \rho_{Al} V_{Al} = \rho_{Fe} V_{Fe}$

Analyze: We use also the volume of a sphere. By substitution,

$$\rho_{Al}\left(\frac{4}{3}\pi r_{Al}^3\right) = \rho_{Fe}\left(\frac{4}{3}\pi(2.00 \text{ cm})^3\right)$$

Now solving for the unknown,

$$r_{Al}^3 = \left(\frac{\rho_{Fe}}{\rho_{Al}}\right)(2.00 \text{ cm})^3 = \left(\frac{7.86 \times 10^3 \text{ kg/m}^3}{2.70 \times 10^3 \text{ kg/m}^3}\right)(2.00 \text{ cm})^3 = 23.3 \text{ cm}^3$$

Taking the cube root, $r_{Al} = 2.86$ cm $\qquad \blacksquare$

Finalize: The aluminum sphere is only 43% larger than the iron one in radius, diameter, and circumference. Volume is proportional to the cube of the linear dimension, so this moderate excess in linear size gives it the $(1.43)(1.43)(1.43) = 2.92$ times larger volume it needs for equal mass.

---

23. One gallon of paint (volume $= 3.78 \times 10^{-3}$ m³) covers an area of 25.0 m². What is the thickness of the fresh paint on the wall?

Solution

Conceptualize: We assume the paint keeps the same volume in the can and on the wall.

Categorize: We model the film on the wall as a rectangular solid, with its volume given by its "footprint" area, which is the area of the wall, multiplied by its thickness t perpendicular to this area and assumed to be uniform.

Analyze: $V = At$ gives $t = \dfrac{V}{A} = \dfrac{3.78 \times 10^{-3} \text{ m}^3}{25.0 \text{ m}^2} = 1.51 \times 10^{-4}$ m ∎

Finalize: The thickness of 1.5 tenths of a millimeter is comparable to the thickness of a sheet of paper, so it is reasonable. The film is many molecules thick.

---

25. (a) At the time of this book's printing, the U.S. national debt is about $10 trillion. If payments were made at the rate of $1 000 per second, how many years would it take to pay off the debt, assuming no interest were charged? (b) A dollar bill is about 15.5 cm long. How many dollar bills, attached end to end, would it take to reach the Moon? The front endpapers give the Earth-Moon distance. *Note:* Before doing these calculations, try to guess at the answers. You may be very surprised.

Solution

(a) **Conceptualize:** $10 trillion is certainly a large amount of money, so even at a rate of $1 000/second, we might guess that it will take a lifetime (~100 years) to pay off the debt.

Categorize: The time interval required to repay the debt will be calculated by dividing the total debt by the rate at which it is repaid.

Analyze: $T = \dfrac{\$10 \text{ trillion}}{\$1000/\text{s}} = \dfrac{\$10 \times 10^{12}}{(\$1000/\text{s})(3.156 \times 10^7 \text{s/yr})} = 317$ yr ∎

Finalize: Our guess was a bit low. $10 trillion really is a lot of money!

(b) **Conceptualize:** We might guess that 10 trillion bills would reach from the Earth to the Moon, and perhaps back again, since our first estimate was low.

Categorize: The number of bills is the distance to the Moon divided by the length of a dollar.

Analyze: $N = \dfrac{D}{\ell} = \dfrac{3.84 \times 10^8 \text{ m}}{0.155 \text{ m}} = 2.48 \times 10^9 \text{ bills}$ ∎

Finalize: Ten trillion dollars is larger than this two-and-a-half billion dollars by four thousand times. The ribbon of bills comprising the debt reaches across the cosmic gulf thousands of times, so again our guess was low. Similar calculations show that the bills could span the distance between the Earth and the Sun ten times. The strip could encircle the Earth's equator nearly 40 000 times. With successive turns wound edge to edge without overlapping, the dollars would cover a zone centered on the equator and about 2.6 km wide.

---

27. Find the order of magnitude of the number of table-tennis balls that would fit into a typical-size room (without being crushed). [In your solution, state the quantities you measure or estimate and the values you take for them.]

Solution

Conceptualize: Since the volume of a typical room is much larger than a Ping-Pong ball, we should expect that a very large number of balls (maybe a million) could fit in a room.

Categorize: Since we are only asked to find an estimate, we do not need to be too concerned about how the balls are arranged. Therefore, to find the number of balls we can simply divide the volume of an average-size living room (perhaps 15 ft × 20 ft × 8 ft) by the volume of an individual Ping-Pong ball.

Analyze: Using the approximate conversion 1 ft = 30 cm, we find

$$V_{\text{Room}} = (15 \text{ ft})(20 \text{ ft})(8 \text{ ft})(30 \text{ cm/ft})^3 \approx 6 \times 10^7 \text{ cm}^3$$

A Ping-Pong ball has a diameter of about 3 cm, so we can estimate its volume as a cube:

$$V_{\text{ball}} = (3 \text{ cm})(3 \text{ cm})(3 \text{ cm}) \approx 30 \text{ cm}^3$$

The number of Ping-Pong balls that can fill the room is

$$N \approx \dfrac{V_{\text{Room}}}{V_{\text{ball}}} \approx 2 \times 10^6 \text{ balls} \sim 10^6 \text{ balls}$$ ∎

Finalize: So a typical room can hold on the order of a million Ping-Pong balls. This problem gives us a sense of how large a quantity "a million" really is.

---

29. To an order of magnitude, how many piano tuners reside in New York City? The physicist Enrico Fermi was famous for asking questions like this one on oral Ph.D. qualifying examinations.

Solution

Conceptualize: Don't reach for the telephone book! Think.

Categorize: Each full-time piano tuner must keep busy enough to earn a living.

Analyze: Assume a total population of 10^7 people. Also, let us estimate that one person in one hundred owns a piano. Assume that in one year a single piano tuner can service about 1 000 pianos (about 4 per day for 250 weekdays), and that each piano is tuned once per year.

Therefore, the number of tuners

$$= \left(\frac{1 \text{ tuner}}{1\ 000 \text{ pianos}} \right) \left(\frac{1 \text{ piano}}{100 \text{ people}} \right) \left(10^7 \text{ people} \right) \sim 100 \text{ tuners} \qquad \blacksquare$$

Finalize: If you did reach for an Internet directory, you would have to count. Instead, have faith in your estimate. Fermi's own ability in making an order-of-magnitude estimate is exemplified by his measurement of the energy output of the first nuclear bomb (the Trinity test at Alamogordo, New Mexico) by observing the fall of bits of paper as the blast wave swept past his station, 14 km away from ground zero.

---

43. Review. A pet lamb grows rapidly, with its mass proportional to the cube of its length. When the lamb's length changes by 15.8%, its mass increases by 17.3 kg. Find the lamb's mass at the end of this process.

Solution

Conceptualize: The little sheep's final mass must be a lot more than 17 kg, so an order of magnitude estimate is 100 kg.

Categorize: When the length changes by 15.8%, the mass changes by a much larger percentage. We will write each of the sentences in the problem as a mathematical equation.

Analyze: Mass is proportional to length cubed: $m = k\ell^3$ where k is a constant. This model of growth is reasonable because the lamb gets thicker as it gets longer, growing in three-dimensional space.

At the initial and final points, $\qquad m_i = k\ell_i^3 \qquad$ and $\qquad m_f = k\ell_f^3$

Length changes by 15.8%: $\qquad$ Long ago you were told that 15.8% **of** ℓ means
$\qquad\qquad\qquad\qquad\qquad\qquad$ 0.158 **times** ℓ.

Thus $\qquad \ell_i + 0.158\ \ell_i = \ell_f \quad$ and $\quad \ell_f = 1.158\ \ell_i$

Mass increases by 17.3 kg: $\qquad m_i + 17.3 \text{ kg} = m_f$

Now we combine the equations using algebra, eliminating the unknowns ℓ_i, ℓ_f, k, and m_i by substitution:

From $\qquad \ell_f = 1.158\,\ell_i$, we have $\ell_f^3 = 1.158^3\,\ell_i^3 = 1.553\,\ell_i^3$

Then $\qquad m_f = k\ell_f^3 = k(1.553)\,\ell_i^3 = 1.553\,k\ell_i^3 = 1.553\,m_i$ and $m_i = m_f/1.553$

Next, $\qquad m_i + 17.3 \text{ kg} = m_f$ becomes $m_f/1.553 + 17.3 \text{ kg} = m_f$

Solving, $\qquad 17.3 \text{ kg} = m_f - m_f/1.553 = m_f(1 - 1/1.553) = 0.356\,m_f$

$\qquad$ and $m_f = 17.3 \text{ kg}/0.356 = 48.6 \text{ kg}$ ∎

Finalize: Our 100-kg estimate was of the right order of magnitude. The 15.8% increase in length produces a 55.3% increase in mass, which is an increase by a factor of 1.553. The writer of this problem was thinking of the experience of a young girl in the Oneida community of New York State. Before the dawn of a spring day she helped with the birth of lambs. She was allowed to choose one lamb as her pet, and braided for it a necklace of straw to distinguish it from the others. Then she went into the house, where her mother had made cocoa with breakfast. Many years later she told this story of widening, overlapping circles of trust and faithfulness, to a group of people working to visualize peace. Over many more years the story is spreading farther.

---

51. The diameter of our disk-shaped galaxy, the Milky Way, is about 1.0×10^5 light-years (ly). The distance to the Andromeda galaxy (Fig. P1.51 in the text), which is the spiral galaxy nearest to the Milky Way, is about 2.0 million ly. If a scale model represents the Milky Way and Andromeda galaxies as dinner plates 25 cm in diameter, determine the distance between the centers of the two plates.

Solution

Conceptualize: Individual stars are fantastically small compared to the distance between them, but galaxies in a cluster are pretty close compared to their sizes.

Categorize: We can say that we solve the problem "by proportion" or by finding and using a scale factor, as if it were a unit conversion factor, between real space and its model.

Analyze: The scale used in the "dinner plate" model is

$$S = \frac{1.0 \times 10^5 \text{ light-years}}{25 \text{ cm}} = 4.00 \times 10^3 \frac{\text{ly}}{\text{cm}}$$

The distance to Andromeda in the dinner plate model will be

$$D = \frac{2.00 \times 10^6 \text{ ly}}{4.00 \times 10^3 \text{ ly/cm}} = 5.00 \times 10^2 \text{ cm} = 5.00$$ ∎

Finalize: Standing at one dinner plate, you can cover your view of the other plate with three fingers held at arm's length. The Andromeda galaxy, called Messier 31, fills this same angle in the field of view that the human race has and will always have.

---

53. A high fountain of water is located at the center of a circular pool as shown in Figure P1.53. A student walks around the pool and measures its circumference to be 15.0 m. Next, the student stands at the edge of the pool and uses a protractor to gauge the angle of elevation of the top of the fountain to be $\phi = 55.0°$. How high is the fountain?

Solution

Conceptualize: Geometry was invented to make indirect distance measurements, such as this one.

Categorize: We imagine a top view to figure the radius of the pool from its circumference. We imagine a straight-on side view to use trigonometry to find the height.

Analyze: Define a right triangle whose legs represent the height and radius of the fountain. From the dimensions of the fountain and the triangle, the circumference is $C = 2\pi r$ and the angle satisfies $\tan \phi = h/r$.

Then by substitution

$$h = r \tan \phi = \left(\frac{C}{2\pi}\right)\tan \phi$$

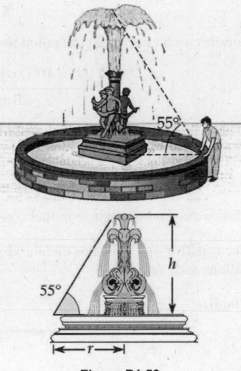

Figure P1.53

Evaluating,

$$h = \frac{15.0\,\text{m}}{2\pi}\tan 55.0° = 3.41\,\text{m}$$

■

Finalize: When we look at a three-dimensional system from a particular direction, we may discover a view to which simple mathematics applies.

---

57. Assume there are 100 million passenger cars in the United States and the average fuel consumption is 20 mi/gal of gasoline. If the average distance traveled by each car is 10 000 mi/yr, how much gasoline would be saved per year if average fuel consumption could be increased to 25 mi/gal?

Solution

Conceptualize: Five miles per gallon is not much for one car, but a big country can save a lot of gasoline.

Categorize: We define an average national fuel consumption rate based upon the total miles driven by all cars combined.

Analyze: In symbols,

$$\text{fuel consumed} = \frac{\text{total miles driven}}{\text{average fuel consumption rate}}$$

or

$$f = \frac{s}{c}$$

For the current rate of 20 mi/gallon we have

$$f = \frac{\left(100 \times 10^{6}\,\text{cars}\right)\left(10^{4}\,(\text{mi/yr})/\text{car}\right)}{20\;\text{mi/gal}} = 5 \times 10^{10}\;\text{gal/yr}$$

Since we consider the same total number of miles driven in each case, at 25 mi/gal we have

$$f = \frac{\left(100 \times 10^{6}\,\text{cars}\right)\left(10^{4}\,(\text{mi/yr})/\text{car}\right)}{25\;\text{mi/gal}} = 4 \times 10^{10}\;\text{gal/yr}$$

Thus we estimate a change in fuel consumption of $\quad \Delta f = -1 \times 10^{10}\;\text{gal/yr}$ ∎

The negative sign indicates that the change is a reduction. It is a fuel savings of ten billion gallons each year.

Finalize: Let's do it!

---

63. The consumption of natural gas by a company satisfies the empirical equation $V = 1.50t + 0.008\,00t^2$, where V is the volume of gas in millions of cubic feet and t is the time in months. Express this equation in units of cubic feet and seconds. Assume a month is 30.0 days.

Solution

Conceptualize: The units of volume and time imply particular combination-units for the coefficients in the equation.

Categorize: We write "millions of cubic feet" as $10^6\;\text{ft}^3$, and use the given units of time and volume to assign units to the equation:

Analyze: $V = \left(1.50 \times 10^6\;\text{ft}^3/\text{mo}\right)t + \left(0.008\,00 \times 10^6\;\text{ft}^3/\text{mo}^2\right)t^2$

To convert the units to seconds, use

$$1 \text{ month} = 30.0 \text{ d}\left(\frac{24 \text{ h}}{1 \text{ d}}\right)\left(\frac{3600 \text{ s}}{1 \text{ h}}\right) = 2.59 \times 10^6 \text{ s, to obtain}$$

$$V = \left(1.50 \times 10^6 \frac{\text{ft}^3}{\text{mo}}\right)\left(\frac{1 \text{ mo}}{2.59 \times 10^6 \text{ s}}\right)t + \left(0.008\,00 \times 10^6 \frac{\text{ft}^3}{\text{mo}^2}\right)\left(\frac{1 \text{ mo}}{2.59 \times 10^6 \text{ s}}\right)^2 t^2$$

$$V = \left(0.579 \text{ ft}^3/\text{s}\right)t + \left(1.19 \times 10^{-9} \text{ ft}^3/\text{s}^2\right)t^2$$

or

$$V = 0.579t + 1.19 \times 10^{-9}t^2, \text{ where } V \text{ is in cubic feet and } t \text{ is in seconds} \qquad \blacksquare$$

Finalize: The coefficient of the first term is the volume rate of flow of gas at the beginning of the month. The second term's coefficient is related to how much the rate of flow increases every second.

2

Motion in One Dimension

EQUATIONS AND CONCEPTS

The **displacement** Δx of a particle moving from position x_i to position x_f equals the final coordinate minus the initial coordinate. *Displacement can be positive, negative or zero.*

$$\Delta x \equiv x_f - x_i \qquad (2.1)$$

Distance traveled is the length of the path followed by a particle and should not be confused with displacement. When $x_f = x_i$, the displacement is zero; however, if the particle leaves x_i, travels along a path, and returns to x_i, the distance traveled will not be zero.

The **average velocity** of an object during a time interval Δt is the ratio of the total displacement Δx to the time interval during which the displacement occurs.

$$v_{x,avg} \equiv \frac{\Delta x}{\Delta t} \qquad (2.2)$$

The **average speed** of a particle is a scalar quantity defined as the ratio of the total distance d traveled to the time required to travel that distance. Average speed has no direction and carries no algebraic sign. *The magnitude of the average velocity is not the average speed; although in certain cases they may be numerically equal.*

$$v_{avg} \equiv \frac{d}{\Delta t} \qquad (2.3)$$

The **instantaneous velocity** v is defined as the limit of the ratio $\Delta x/\Delta t$ as Δt approaches zero. This limit is called the derivative of x with respect to t. The instantaneous velocity at any time is the slope of the position-time graph at that time. As illustrated in the figure, the slope can be positive, negative, or zero.

$$v_x \equiv \lim_{\Delta t \to 0} \frac{\Delta x}{\Delta t} = \frac{dx}{dt} \qquad (2.5)$$

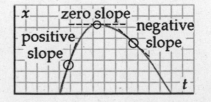

12

The **instantaneous speed** is the magnitude of the instantaneous velocity.

The **average acceleration** of an object is defined as the ratio of the change in velocity to the time interval during which the change in velocity occurs. Equation 2.9 gives the average acceleration of a particle in one-dimensional motion along the x axis.

$$a_{x,avg} \equiv \frac{\Delta v_x}{\Delta t} = \frac{v_{xf} - v_{xi}}{t_f - t_i} \qquad (2.9)$$

The **instantaneous acceleration** a is defined as the limit of the ratio $\Delta v_x / \Delta t$ as Δt approaches zero. This limit is the derivative of the velocity along the x direction with respect to time. *A negative acceleration does not necessarily imply a decreasing speed. Acceleration and velocity are not always in the same direction.*

$$a_x \equiv \lim_{\Delta t \to 0} \frac{\Delta v_x}{\Delta t} = \frac{dv_x}{dt} \qquad (2.10)$$

The **acceleration** can also be expressed as the second derivative of the position with respect to time. This is shown in Equation 2.12 for the case of a particle in one-dimensional motion along the x axis.

$$a_x = \frac{d^2 x}{dt^2} \qquad (2.12)$$

The **kinematic equations,** (2.13–2.17), can be used to describe one-dimensional motion along the x axis with constant acceleration. Note that each equation shows a different relationship among physical quantities: initial velocity, final velocity, acceleration, time, and position.

Remember, the relationships stated in Equations 2.13–2.17 are true only for cases in which the acceleration is constant.

$$v_{xf} = v_{xi} + a_x t \qquad (2.13)$$

$$v_{x,avg} = \frac{v_{xi} + v_{xf}}{2} \qquad (2.14)$$

$$x_f = x_i + \tfrac{1}{2}\left(v_{xi} + v_{xf}\right)t \qquad (2.15)$$

$$x_f = x_i + v_{xi} t + \tfrac{1}{2} a_x t^2 \qquad (2.16)$$

$$v_{xf}^2 = v_{xi}^2 + 2a_x \left(x_f - x_i\right) \qquad (2.17)$$

A **freely falling object** is any object moving under the influence of the gravitational force alone. Equations 2.13–2.17 can be modified to describe the motion of freely falling objects by denoting the motion to be along the y axis (defining "up" as positive) and setting $a_y = -g$. *A freely falling object experiences an acceleration that is directed downward regardless of the direction or magnitude of its actual motion.*

$$v_{yf} = v_{yi} - gt$$

$$y_f = y_i + \tfrac{1}{2}(v_{yi} + v_{yf})t$$

$$v_{y,avg} = \tfrac{1}{2}(v_{yi} + v_{yf})$$

$$y_f = y_i + v_{yi} t - \tfrac{1}{2} g t^2$$

$$v_{yf}^2 = v_{yi}^2 - 2g(y_f - y_i)$$

SUGGESTIONS, SKILLS, AND STRATEGIES

Organize your problem-solving by considering each step of the **Conceptualize**, **Categorize**, **Analyze**, and **Finalize** protocol described in your textbook and implemented in the solution to problems in this Student Solutions Manual and Study Guide. Refer to the step-by-step description of the General Problem-Solving Strategy following Section (2.8) of your textbook, and in the Preface to this Manual.

REVIEW CHECKLIST

- For each of the following pairs of terms, define each quantity and state how each is related to the other member of the pair: distance and displacement; instantaneous and average velocity; speed and instantaneous velocity; instantaneous and average acceleration. (Sections 2.1, 2.2, 2.3 and 2.4)

- Construct a graph of position versus time (given a function such as $x = 5 + 3t - 2t^2$) for a particle in motion along a straight line. From this graph, you should be able to determine the value of the average velocity between two points t_1 and t_2 and the instantaneous velocity at a given point. Average velocity is the slope of the chord between the two points and the instantaneous velocity at a given time is the slope of the tangent to the graph at that time. (Section 2.2)

- Be able to interpret graphs of one-dimensional motion showing position vs. time, velocity vs. time, and acceleration vs. time. (Section 2.5)

- Apply the equations of kinematics to any situation where the motion occurs under constant acceleration. (Section 2.6)

- Describe what is meant by a freely falling body (one moving under the influence of gravity—where air resistance is neglected). Recognize that the equations of kinematics apply directly to a freely falling object and that the acceleration is then given by $a = -g$ (where $g = 9.80$ m/s^2). (Section 2.7)

ANSWER TO AN OBJECTIVE QUESTION

13. A student at the top of a building of height h throws one ball upward with a speed of v_i and then throws a second ball downward with the same initial speed $|v_i|$. Just before it reaches the ground, is the final speed of the ball thrown upward (a) larger, (b) smaller, or (c) the same in magnitude, compared with the final speed of the ball thrown downward?

Answer (c) They are the same, if the balls are in free fall with no air resistance. After the first ball reaches its apex and falls back downward past the student, it will have a downward velocity with a magnitude equal to v_i. This velocity is the same as the initial velocity of the second ball, so after they fall through equal heights their impact speeds will also be the same. By contrast, the balls are in flight for very different time intervals.

ANSWERS TO SELECTED CONCEPTUAL QUESTIONS

1. If the average velocity of an object is zero in some time interval, what can you say about the displacement of the object for that interval?

Answer The displacement is **zero**, since the displacement is proportional to average velocity.

□   □   □   □

9. Two cars are moving in the same direction in parallel lanes along a highway. At some instant, the velocity of car A exceeds the velocity of car B. Does that mean that the acceleration of A is greater than that of B? Explain.

Answer No. If Car A has been traveling with cruise control, its velocity may be high (say 60 mi/h = 27 m/s), but its acceleration will be close to zero. If Car B is just pulling onto the highway, its velocity is likely to be low (15 m/s), but its acceleration will be high.

□   □   □   □

SOLUTIONS TO SELECTED END-OF-CHAPTER PROBLEMS

1. The position versus time for a certain particle moving along the x axis is shown in Figure P2.1. Find the average velocity in the time intervals (a) 0 to 2 s, (b) 0 to 4 s, (c) 2 s to 4 s, (d) 4 s to 7 s, and (e) 0 to 8 s.

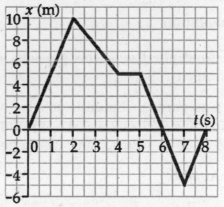

Figure P2.1

Solution

Conceptualize: We must think about how x is changing with t in the graph.

Categorize: The average velocity is the slope of a secant line drawn into the graph between specified points.

Analyze: On this graph, we can tell positions to two significant figures:

(a) $x = 0$ at $t = 0$ and $x = 10$ m at $t = 2$ s:

$$v_{x,avg} = \frac{\Delta x}{\Delta t} = \frac{10 \text{ m} - 0}{2 \text{ s} - 0} = 5.0 \text{ m/s}$$ ■

(b) $x = 5.0$ m at $t = 4$ s:

$$v_{x,avg} = \frac{\Delta x}{\Delta t} = \frac{5.0 \text{ m} - 0}{4 \text{ s} - 0} = 1.2 \text{ m/s}$$ ■

(c) $v_{x,avg} = \frac{\Delta x}{\Delta t} = \frac{5.0 \text{ m} - 10 \text{ m}}{4 \text{ s} - 2 \text{ s}} = -2.5 \text{ m/s}$ ■

(d) $v_{x,avg} = \dfrac{\Delta x}{\Delta t} = \dfrac{-5.0 \text{ m} - 5.0 \text{ m}}{7 \text{ s} - 4 \text{ s}} = -3.3 \text{ m/s}$ ∎

(e) $v_{x,avg} = \dfrac{\Delta x}{\Delta t} = \dfrac{0.0 \text{ m} - 0.0 \text{ m}}{8 \text{ s} - 0 \text{ s}} = 0 \text{ m/s}$ ∎

Finalize: The average velocity is the slope, not necessarily of the graph line itself, but of a secant line cutting across the graph between specified points. The slope of the graph line itself is the instantaneous velocity, found, for example, in the following problem 5 part (b).

Note with care that the change in a quantity is defined as the final value minus the original value of the quantity. We use this "final-first" definition so that a positive change will describe an increase, and a negative value for change represents the later value being less than the earlier value.

———————————

3.   A person walks first at a constant speed of 5.00 m/s along a straight line from point Ⓐ to point Ⓑ and then back along the line from Ⓑ to Ⓐ at a constant speed of 3.00 m/s. (a) What is her average speed over the entire trip? (b) What is her average velocity over the entire trip?

Solution

Conceptualize: This problem lets you think about the distinction between speed and velocity.

Categorize: Speed is positive whenever motion occurs, so the average speed must be positive. Velocity we take as positive for motion to the right and negative for motion to the left, so its average value can be positive, negative, or zero.

Analyze:

(a)   The average speed during any time interval is equal to the total distance of travel divided by the total time:

$$\text{average speed} = \frac{\text{total distance}}{\text{total time}} = \frac{d_{AB} + d_{BA}}{t_{AB} + t_{BA}}$$

But $d_{AB} = d_{BA}, \quad t_{AB} = d/v_{AB}, \quad \text{and} \quad t_{BA} = d/v_{BA}$

so $\text{average speed} = \dfrac{d + d}{\left(d/v_{AB}\right) + \left(d/v_{BA}\right)} = \dfrac{2\left(v_{AB}\right)\left(v_{BA}\right)}{v_{AB} + v_{BA}}$

and $\text{average speed} = 2\left[\dfrac{(5.00 \text{ m/s})(3.00 \text{ m/s})}{5.00 \text{ m/s} + 3.00 \text{ m/s}}\right] = 3.75 \text{ m/s}$ ∎

(b) The average velocity during any time interval equals total displacement divided by elapsed time.

$$v_{x,avg} = \frac{\Delta x}{\Delta t}$$

Since the walker returns to the starting point, $\Delta x = 0$ and $v_{x,avg} = 0$ ∎

Finalize: The velocity can be thought to average out to zero because it has a higher positive value for a short time interval and a lower negative value for a longer time.

---

5. A position–time graph for a particle moving along the x axis is shown in Figure P2.5. (a) Find the average velocity in the time interval $t = 1.50$ s to $t = 4.00$ s. (b) Determine the instantaneous velocity at $t = 2.00$ s by measuring the slope of the tangent line shown in the graph. (c) At what value of t is the velocity zero?

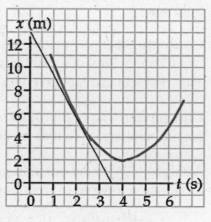

Figure P2.5

Solution

Conceptualize: We will have to distinguish between average and instantaneous velocities.

Categorize: For average velocity, we find the slope of a secant line running across the graph between the 1.5-s and 4-s points. Then for instantaneous velocities we think of slopes of tangent lines, which means the slope of the graph itself at a point.

Analyze:

(a) From the graph: At $t_1 = 1.5$ s, $x = x_1 = 8.0$ m

At $t_2 = 4.0$ s $x = x_2 = 2.0$ m

Therefore, $v_{1 \to 2avg} = \dfrac{\Delta x}{\Delta t} = \dfrac{2.0 \text{ m} - 8.0 \text{ m}}{4.0 \text{ s} - 1.5 \text{ s}} = -2.4$ m/s   ∎

(b) Choose two points along a line which is tangent to the curve at $t = 2.0$ s. We will use the two points ($t_i = 0.0$ s, and $x_i = 13.0$ m) and ($t_f = 3.5$ s, $x_f = 0.0$ m). Instantaneous velocity equals the slope of the tangent line,

so $v_x = \dfrac{x_f - x_i}{t_f - t_i} = \dfrac{0.0 \text{ m} - 13.0 \text{ m}}{3.5 \text{ s} - 0.0 \text{ s}} = -3.7$ m/s   ∎

The negative sign shows that the **direction** of v_x is along the negative x direction.   ∎

(c) The velocity will be zero when the slope of the tangent line is zero. This occurs for the point on the graph where x has its minimum value.

Therefore, $v = 0$ at $t = 4.0$ s   ∎

Finalize: Try moving your hand to mimic the motion graphed. Start a meter away from a motion detector defined to be at $x = 0$, moving toward it rapidly. Then slow down your motion smoothly, coming to rest at distance 20 cm at time 4 s, reversing your motion and moving away, slowly at first and then more rapidly.

---

17.   A particle moves along the x axis according to the equation $x = 2.00 + 3.00t - 1.00t^2$, where x is in meters and t is in seconds. At $t = 3.00$ s, find (a) the position of the particle, (b) its velocity, and (c) its acceleration.

Solution

Conceptualize: A mathematical function can be specified as a table of values, a graph, or a formula. Previous problems have displayed position as a function of time with a graph. This problem displays $x(t)$ with an equation.

Categorize: To find position we simply evaluate the given expression. To find velocity we differentiate it. To find acceleration we take a second derivative.

Analyze: With the position given by $x = 2.00 + 3.00t - t^2$, we can use the rules for differentiation to write expressions for the velocity and acceleration as functions of time:

$$v_x = \frac{dx}{dt} = \frac{d}{dt}\left(2 + 3t - t^2\right) = 3 - 2t \qquad \text{and} \qquad a_x = \frac{dv}{dt} = \frac{d}{dt}(3 - 2t) = -2$$

Now we can evaluate x, v, and a at $t = 3.00$ s.

(a) $x = 2.00 + 3.00(3.00) - (3.00)^2 = 2.00$ m ∎

(b) $v_x = 3.00 - 2(3.00) = -3.00$ m/s ∎

(c) $a_x = -2.00$ m/s^2 ∎

Finalize: The operation of taking a time derivative corresponds physically to finding out how fast a quantity is changing—to finding its rate of change.

---

23.   An object moving with uniform acceleration has a velocity of 12.0 cm/s in the positive x direction when its x coordinate is 3.00 cm. If its x coordinate 2.00 s later is −5.00 cm, what is its acceleration?

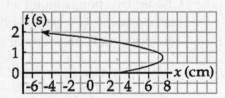

(acceleration is to the left)

Solution

Conceptualize: Study the graph. Move your hand to imitate the motion, first rapidly to the right, then slowing down, stopping, turning around, and speeding up to move to the left faster than before.

Categorize: The velocity is always changing; there is always nonzero acceleration and the problem says it is constant. So we can use one of the set of equations describing constant-acceleration motion.

Analyze:

Take the initial point to be the moment when $x_i = 3.00$ cm

and $v_{xi} = 12.0$ cm/s

Also, at $t = 2.00$ s $x_f = -5.00$ cm

Once you have classified the object as a particle moving with constant acceleration and have the standard set of four equations in front of you, how do you choose which equation to use? Make a list of all of the six symbols in the equations: x_i, x_f, v_{xi}, v_{xf}, a_x, and t. On the list fill in values as above, showing that x_i, x_f, v_{xi}, and t are known. Identify a_x as the unknown. Choose an equation involving only one unknown and the knowns. That is, choose an equation *not* involving v_{xf}. Thus we choose the kinematic equation

$$x_f = x_i + v_{xi}t + \tfrac{1}{2}a_x t^2$$

and solve for a: $a_x = \dfrac{2\left[x_f - x_i - v_{xi}t\right]}{t^2}$

We substitute: $a = \dfrac{2[-5.00 \text{ cm} - 3.00 \text{ cm} - (12.0 \text{ cm/s})(2.00 \text{ s})]}{(2.00 \text{ s})^2}$

and calculate: $a = -16.0$ cm/s^2 ∎

Finalize: Think with care about how slowing down in motion to the right, turning around, and speeding up in motion to the left all exemplify acceleration toward the left. The acceleration really is the same (16-cm/s-of-velocity-change-toward-the-left-in-every-second) throughout the motion, and notably including the point when the object is momentarily at rest.

─────────────────────────

24.   In Example 2.7, we investigated a jet landing on an aircraft carrier. In a later maneuver, the jet comes in for a landing on solid ground with a speed of 100 m/s and its acceleration can have a maximum magnitude of 5.00 m/s^2 as it comes to rest. (a) From the instant the jet touches the runway, what is the minimum time interval needed before it can come to rest? (b) Can this jet land at a small tropical island airport where the runway is 0.800 km long? (c) Explain your answer.

Solution

Conceptualize: We think of the plane moving with maximum-size backward acceleration throughout the landing…

Categorize: …so the acceleration is constant, the stopping time a minimum, and the stopping distance as short as it can be.

Analyze: The negative acceleration of the plane as it lands can be called deceleration, but it is simpler to use the single general term acceleration for all rates of velocity change.

(a)   The plane can be modeled as a particle under constant acceleration:

$$a_x = -5.00 \text{ m/s}^2$$

Given $v_{xi} = 100$ m/s and $v_{xf} = 0$, we use the equation $v_{xf} = v_{xi} + a_x t$

and solve for t: $t = \dfrac{v_{xf} - v_{xi}}{a_x} = \dfrac{0 - 100 \text{ m/s}}{-5.00 \text{ m/s}^2} = 20.0 \text{ s}$ ∎

(b)   Find the required stopping distance and compare this to the length of the runway. Taking x_i to be zero, we get

$$v_{xf}^2 = v_{xi}^2 + 2a_x(x_f - x_i)$$

or $\Delta x = x_f - x_i = \dfrac{v_{xf}^2 - v_{xi}^2}{2a_x} = \dfrac{0 - \left(100 \text{ m/s}\right)^2}{2\left(-5.00 \text{ m/s}^2\right)} = 1\ 000 \text{ m}$

(c)   The stopping distance is greater than the length of the runway; the plane **cannot land**.   ∎

Finalize: From the list of four standard equations about motion with constant acceleration, we can usually choose one that contains the unknown and the given information, to solve that part of the problem directly.

---

25.   Colonel John P. Stapp, USAF, participated in studying whether a jet pilot could survive emergency ejection. On March 19, 1954, he rode a rocket-propelled sled that moved down a track at a speed of 632 mi/h. He and the sled were safely brought to rest in 1.40 s. Determine (a) the negative acceleration he experienced and (b) the distance he traveled during this negative acceleration.

Solution

Conceptualize: We estimate the acceleration as between $-10g$ and $-100g$: that is, between -100 m/s^2 and -1000 m/s^2. We have already chosen the straight track as the x axis and the direction of travel as positive. We expect the stopping distance to be on the order of 100 m.

Categorize: We assume the acceleration is constant. We choose the initial and final points 1.40 s apart, bracketing the slowing-down process. Then we have a straightforward problem about a particle under constant acceleration.

Analyze: $v_{xi} = 632 \text{ mi/h} = 632 \text{ mi/h}\left(\dfrac{1609 \text{ m}}{1 \text{ mi}}\right)\left(\dfrac{1 \text{ h}}{3600 \text{ s}}\right) = 282 \text{ m/s}$

(a)   Taking $v_{xf} = v_{xi} + a_x t$ with $v_{xf} = 0$,

$$a_x = \dfrac{v_{xf} - v_{xi}}{t} = \dfrac{0 - 282 \text{ m/s}}{1.40 \text{ s}} = -202 \text{ m/s}^2$$ ∎

This has a magnitude of approximately $20g$.

(b) $x_f - x_i = \frac{1}{2}(v_{xi} + v_{xf})t = \frac{1}{2}(282 \text{ m/s} + 0)(1.40 \text{ s}) = 198 \text{ m}$ ∎

Finalize: While $x_f - x_i$, v_{xi}, and t are all positive, a_x is negative as expected. Our answers for a_x and for the distance agree with our order-of-magnitude estimates. For many years Colonel Stapp held the world land speed record.

––––––––––––––––––––

40.   A baseball is hit so that it travels straight upward after being struck by the bat. A fan observes that it takes 3.00 s for the ball to reach its maximum height. Find (a) the ball's initial velocity and (b) the height it reaches.

Solution

Conceptualize: The initial speed of the ball is probably somewhat greater than the speed of the pitch, which might be about 60 mi/h (~30 m/s), so an initial upward velocity off the bat of somewhere between 20 and 100 m/s would be reasonable. We also know that the length of a ball field is about 300 ft (~100 m), and a pop-fly usually does not go higher than this distance, so a maximum height of 10 to 100 m would be reasonable for the situation described in this problem.

Categorize: Since the ball's motion is entirely vertical, we can use the equations for free fall to find the initial velocity and maximum height from the elapsed time.

Analyze: After leaving the bat, the ball is in free fall for $t = 3.00$ s and has constant acceleration $a_y = -g = -9.80$ m/s^2.

Solve the equation $v_{yf} = v_{yi} + a_y t$ with $a_y = -g$ to obtain v_{yi} with $v_{yf} = 0$ when the ball reaches its maximum height.

(a) $v_{yi} = v_{yf} + gt = 0 + (9.80 \text{ m/s}^2)(3.00 \text{ s}) = 29.4$ m/s (upward)   ∎

(b)   The maximum height is $y_f = v_{yi} t - \frac{1}{2} g t^2$:

$y_f = (29.4 \text{ m/s})(3.00 \text{ s}) - \frac{1}{2}(9.80 \text{ m/s}^2)(3.00 \text{ s})^2 = 44.1$ m   ∎

Finalize: The calculated answers seem reasonable since they lie within our expected ranges, and they have the correct units and direction. We say that the ball is in free fall in its upward motion as well as in its subsequent downward motion and at the moment when its instantaneous velocity is zero at the top. On the other hand, it is not in free fall when it is in contact with the bat or with the catcher's glove.

––––––––––––––––––––

43.   A student throws a set of keys vertically upward to her sorority sister, who is in a window 4.00 m above. The second student catches the keys 1.50 s later. (a) With what initial velocity were the keys thrown? (b) What was the velocity of the keys just before they were caught?

Solution

Conceptualize: We do not need to know in advance whether the keys are moving up or have turned around to be moving down when the second woman catches them. The answer to part (b) will tell us which it is.

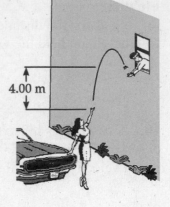

Categorize: We model the keys as a particle under the constant free-fall acceleration.

Analyze: Take the first student's position to be $y_i = 0$
and we are given that at $t = 1.50$ s $y_f = 4.00$ m
and $a_y = -9.80$ m/s^2

(a)   We choose the equation $y_f = y_i + v_{yi}t + \frac{1}{2}a_yt^2$ to connect the data and the unknown.

We solve: $$v_{yi} = \frac{y_f - y_i - \frac{1}{2}a_yt^2}{t}$$

and substitute: $$v_{yi} = \frac{4.00 \text{ m} - \frac{1}{2}\left(-9.80/s^2\right)(1.50 \text{ s})^2}{1.50 \text{ s}} = 10.0 \text{ m/s}$$ ∎

(b)   The velocity at any time $t > 0$ is given by $v_{yf} = v_{yi} + a_yt$
Therefore, at $t = 1.50$ s,

$$v_{yf} = 10.0 \text{ m/s} - (9.80 \text{ m/s}^2)(1.50 \text{ s}) = -4.68 \text{ m/s}$$ ∎

The negative sign means that the keys are moving **downward** just before they are caught.

Finalize: The 'initial' point is really just after the keys leave the first student's hand and the 'final' point is just before the second woman catches them. Then the hands do not give the keys some unknown acceleration during the motion we consider. The acceleration between these points is the known acceleration caused by the planet's gravitation.

---

45. A daring ranch hand sitting on a tree limb wishes to drop vertically onto a horse galloping under the tree. The constant speed of the horse is 10.0 m/s, and the distance from the limb to the level of the saddle is 3.00 m. (a) What must be the horizontal distance between the saddle and limb when the ranch hand makes his move? (b) For what time interval is he in the air?

Solution

Conceptualize: The man will be a particle in free fall starting from rest. The horse moves with constant velocity.

Categorize: Both horse and man have constant accelerations: they are g downward for the man and 0 for the horse.

Analyze: We choose to do part (b) first.

(b) Consider the vertical motion of the man after leaving the limb (with $v_i = 0$ at $y_i = 3.00$ m) until reaching the saddle (at $y_f = 0$).

Modeling the man as a particle under constant acceleration, we find his time of fall from $y_f = y_i + v_{yi}t + \frac{1}{2}a_y t^2$

When $v_i = 0$, $t = \sqrt{\dfrac{2(y_f - y_i)}{a_y}} = \sqrt{\dfrac{2(0 - 3.00 \text{ m})}{-9.80 \text{ m/s}^2}} = 0.782$ s   ∎

(a) During this time interval, the horse is modeled as a particle under constant velocity in the horizontal direction.

$$v_{xi} = v_{xf} = 10.0 \text{ m s}$$

so $x_f - x_i = v_{xi}t = (10.0 \text{ m/s})(0.782 \text{ s}) = 7.82$ m

and the ranch hand must let go when the horse is 7.82 m from the tree.   ∎

Finalize: Visualizing the motions, starting at different points and ending at the same point, guided us to make our first step computation of the time interval common to both motions.

————————

47. Automotive engineers refer to the time rate of change of acceleration as the "jerk." Assume an object moves in one dimension such that its jerk J is constant. (a) Determine expressions for its acceleration $a_x(t)$, velocity $v_x(t)$, and position $x(t)$, given that its initial acceleration, velocity, and position are a_{xi}, v_{xi}, and x_i, respectively. (b) Show that $a_x^2 = a_{xi}^2 + 2J(v_x - v_{xi})$.

Solution

Conceptualize: Steadily changing force acting on an object can make it move for a while with steadily changing acceleration, which means with constant jerk.

Categorize: This is a derivation problem. We start from basic definitions.

Analyze: We are given $J = da_x/dt = $ constant, so we know that $da_x = J dt$.

(a) Integrating from the 'initial' moment when we know the acceleration to any later moment,

$$\int_{a_{ix}}^{a_x} da = \int_0^t J dt \qquad a_x - a_{ix} = J(t - 0)$$

Therefore, $a_x = Jt + a_{xi}$ ∎

From $a_x = dv_x/dt$, $dv_x = a_x\,dt$

Integration between the same two points tells us the velocity as a function of time:

$$\int_{v_{xi}}^{v_x} dv_x = \int_0^t a_x\,dt = \int_0^t (a_{xi} + Jt)dt$$

$$v_x - v_{xi} = a_{xi}t + \tfrac{1}{2}Jt^2 \qquad v_x = v_{xi} + a_{xi}t + \tfrac{1}{2}Jt^2 \qquad ∎$$

From $v_x = dx/dt$, $dx = v_x\,dt$. Integrating a third time gives us $x(t)$:

$$\int_{x_i}^{x} dx = \int_0^t v_x\,dt = \int_0^t (v_{xi} + a_{xi}t + \tfrac{1}{2}Jt^2)\,dt$$

$$x - x_i = v_{xi}t + \tfrac{1}{2}a_{xi}t^2 + \tfrac{1}{6}Jt^3$$

and $x = \tfrac{1}{6}Jt^3 + \tfrac{1}{2}a_{xi}t^2 + v_{xi}t + x_i$ ∎

(b) Squaring the acceleration, $a_x^2 = (Jt + a_{xi})^2 = J^2t^2 + a_{xi}^2 + 2Ja_{xi}t$

Rearranging, $a_x^2 = a_{xi}^2 + 2J(\tfrac{1}{2}Jt^2 + a_{xi}t)$

The expression for v_x was $v_x = \tfrac{1}{2}Jt^2 + a_{xi}t + v_{xi}$

So $(v_x - v_{xi}) = \tfrac{1}{2}Jt^2 + a_{xi}t$

and by substitution $a_x^2 = a_{xi}^2 + 2J(v_x - v_{xi})$ ∎

Finalize: Our steps here have been parallel to one way of deriving the equations for constant acceleration motion. In fact, if we put in $J = 0$ our equations turn into standard equations for motion with a constant.

---

53. An inquisitive physics student and mountain climber climbs a 50.0-m-high cliff that overhangs a calm pool of water. He throws two stones vertically downward, 1.00 s apart, and observes that they cause a single splash. The first stone has an initial speed of 2.00 m/s. (a) How long after release of the first stone do the two stones hit the water? (b) What initial velocity must the second stone have if the two stones are to hit the water simultaneously? (c) What is the speed of each stone at the instant the two stones hit the water?

Solution

Conceptualize: The different nonzero original speeds of the two stones do not affect their accelerations, which have the same value g downward. This is a (pair of) free-fall problem(s).

Categorize: Equations chosen from the standard constant-acceleration set describe each stone separately, but look out for having to solve a quadratic equation.

Analyze: We set $y_i = 0$ at the top of the cliff, and find the time interval required for the first stone to reach the water using the particle under constant acceleration model:

$$y_f = y_i + v_{yi}t + \tfrac{1}{2}a_y t^2$$

or in quadratic form, $-\tfrac{1}{2}a_y t^2 - v_{yi}t + y_f - y_i = 0$

(a)   If we take the direction downward to be negative,

$$y_f = -50.0 \text{ m} \quad v_{yi} = -2.00 \text{ m/s} \quad \text{and} \quad a_y = -9.80 \text{ m/s}^2$$

Substituting these values into the equation, we find

$$(4.90 \text{ m/s}^2)t^2 + (2.00 \text{ m/s})t - 50.0 \text{ m} = 0$$

Use the quadratic formula. The stone reaches the pool after it is thrown, so time must be positive and only the positive root describes the physical situation:

$$t = \frac{-2.00 \text{ m/s} \pm \sqrt{(2.00 \text{ m/s})^2 - 4(4.90 \text{ m/s}^2)(-50.0 \text{ m})}}{2(4.90 \text{ m/s}^2)} = 3.00 \text{ s} \qquad \blacksquare$$

(b)   For the second stone, the time of travel is $t = 3.00 \text{ s} - 1.00 \text{ s} = 2.00 \text{ s}$.

Since $y_f = y_i + v_{yi}t + \tfrac{1}{2}a_y t^2$,

$$v_{yi} = \frac{(y_f - y_i) - \tfrac{1}{2}a_y t^2}{t} = \frac{-50.0 \text{ m} - \tfrac{1}{2}(-9.80 \text{ m/s}^2)(2.00 \text{ s})^2}{2.00 \text{ s}} = -15.3 \text{ m/s} \qquad \blacksquare$$

The negative value indicates the downward direction of the initial velocity of the second stone.

(c)   For the first stone,

$$v_{1f} = v_{1i} + a_1 t_1 = -2.00 \text{ m/s} + (-9.80 \text{ m/s}^2)(3.00 \text{ s})$$

$$v_{1f} = -31.4 \text{ m/s} \qquad \blacksquare$$

For the second stone, $v_{2f} = v_{2i} + a_2 t_2 = -15.3 \text{ m/s} + (-9.80 \text{ m/s}^2)(2.00 \text{ s})$

$$v_{2f} = -34.8 \text{ m/s} \qquad \blacksquare$$

Finalize: Make sure you know that the solution to $ax^2 + bx + c = 0$ is given by the "quadratic formula" $x = \dfrac{-b \pm \sqrt{b^2 - 4ac}}{2a}$. The equation has two solutions that have to be considered. We need the quadratic formula when an equation has a term containing the unknown to the second power, a term containing the unknown to the first power, and a constant term containing the unknown to the zeroth power. In this problem, the other root of the quadratic equation is -3.40 s. It means that the student did not need to throw the first stone down into the pool at time zero. The stone could have been burped up by a moat monster in the pool, who gave it an upward velocity of 31.4 m/s, at 3.40 s before the stone then passes the student on the way down. In this different story the stone has the same motion from cliff to pool.

61. Kathy tests her new sports car by racing with Stan, an experienced racer. Both start from rest, but Kathy leaves the starting line 1.00 s after Stan does. Stan moves with a constant acceleration of 3.50 m/s² while Kathy maintains an acceleration of 4.90 m/s². Find (a) the time at which Kathy overtakes Stan, (b) the distance she travels before she catches him, and (c) the speeds of both cars at the instant Kathy overtakes Stan.

Solution

Conceptualize: The two racers travel equal distances between the starting line and the overtake point, but their travel times and their speeds at the overtake point are different.

Categorize: We have constant-acceleration equations to apply to the two cars separately.

Analyze:

(a)   Let the times of travel for Kathy and Stan be t_K and t_S where $t_S = t_K + 1.00$ s.

Both start from rest ($v_{xiK} = v_{xiS} = 0$), so the expressions for the distances traveled are

$$x_K = \tfrac{1}{2} a_{x,K} t_K^2 = \tfrac{1}{2}(4.90 \text{ m/s}^2)t_K^2$$

and $$x_s = \tfrac{1}{2} a_{x,s} t_s^2 = \tfrac{1}{2}(3.50 \text{ m/s}^2)(t_K + 1.00 \text{ s})^2$$

When Kathy overtakes Stan, the two distances will be equal. Setting $x_K = x_s$ gives

$$\tfrac{1}{2}(4.90 \text{ m/s}^2)t_K^2 = \tfrac{1}{2}(3.50 \text{ m/s}^2)(t_K + 1.00 \text{ s})^2$$

This we simplify and write in the standard form of a quadratic as

$$t_K^2 - 5.00\, t_K s - 2.50 \text{ s}^2 = 0$$

We solve using the quadratic formula $t = \dfrac{-b \pm \sqrt{b^2 - 4ac}}{2a}$ to find

$$t_K = \frac{5 \pm \sqrt{5^2 - 4(1)(-2.5)}}{2(1)} = \frac{5 + \sqrt{35}}{2} = 5.46 \text{ s} \qquad \blacksquare$$

Only the positive root makes sense physically, because the overtake point must be after the starting point in time.

(b)   Use the equation from part (a) for distance of travel,

$$x_K = \tfrac{1}{2} a_{x,K} t_K^2 = \tfrac{1}{2}(4.90 \text{ m/s}^2)(5.46 \text{ s})^2 = 73.0 \text{ m} \qquad \blacksquare$$

(c)   Remembering that $v_{xi,K} = v_{xi,S} = 0$, the final velocities will be:

$$v_{xf,K} = a_{x,K} t_K = (4.90 \text{ m/s}^2)(5.46 \text{ s}) = 26.7 \text{ m/s} \qquad \blacksquare$$

$$v_{xf,S} = a_{x,S} t_S = (3.50 \text{ m/s}^2)(6.46 \text{ s}) = 22.6 \text{ m/s} \qquad \blacksquare$$

Finalize: Triple subscripts! You may find it easiest to keep track of all the parameters describing the accelerated motion of two objects by making a table of all the symbols

$$x_{iK} \quad x_K \quad v_{xiK} \quad v_{xfK} \quad a_{xK} \quad x_{iS} \quad x_S \quad v_{xiS} \quad v_{xfS} \quad a_{xS}$$

Think of the particular meaning of each. Show the known values in the listing at the start and fill in the unknowns as you go.

---

63. Two objects, A and B, are connected by hinges to a rigid rod that has a length L. The objects slide along perpendicular guide rails, as shown in Figure P2.63. Assume object A slides to the left with a constant speed v. (a) Find the velocity v_B of object B as a function of the angle θ. (b) Describe v_B relative to v: is v_B always smaller than v, larger than v, the same as v, or does it have some other relationship?

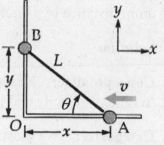

Figure P2.63

Solution

Conceptualize: Imitate the motion with your ruler, guiding its ends along sides of a sheet of paper. B starts moving very rapidly and then slows down, but not with constant acceleration.

Categorize: We translate from a pictorial representation through a geometric model to a mathematical representation by observing that the distances x and y are always related by $x^2 + y^2 = L^2$.

Analyze: (a) Differentiating this equation with respect to time, we have

$$2x\frac{dx}{dt} + 2y\frac{dy}{dt} = 0$$

Now the unknown velocity of B is

$$\frac{dy}{dt} = v_B$$

and

$$\frac{dx}{dt} = -v$$

So the differentiated equation becomes

$$\frac{dy}{dt} = -\frac{x}{y}\left(\frac{dx}{dt}\right) = -\left(\frac{x}{y}\right)(-v) = v_B$$

But

$$\frac{y}{x} = \tan\theta$$

so

$$v_B = \left(\frac{1}{\tan\theta}\right)v \qquad \blacksquare$$

(b) We assume that θ starts from zero. At this instant $1/\tan\theta$ is infinite, and the velocity of B is infinitely larger than that of A. As θ increases, the velocity of object B decreases, becoming equal to v when $\theta = 45°$. After that instant, B continues to slow down with non-constant acceleration, coming to rest as θ goes to 90°.

Finalize: The definition of velocity as the time derivative of position is always true. Differentiation is an operation you can always do to both sides of an equation. It is perhaps a surprise that the value of L does not affect the answer.

---

65. In a women's 100-m race, accelerating uniformly, Laura takes 2.00 s and Healan 3.00 s to attain their maximum speeds, which they each maintain for the rest of the race. They cross the finish line simultaneously, both setting a world record of 10.4 s. (a) What is the acceleration of each sprinter? (b) What are their respective maximum speeds? (c) Which sprinter is ahead at the 6.00-s mark, and by how much? (d) What is the maximum distance by which Healan is behind Laura and at what time does that occur?

Solution

Conceptualize: Healan spends more time speeding up, so she must speed up to a higher 'terminal' velocity and just catch up to Laura at the finish line.

Categorize: We must take the motion of each athlete apart into two sections, one with constant nonzero acceleration and one with constant velocity, in order to apply our standard equations.

Analyze:

(a) Laura moves with constant positive acceleration a_L for 2.00 s, then with constant speed (zero acceleration) for 8.40 s, covering a distance of

$$x_{L1} + x_{L2} = 100 \text{ m}$$

The two component distances are $x_{L1} = \frac{1}{2}a_L(2.00 \text{ s})^2$

and $x_{L2} = v_L(8.40 \text{ s})$

where v_L is her maximum speed $v_L = 0 + a_L(2.00 \text{ s})$

By substitution $\frac{1}{2}a_L(2.00 \text{ s})^2 + a_L(2.00 \text{ s})(8.40 \text{ s}) = 100 \text{ m}$

so we solve to find $a_L = 5.32 \text{ m/s}^2$ ∎

Similarly, for Healan $x_{H1} + x_{H2} = 100 \text{ m}$

with $x_{H1} = \frac{1}{2}a_H(3.00 \text{ s})^2$ $x_{H2} = v_H(7.40 \text{ s})$ $v_H = a_H(3.00 \text{ s})$

$$\frac{1}{2}a_H(3.00 \text{ s})^2 + a_H(3.00 \text{ s})(7.40 \text{ s}) = 100 \text{ m}$$
$$a_H = 3.75 \text{ m/s}^2$$ ∎

(b) Their speeds after accelerating are

$$v_L = a_L(2.00 \text{ s}) = (5.32 \text{ m/s}^2)(2.00 \text{ s}) = 10.6 \text{ m/s}$$ ∎

and $v_H = a_H(3.00 \text{ s}) = (3.75 \text{ m/s}^2)(3.00 \text{ s}) = 11.2 \text{ m/s}$ ∎

(c) In the first 6.00 s, Laura covers a distance

$$\frac{1}{2}a_L(2.00 \text{ s})^2 + v_L(4.00 \text{ s}) = \frac{1}{2}(5.32 \text{ m/s}^2)(2.00 \text{ s})^2 + (10.6 \text{ m/s})(4.00 \text{ s}) = 53.2 \text{ m}$$

and Healan has run a distance

$$\frac{1}{2}a_H(3.00 \text{ s})^2 + v_H(3.00 \text{ s}) = \frac{1}{2}(3.75 \text{ m/s}^2)(3.00 \text{ s})^2 + (11.2 \text{ m/s})(3.00 \text{ s}) = 50.6 \text{ m}$$

So Laura is ahead by $53.2 \text{ m} - 50.6 \text{ m} = 2.63 \text{ m}$ ∎

(d) Between 0 and 2 s Laura has the higher acceleration, so the distance between the runners increases. After 3 s, Healan's speed is higher, so the distance between them decreases. The distance between them is momentarily staying constant at its maximum value when they have equal speeds. This instant t_m is when Healan, still accelerating, has speed 10.6 m/s:

$$10.6 \text{ m/s} = (3.75 \text{ m/s}^2)\, t_m \qquad t_m = 2.84 \text{ s.} \qquad \blacksquare$$

Evaluating their positions again at this moment gives

Laura: $x = (1/2)(5.32)(2.00)^2 + (10.6)(0.840) = 19.6 \text{ m}$

Healan: $x = (1/2)(3.75)(2.84)^2 = 15.1 \text{ m}$

and shows that Laura is ahead by 19.6 m – 15.1 m = 4.47 m $\blacksquare$

Finalize: We carried out all of the same analysis steps for both of the runners. At the very end of part (c) we compared their positions. Do you think that 53.2 minus 50.6 should give 2.6 precisely, or 2.6 with the next digit unknown? Our method is to store all intermediate results in calculator memory with many digits, and to write down the three-digit final answer only after an unbroken chain of calculations with no "rounding off." Then we never pay attention to arguments about just when or how many intermediate results should be rounded off, and we never have to retype numbers into the calculator.

3

Vectors

EQUATIONS AND CONCEPTS

The **location of a point P in a plane** can be specified by either Cartesian coordinates, x and y, or polar coordinates, r and θ. *If one set of coordinates is known, values for the other set can be calculated.*

$$x = r\cos\theta \qquad (3.1)$$

$$y = r\sin\theta \qquad (3.2)$$

$$\tan\theta = \frac{y}{x} \qquad (3.3)$$

$$r = \sqrt{x^2 + y^2} \qquad (3.4)$$

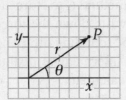

The **commutative law of addition** states that when two or more vectors are added, the sum is independent of the order of addition. To add vector $\vec{A}$ to vector $\vec{B}$ using the graphical method, first construct $\vec{A}$ and then draw $\vec{B}$ such that the tail of $\vec{B}$ starts at the head of $\vec{A}$. The sum of $\vec{A} + \vec{B}$ is the vector that completes the triangle by connecting the tail of $\vec{A}$ to the head of $\vec{B}$.

$$\vec{A} + \vec{B} = \vec{B} + \vec{A} \qquad (3.5)$$

$$\vec{A} + \vec{B} = \vec{B} + \vec{A}$$

The **associative law of addition** states that when three or more vectors are added the sum is independent of the way in which the individual vectors are grouped.

$$\vec{A} + (\vec{B} + \vec{C}) = (\vec{A} + \vec{B}) + \vec{C} \qquad (3.6)$$

In the **graphical or geometric method** of vector addition, the vectors to be added (or subtracted) are represented by arrows connected head-to-tail in any order. The resultant or sum is the vector which joins the tail of the first vector to the head of the last vector. The length of each arrow must be proportional to the magnitude of the corresponding vector and must be along the

$$\vec{R} = \vec{A} + \vec{B} + \vec{C} + \vec{D}$$

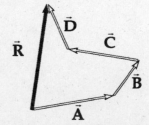

30

direction which makes the proper angle relative to the others. *When two or more vectors are to be added, all of them must represent the same physical quantity—that is, have the same units.*

The **operation of vector subtraction** utilizes the definition of the negative of a vector. The vector $(-\vec{A})$ has a magnitude equal to the magnitude of $\vec{A}$, but acts or points along a direction opposite the direction of $\vec{A}$. *The negative of vector $\vec{A}$ is defined as the vector that when added to $\vec{A}$ gives zero for the vector sum.*

$$\vec{A} - \vec{B} = \vec{A} + (-\vec{B}) \tag{3.7}$$

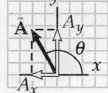

The **rectangular components** of a vector are the projections of the vector onto the respective coordinate axes. As illustrated in the figures, the projection of $\vec{A}$ onto the x axis is the x component of $\vec{A}$; and the projection of $\vec{A}$ onto the y axis is the y component of $\vec{A}$. *The angle θ is measured counterclockwise relative to the positive x axis and the algebraic sign of the components will depend on the value of θ.*

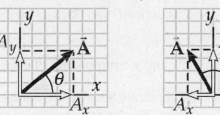

$$A_x = A \cos \theta \tag{3.8}$$

$$A_y = A \sin \theta \tag{3.9}$$

The **magnitude of $\vec{A}$ and the angle**, θ, which the vector makes with the positive x axis can be determined from the values of the x and y components of $\vec{A}$.

$$A = \sqrt{A_x^2 + A_y^2} \tag{3.10}$$

$$\theta = \tan^{-1}\left(\frac{A_y}{A_x}\right) \tag{3.11}$$

Unit vectors are dimensionless and have a magnitude of exactly 1. A vector $\vec{A}$ lying in the xy plane, having rectangular components A_x and A_y, can be expressed in unit vector notation. *Unit vectors specify the directions of the vector components.*

$$\vec{A} = A_x \hat{\mathbf{i}} + A_y \hat{\mathbf{j}} \tag{3.12}$$

The **resultant, $\vec{R}$,** of adding two vectors $\vec{A}$ and $\vec{B}$ can be expressed in terms of the components of the two vectors.

$$\vec{R} = \left(A_x + B_x\right)\hat{\mathbf{i}} + \left(A_y + B_y\right)\hat{\mathbf{j}} \tag{3.14}$$

The **magnitude and direction** of a resultant vector can be determined from the values of the components of the vectors in the sum.

$$R = \sqrt{R_x{}^2 + R_y{}^2}$$

$$= \sqrt{\left(A_x + B_x\right)^2 + \left(A_y + B_y\right)^2} \qquad (3.16)$$

$$\tan\theta = \frac{R_y}{R_x} = \frac{A_y + B_y}{A_x + B_x} \qquad (3.17)$$

SUGGESTIONS, SKILLS, AND STRATEGIES

When two or more vectors are to be added, the following step-by-step procedure is recommended:

- Select a coordinate system.

- Draw a sketch of the vectors to be added (or subtracted), with a label on each vector.

- Find the x and y components of each vector.

- Find the algebraic sum of the components of the individual vectors in both the x and y directions. These sums are the components of the resultant vector.

- Use the Pythagorean theorem to find the magnitude of the resultant vector.

- Use a suitable trigonometric function to find the angle the resultant vector makes with the x axis.

REVIEW CHECKLIST

You should be able to:

- Locate a point in space using both rectangular coordinates and polar coordinates. (Section 3.1)

- Use the graphical method for addition and subtraction of vectors. (Section 3.3)

- Resolve a vector into its rectangular components. Determine the magnitude and direction of a vector from its rectangular components. Use unit vectors to express any vector in unit vector notation. (Section 3.4)

- Determine the magnitude and direction of a resultant vector in terms of the components of individual vectors which have been added or subtracted. Express the resultant vector in unit vector notation. (Section 3.4)

ANSWER TO AN OBJECTIVE QUESTION

11. Vector $\vec{A}$ lies in the *xy* plane. Both of its components will be negative if it points from the origin into which quadrant? (a) the first quadrant (b) the second quadrant (c) the third quadrant (d) the fourth quadrant (e) the second or fourth quadrant.

Answer The vector $\vec{A}$ will have both rectangular components negative in the third quadrant, when the angle of $\vec{A}$ from the *x* axis is between π rad (180°) and $3\pi/2$ rad (270°). Thus the answer is (c).

□   □   □   □

ANSWER TO A CONCEPTUAL QUESTION

3. Is it possible to add a vector quantity to a scalar quantity? Explain.

Answer Vectors and scalars are distinctly different and cannot be added to each other. Remember that a vector defines a quantity **in a certain direction**, while a scalar only defines a quantity with no associated direction. It makes no sense to add a football uniform number to a number of apples, and similarly it makes no sense to add a number of apples to a wind velocity.

□   □   □   .□

SOLUTIONS TO SELECTED END-OF-CHAPTER PROBLEMS

1. The polar coordinates of a point are $r = 5.50$ m and $\theta = 240°$. What are the Cartesian coordinates of this point?

Solution

Conceptualize: The diagram helps to visualize the *x* and *y* coordinates as about −3 m and −5 m respectively.

Categorize: Trigonometric functions will tell us the coordinates directly.

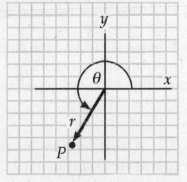

Analyze: When the polar coordinates (r, θ) of a point P are known, the cartesian coordinates are found as

$$x = r \cos \theta \quad \text{and} \quad y = r \sin \theta$$

here $x = (5.50 \text{ m}) \cos 240° = -2.75$ m

$y = (5.50 \text{ m}) \sin 240° = -4.76$ m   ■

Finalize: The coordinates are both negative because the point is in the third quadrant. The y coordinate is larger in magnitude because the point is closer to the y axis than to the x axis.

---

3. A fly lands on one wall of a room. The lower left-hand corner of the wall is selected as the origin of a two-dimensional Cartesian coordinate system. If the fly is located at the point having coordinates (2.00, 1.00) m, (a) how far is it from the origin? (b) What is its location in polar coordinates?

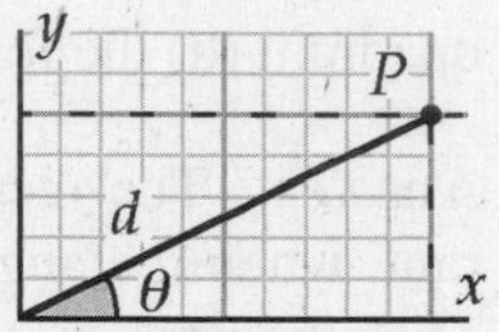

Solution

Conceptualize: The diagram helps to visualize the distance d from the corner as more than 2 m but much less than 3 m, and the angle coordinate as between 0 and 45°.

Categorize: The Pythagorean theorem and the definition of the tangent will be the starting points for our calculation.

Analyze:

(a) Take the wall as the xy plane so that the coordinates are $x = 2.00$ m and $y = 1.00$ m; and the fly is located at point P. The distance between two points in the xy plane is

$$d = \sqrt{(x_2 - x_1)^2 + (y_2 - y_1)^2}$$

so here $$d = \sqrt{(2.00\text{ m} - 0)^2 + (1.00\text{ m} - 0)^2} = 2.24\text{ m}$$ ■

(b) From the figure, $\theta = \tan^{-1}\left(\dfrac{y}{x}\right) = \tan^{-1}\left(\dfrac{1.00}{2.00}\right) = 26.6°$ and $r = d$

Therefore, the polar coordinates of the point P are (2.24 m, 26.6°) ■

Finalize: To find the angle we could have used the inverse sine of (1 m/2.26 m) or the inverse cosine of (2 m/2.26 m). But it is standard to use the tangent. Then a mistake in calculating the radial coordinate cannot also affect the angle coordinate.

---

7. A surveyor measures the distance across a straight river by the following method (Fig. P3.7). Starting directly across from a tree on the opposite bank, she walks $d = 100$ m along the riverbank to establish a baseline. Then she sights across to the tree. The angle from her baseline to the tree is $\theta = 35.0°$. How wide is the river?

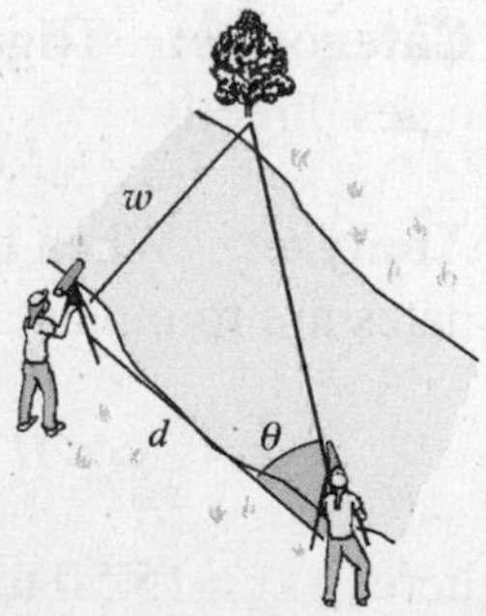

Figure P3.7

Solution

Conceptualize: Make a sketch of the area as viewed from above. Assume the river banks are straight and parallel, show the location of the tree, and the original location of the surveyor.

Categorize: The width *w* between tree and surveyor must be perpendicular to the river banks, as the surveyor chooses to start "directly across" from the tree. Now draw the 100-meter baseline, *d*, and the line showing the line of sight to the tree. We have a right triangle, so the definition of a trigonometric function will give us the answer.

Analyze: From the figure, $\tan \theta = \dfrac{w}{d}$

and $\quad w = d \tan \theta = (100 \text{ m}) \tan 35.0° = 70.0 \text{ m}$ ∎

Finalize: Observe that just the definitions of the sine, cosine, and tangent are sufficient for our applications in physics. For our applications, you will likely never have to use such theorems as the law of cosines or the law of sines.

---

9. *Why is the following situation impossible?* A skater glides along a circular path. She defines a certain point on the circle as her origin. Later on, she passes through a point at which the distance she has traveled along the path from the origin is smaller than the magnitude of her displacement vector from the origin.

Solution

Conceptualize: Draw a diagram of the skater's path. It needs to be the view from a hovering helicopter to see the circular path as circular in shape. To start with a concrete example, we have chosen to draw motion *ABC* around one half of a circle of radius 5 m.

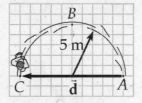

Categorize: In solving this problem we must contrast displacement with distance traveled.

Analyze: The displacement, shown as $\vec{\mathbf{d}}$ in the diagram, is the straight-line change in position from starting point *A* to finish *C*. In the specific case we have chosen to draw, it lies along a diameter of the circle. Its magnitude is $\left|\vec{\mathbf{d}}\right| = \left|-10.0\hat{\mathbf{i}}\right| = 10.0 \text{ m}$

The distance skated is greater than the straight-line displacement. The distance follows the curved path of the semicircle (*ABC*). Its length is half of the circumference: $s = \frac{1}{2}(2\pi r) = 5.00\pi \text{ m} = 15.7 \text{ m}$

A straight line is the shortest distance between two points. For any nonzero displacement, less or more than across a semicircle, the distance along the path will be greater than the displacement magnitude. If the skater completes one or several full revolutions, her displacement will be zero but the distance traveled will be one or several times the circumference of the circle. ∎

Finalize: There is nothing special about a circle—it could be any curve. For any motion involving change in direction, the distance traveled is always greater than the magnitude of the displacement.

---

11. The displacement vectors $\vec{A}$ and $\vec{B}$ shown in Figure P3.11 both have magnitudes of 3.00 m. The direction of vector $\vec{A}$ is $\theta = 30.0°$. Find graphically (a) $\vec{A} + \vec{B}$, (b) $\vec{A} - \vec{B}$, (c) $\vec{B} - \vec{A}$, and (d) $\vec{A} - 2\vec{B}$. Report all angles counterclockwise from the positive x axis.

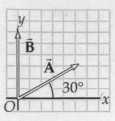

Figure P3.11

Solution

Conceptualize: Visualize adding vectors as playing golf. You string along one displacement after another, each beginning where the last ended. Visualize the negative of a vector as an equal-magnitude vector in the opposite direction. Visualize $-2\vec{B}$ as having twice the length of $\vec{B}$ and being in the opposite direction.

Categorize: We must draw with protractor and ruler to construct the additions. Then we must measure with ruler and protractor to read the answers.

Analyze: To find these vectors graphically, we draw each set of vectors as in the following drawings. We have a free choice of starting point and of scale. When subtracting vectors, we use $\vec{A} - \vec{B} = \vec{A} + (-\vec{B})$.

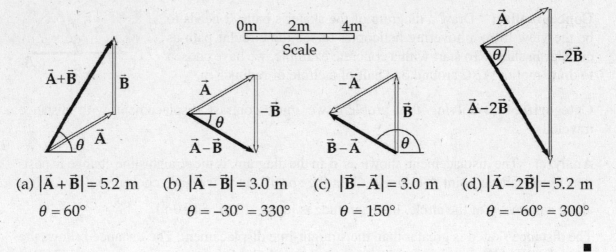

(a) $|\vec{A} + \vec{B}| = 5.2$ m   (b) $|\vec{A} - \vec{B}| = 3.0$ m   (c) $|\vec{B} - \vec{A}| = 3.0$ m   (d) $|\vec{A} - 2\vec{B}| = 5.2$ m

$\theta = 60°$ $\theta = -30° = 330°$ $\theta = 150°$ $\theta = -60° = 300°$

■

Finalize: A picture of reasonably small size gives us a result with only about two-significant-digit precision.

─────────────────

13. A roller-coaster car moves 200 ft horizontally and then rises 135 ft at an angle of 30.0° above the horizontal. It next travels 135 ft at an angle of 40.0° downward. What is its displacement from its starting point? Use graphical techniques.

Solution

Conceptualize: We will draw a side view picture.

Categorize: A vector is represented by an arrow with length representing its magnitude to a conveniently chosen scale. When adding vectors graphically, the directions of the vectors must be maintained as they start from different points in sequence. They are connected "head-to-tail."

Analyze: Your sketch should look like the one to the right. You will probably only be able to obtain a measurement to one or two significant figures.

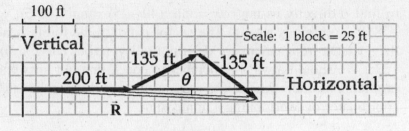

Figure P3.13

The distance R and the angle θ can be measured to give, upon use of your scale factor, the values of:

$$\vec{R} = 4.2 \times 10^2 \text{ ft at } 3° \text{ below the horizontal.}$$ ∎

Finalize: Here the drawing looks like a photographic side view of the roller coaster frame and track. In problems about adding other kinds of vectors, a sketch may not seem so real, but it will still be useful as an adjunct to a calculation about vector addition, and the sketch need not be measured to scale to be useful.

---

15. A vector has an x component of -25.0 units and a y component of 40.0 units. Find the magnitude and direction of this vector.

Solution

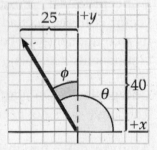

Conceptualize: First we should visualize the vector either in our mind or with a sketch. Since the hypotenuse of the right triangle must be greater than either the x or y components that form the legs, we can estimate the magnitude of the vector to be about 50 units. The direction of the vector can be quantified as the angle θ that appears to be about 120° counterclockwise from the +x axis.

Categorize: We use geometry and trigonometry to obtain a more precise result.

Analyze: The magnitude can be found by the Pythagorean theorem:

$$r = \sqrt{x^2 + y^2} \qquad r = \sqrt{(-25.0 \text{ units})^2 + (40.0 \text{ units})^2} = 47.2 \text{ units}$$ ∎

We observe that $\tan \phi = x/y$ (if we consider x and y to both be positive).

$$\phi = \tan^{-1}\left(\frac{x}{y}\right) = \tan^{-1}\left(\frac{25.0 \text{ units}}{40.0 \text{ units}}\right) = \tan^{-1}(0.625) = 32.0°$$

The angle from the +x axis can be found by adding 90° to ϕ.

$$\theta = \phi + 90° = 122°$$

Finalize: Our calculated results agree with our graphical estimates. Look out! If you tried to find θ directly as $\tan^{-1}y/x = \tan^{-1}(40/-25) = \tan^{-1}(-1.6)$, your calculator would report $-58.0°$, and this is wrong! The vector is not in the fourth quadrant. It is not at 58° below the +x axis. Your calculator does not tell you, but $\tan^{-1}(-1.6)$ has another value, namely 122°. We should always remember to check that our answers make sense, especially for problems like this where it is easy to mistakenly calculate the wrong angle by confusing coordinates or overlooking a minus sign.

The direction angle of a vector can generally be specified in more than one way, and we may choose a notation that is most convenient for the given problem. If compass directions were stated in this question, we could have reported the vector angle to be 32.0° west of north or at a compass heading of 328° clockwise from north.

---

19. Obtain expressions in component form for the position vectors having the polar coordinates (a) 12.8 m, 150°; (b) 3.30 cm, 60.0°; and (c) 22.0 in., 215°.

Solution

Conceptualize: Position vector (a) will have a large negative x component because the vector is in the second quadrant, a bit above the $-x$ axis. (b) is in the first quadrant and its y component will be larger than the x component. (c) is in the third quadrant with both components negative.

Categorize: Do not think of $\sin\theta = $ opposite/hypotenuse, but jump right to $y = R\sin\theta$. The angle does not need to fit inside a triangle.

Analyze: We find the x and y components of each vector using $x = R\cos\theta$ and $y = R\sin\theta$. In unit vector notation, $\vec{R} = R_x\hat{i} + R_y\hat{j}$

(a) $x = (12.8\text{ m})\cos 150° = -11.1\text{ m}$ and $y = (12.8\text{ m})\sin 150° = 6.40\text{ m}$

so $\vec{R} = \left(-11.1\hat{i} + 6.40\hat{j}\right)\text{ m}$ ∎

(b) $x = (3.30\text{ cm})\cos 60.0° = 1.65\text{ cm}$ and $y = (3.30\text{ cm})\sin 60.0° = 2.86\text{ cm}$

so $\vec{R} = \left(1.65\hat{i} + 2.86\hat{j}\right)\text{ cm}$ ∎

(c) $x = (22.0\text{ in.})\cos 215° = -18.0\text{ in.}$ and $y = (22.0\text{ in.})\sin 215° = -12.6\text{ in.}$

so $\vec{R} = \left(-18.0\hat{i} - 12.6\hat{j}\right)\text{ in.}$ ∎

Finalize: We check each answer against our expectations in the Conceptualize step and find consistency. Even without drawing on paper, we are thinking in pictorial terms.

---

23. Consider the two vectors $\vec{A} = 3\hat{i} - 2\hat{j}$ and $\vec{B} = -\hat{i} - 4\hat{j}$. Calculate (a) $\vec{A} + \vec{B}$, (b) $\vec{A} - \vec{B}$, (c) $|\vec{A} + \vec{B}|$, (d) $|\vec{A} - \vec{B}|$, and (e) the directions of $\vec{A} + \vec{B}$ and $\vec{A} - \vec{B}$.

Solution

Conceptualize: It would be good to sketch the vectors and their sum and difference as we did in problem 11, but …

Categorize: …we can get answers in unit-vector form just by doing calculations with each term labeled with an $\hat{i}$ or a $\hat{j}$. There are in a sense only two vectors to calculate, since parts (c), (d), and (e) just ask about the magnitudes and directions of the answers to (a) and (b).

Analyze: Use the property of vector addition that states that the components of $\vec{R} = \vec{A} + \vec{B}$ are computed as $R_x = A_x + B_x$ and $R_y = A_y + B_y$

(a) $\vec{A} + \vec{B} = \left(3\hat{i} - 2\hat{j}\right) + \left(-\hat{i} - 4\hat{j}\right) = 2.00\hat{i} - 6.00\hat{j}$ ■

(b) $\vec{A} - \vec{B} = \left(3\hat{i} - 2\hat{j}\right) - \left(-\hat{i} - 4\hat{j}\right) = 4.00\hat{i} + 2.00\hat{j}$ ■

For the vector $\vec{R} = R_x\hat{i} + R_y\hat{j}$ the magnitude is $|\vec{R}| = \sqrt{R_x^2 + R_y^2}$ so

(c) $|\vec{A} + \vec{B}| = \sqrt{2^2 + (-6)^2} = 6.32$ and   ■

(d) $|\vec{A} - \vec{B}| = \sqrt{4^2 + 2^2} = 4.47$ ■

The direction of a vector relative to the positive x axis is $\theta = \tan^{-1}\left(R_y / R_x\right)$

(e) For $\vec{A} + \vec{B}$, $\theta = \tan^{-1}(-6/2)$ in the fourth quadrant $= -71.6° = 288°$ ■

For $\vec{A} - \vec{B}$, $\theta = \tan^{-1}(2/4)$ in the first quadrant $= 26.6°$ ■

Finalize: The unit-vector notation was invented for brevity and convenience. Use it yourself whenever you have a choice.

---

25. Your dog is running around the grass in your back yard. He undergoes successive displacements 3.50 m south, 8.20 m northeast, and 15.0 m west. What is the resultant displacement?

Solution

Conceptualize: On a morning after a raccoon has wandered through the back yard, your faithful dog is running around with his nose down, stringing up crime-scene tape and chalking in figures on the ground. He will end up somewhat northwest of his starting point.

Categorize: We use the unit-vector addition method. It is just as easy to add three displacements as to add two.

Analyze: Take the direction east to be along $+\hat{\mathbf{i}}$. The three displacements can be written as:

$$\vec{\mathbf{d}}_1 = -3.50\hat{\mathbf{j}} \text{ m}$$

$$\vec{\mathbf{d}}_2 = (8.20 \cos 45.0°)\hat{\mathbf{i}} \text{ m} + (8.20 \sin 45.0°)\hat{\mathbf{j}} \text{ m} = 5.80\hat{\mathbf{i}} \text{ m} + 5.80\hat{\mathbf{j}} \text{ m}$$

and $\vec{\mathbf{d}}_3 = -15.0\hat{\mathbf{i}}$ m

The resultant is $\vec{\mathbf{R}} = \vec{\mathbf{d}}_1 + \vec{\mathbf{d}}_2 + \vec{\mathbf{d}}_3 = (0 + 5.80 - 15.0)\hat{\mathbf{i}}$ m $+ (-3.50 + 5.80 + 0)\hat{\mathbf{j}}$ m

or $\vec{\mathbf{R}} = \left(-9.20\hat{\mathbf{i}} + 2.30\hat{\mathbf{j}}\right)$ m   ∎

The magnitude of the resultant displacement is

$$|\vec{\mathbf{R}}| = \sqrt{R_x^2 + R_y^2} = \sqrt{(-9.20 \text{ m})^2 + (2.30 \text{ m})^2} = 9.48 \text{ m}$$ ∎

The direction of the resultant vector is given by

$$\tan^{-1}\left(\frac{R_y}{R_x}\right) = \tan^{-1}\left(\frac{2.30}{-9.20}\right) \text{ in the second quadrant} = -14.0° + 180° = 166°$$ ∎

Finalize: Note again that your calculator gives you no hint of whether the direction angle is in the fourth quadrant with x component positive and y component negative, at $360° - 14° = 346°$; or in the second quadrant with x negative and y positive, at $180° - 14° = 166°$. Your observation that the x component is negative and the y component is positive diagnoses the resultant as in the second quadrant.

---

33. The vector $\vec{\mathbf{A}}$ has x, y, and z components of 8.00, 12.0, and −4.00 units, respectively. (a) Write a vector expression for $\vec{\mathbf{A}}$ in unit-vector notation. (b) Obtain a unit-vector expression for a vector $\vec{\mathbf{B}}$ one fourth the length of $\vec{\mathbf{A}}$ pointing in the same direction as $\vec{\mathbf{A}}$. (c) Obtain a unit-vector expression for a vector $\vec{\mathbf{C}}$ three times the length of $\vec{\mathbf{A}}$ pointing in the direction opposite the direction of $\vec{\mathbf{A}}$.

Solution

Conceptualize: Hold your fingertip at the center of the front edge of your study desk, defined as point O. Move your finger 8 cm to the right, then 12 cm vertically up, and then 4 cm horizontally away from you. Its location relative to the starting point represents position vector $\vec{A}$. Move three fourths of the way straight back toward O. Now your fingertip is at the location of $\vec{B}$. Now move your finger 50 cm straight through O, through your left thigh, and down toward the floor. Its position vector now is $\vec{C}$.

Categorize: We use unit-vector notation throughout. There is no adding to do here, but just multiplication of a vector by two different scalars.

Analyze:

(a) $\vec{A} = A_x\hat{i} + A_y\hat{j} + A_z\hat{k}$ $\vec{A} = 8.00\hat{i} + 12.0\hat{j} - 4.00\hat{k}$ ∎

(b) $\vec{B} = \vec{A}/4$ $\vec{B} = 2.00\hat{i} + 3.00\hat{j} - 1.00\hat{k}$ ∎

(c) $\vec{C} = -3\vec{A}$ $\vec{C} = -24.0\hat{i} - 36.0\hat{j} + 12.0\hat{k}$ ∎

Finalize: If your finger got lost in the Conceptualize step, trace out now from the center of the front edge of your desktop 24 cm left, 36 cm down, and 12 cm horizontally away from the front of the desk. That gets you to the point with position vector $\vec{C}$. Think of this as a very straightforward problem. You will frequently encounter multiplication of a vector by a scalar, as in the relationship of velocity and momentum and the relationship of total force and acceleration. Later in the course you will study two different ways of forming the product of one vector with another vector.

---

35. Vector $\vec{A}$ has a negative x component 3.00 units in length and a positive y component 2.00 units in length. (a) Determine an expression for $\vec{A}$ in unit-vector notation. (b) Determine the magnitude and direction of $\vec{A}$. (c) What vector $\vec{B}$ when added to $\vec{A}$ gives a resultant vector with no x component and a negative y component 4.00 units in length?

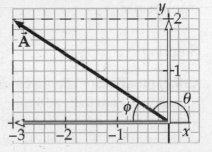

Solution

Conceptualize: The component description of $\vec{A}$ is just restated to constitute the answer to part (a). The vector looks roughly 4 units long at $180° - 35°$. And $\vec{B}$ should clearly be $+3\hat{i} - 6\hat{j}$.

Categorize: You build your faith in the unit-vector notation method, and your understanding, when you use it to confirm by calculation results that you can guess in advance.

Analyze:

(a) We are given $A_x = -3.00$ units   and $A_y = 2.00$ units

so $\vec{A} = A_x\hat{i} + A_y\hat{j} = \left(-3.00\hat{i} + 2.00\hat{j}\right)$ units   ∎

(b) The magnitude is $|\vec{A}| = \sqrt{A_x^2 + A_y^2} = \sqrt{(-3.00)^2 + (2.00)^2} = 3.61$ units   ∎

and the direction is in the second quadrant, described by

$$\tan\phi = \left|\frac{A_y}{A_x}\right| = \left|\frac{2.00}{-3.00}\right| = 0.667 \quad \text{so} \quad \phi = 33.7° \quad \text{(relative to the } -x \text{ axis)}$$

Thus $\vec{A}$ is at $\theta = 180° - \phi = 146°$ ∎

(c) We are given that $\vec{R} = \left(0\hat{i} - 4.00\hat{j}\right)$ units $= \vec{A} + \vec{B} = \left(-3.00\hat{i} + 2.00\hat{j}\right)$ units $+ \vec{B}$

so $\vec{B} = \vec{R} - \vec{A} = \left[(3+0)\hat{i} + (-4-2)\hat{j}\right]$ units $= \left(+3.00\hat{i} - 6.00\hat{j}\right)$ units   ∎

Finalize: Our calculations are as brief as possible and agree with our preliminary estimates. Brace yourself for continued successes.

---

36. Three displacement vectors of a croquet ball are shown in Figure P3.36, where $|\vec{A}| = 20.0$ units, $|\vec{B}| = 40.0$ units, and $|\vec{C}| = 30.0$ units. Find (a) the resultant in unit-vector notation and (b) the magnitude and direction of the resultant displacement.

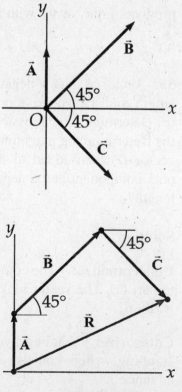

Solution

Conceptualize: The given diagram shows the vectors individually, but not their addition. The second diagram represents a map view of the motion of the ball. From it, the magnitude of the resultant $\vec{R}$ should be about 60 units. Its direction is in the first quadrant, at something like 30° from the x axis. Its x component appears to be about 50 units and its y component about 30 units.

Categorize: According to the definition of a displacement, we ignore any departure from straightness of the actual path of the ball. We model each of the three motions as straight. The simplified problem is solved by straightforward application of the component method of vector addition. It works for adding two, three, or any number of vectors.

Figure P3.36

Analyze:

(a) We find the two components of each of the three vectors

$A_x = (20.0 \text{ units}) \cos 90° = 0$ and $A_y = (20.0 \text{ units}) \sin 90° = 20.0 \text{ units}$

$B_x = (40.0 \text{ u}) \cos 45° = 28.3 \text{ units}$ and $B_y = (40.0 \text{ u}) \sin 45° = 28.3 \text{ units}$

$C_x = (30.0 \text{ u}) \cos 315° = 21.2 \text{ units}$ and $C_y = (30.0 \text{ u}) \sin 315° = -21.2 \text{ units}$

Now adding, $R_x = A_x + B_x + C_x = (0 + 28.3 + 21.2) \text{ units} = 49.5 \text{ units}$

and $R_y = A_y + B_y + C_y = (20 + 28.3 - 21.2) \text{ units} = 27.1 \text{ units}$

so $\vec{R} = 49.5\hat{i} + 27.1\hat{j} \text{ units}$ ∎

(b) $|\vec{R}| = \sqrt{R_x^2 + R_y^2} = \sqrt{(49.5 \text{ units})^2 + (27.1 \text{ units})^2} = 56.4 \text{ units}$ ∎

$\theta = \tan^{-1}\left(\dfrac{R_y}{R_x}\right) = \tan^{-1}\left(\dfrac{27.1}{49.5}\right) = 28.7°$ ∎

Finalize: The approximate values we guessed in the conceptualize step (60 units at $30° \approx (50\hat{i} + 30\hat{j})$ units) agree with the computed values precise to three digits (56.4 units at $28.7° = (49.5\hat{i} + 27.1\hat{j})$ units). Perhaps the greatest usefulness of the diagram is checking positive and negative signs for each component of each vector. The y component of $\vec{C}$ is negative, for example, because the vector is downward. For each vector it is good to check against the diagram whether the y component is less than, equal to, or greater than the x component. If your calculator is set to radians instead of degrees, the diagram can rescue you.

---

39. A man pushing a mop across a floor causes it to undergo two displacements. The first has a magnitude of 150 cm and makes an angle of 120° with the positive x axis. The resultant displacement has a magnitude of 140 cm and is directed at an angle of 35.0° to the positive x axis. Find the magnitude and direction of the second displacement.

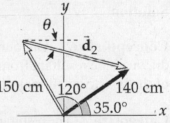

Solution

Conceptualize: The diagram suggests that $\vec{d}_2$ is about 180 cm at 20° south of east, or 180 cm at 340°.

Categorize: We will use the component method for a precise answer. We already know the total displacement, so the algebra of solving a vector equation will guide us to do a subtraction.

Analyze: The total displacement of the mop is $\vec{R} = \vec{d}_1 + \vec{d}_2$

Then $\vec{d}_2 = \vec{R} - \vec{d}_1$

substituting, $\quad\vec{\mathbf{d}}_2 = (140 \text{ cm at } 35°) - (150 \text{ cm at } 120°)$

identifying the meaning of the negative sign gives

$$\vec{\mathbf{d}}_2 = (140 \text{ cm at } 35°) + (150 \text{ cm at } 300°)$$

In components, $\quad\vec{\mathbf{d}}_2 = \left[140 \cos 35° \,\hat{\mathbf{i}} + 140 \sin 35\hat{\mathbf{j}} + 150 \cos 300° \,\hat{\mathbf{i}} + 150 \sin 300° \,\hat{\mathbf{j}} \right]$

evaluating, $\quad\vec{\mathbf{d}}_2 = \left[(115+75)\hat{\mathbf{i}} + (80.3-130)\hat{\mathbf{j}} \right] \text{cm} \quad$ or $\quad \vec{\mathbf{d}}_2 = \left[190\hat{\mathbf{i}} - 49.6\hat{\mathbf{j}} \right] \text{cm}$

Finding magnitude and direction,

$$\vec{\mathbf{d}}_2 = \sqrt{(190 \text{ cm})^2 + (49.6 \text{ cm})^2} \text{ at } \tan^{-1}\left(\frac{49.6 \text{ cm}}{190 \text{ cm}} \right) \text{ from the } x \text{ axis in the fourth quadrant}$$

so at last $\quad\vec{\mathbf{d}}_2 = 196 \text{ cm at } (360° - 14.7°) = 196 \text{ cm at } 345°$ ■

Finalize: To the very limited precision of our not-to-scale sketch, we can claim agreement between it and the calculated value for the second displacement. If the problem had not mentioned finding the magnitude and direction of $\vec{\mathbf{d}}_2$ we could have stopped at $\vec{\mathbf{d}}_2 = [190\hat{\mathbf{i}} - 49.6\hat{\mathbf{j}}]$cm. A vector is completely specified by its components.

---

47. A person going for a walk follows the path shown in Figure P3.47. The total trip consists of four straight-line paths. At the end of the walk, what is the person's resultant displacement measured from the starting point?

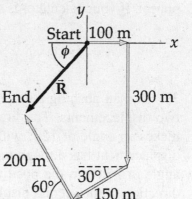

Figure P3.47

Solution

Conceptualize: On our version of the diagram we have drawn in the resultant from the tail of the first arrow to the head of the last arrow. It looks like about 200 m at 180° + 60°.

Categorize: The resultant displacement $\vec{\mathbf{R}}$ is equal to the sum of the four individual displacements, $\vec{\mathbf{R}} = \vec{\mathbf{d}}_1 + \vec{\mathbf{d}}_2 + \vec{\mathbf{d}}_3 + \vec{\mathbf{d}}_4$

Analyze: We translate from the pictorial representation to a mathematical representation by writing the individual displacements in unit-vector notation:

$$\vec{\mathbf{d}}_1 = 100\hat{\mathbf{i}} \text{ m}$$

$$\vec{\mathbf{d}}_2 = -300\hat{\mathbf{j}} \text{ m}$$

$$\vec{\mathbf{d}}_3 = (-150 \cos 30°)\hat{\mathbf{i}} \text{ m} + (-150 \sin 30°)\hat{\mathbf{j}} \text{ m} = -130\hat{\mathbf{i}} \text{ m} - 75\hat{\mathbf{j}} \text{ m}$$

$$\vec{\mathbf{d}}_4 = (-200 \cos 60°)\hat{\mathbf{i}} \text{ m} + (200 \sin 60°)\hat{\mathbf{j}} \text{ m} = -100\hat{\mathbf{i}} \text{ m} + 173\hat{\mathbf{j}} \text{ m}$$

Summing the components together, we find

$$R_x = d_{1x} + d_{2x} + d_{3x} + d_{4x} = (100 + 0 - 130 - 100)\text{m} = -130 \text{ m}$$

$$R_y = d_{1y} + d_{2y} + d_{3y} + d_{4y} = (0 - 300 - 75 + 173)\text{m} = -202 \text{ m}$$

so altogether $\quad \vec{R} = -130\hat{i} \text{ m} - 202\hat{j} \text{ m}$

Its magnitude is $\quad |\vec{R}| = \sqrt{R_x^2 + R_y^2} = \sqrt{(-130 \text{ m})^2 + (-202 \text{ m})^2} = 240 \text{ m}$ ■

We calculate the angle $\quad \phi = \tan^{-1}\left(\dfrac{R_y}{R_x}\right) = \tan^{-1}\left(\dfrac{-202}{-130}\right) = 57.2°$

A Pitfall Prevention in the text explains why this angle does not specify the resultant's direction in standard form. The resultant points into the third quadrant instead of the first quadrant. The angle counterclockwise from the +x axis is $\quad \theta = 57.2° + 180° = 237°$ ■

Finalize. We have the preliminary estimate 200 km at 240° that is consistent with two ways of writing the calculated total displacement, $\vec{R} = -130\hat{i} \text{ m} - 202\hat{j} \text{ m} = 240 \text{ m at } 237°$

4

Motion in Two Dimensions

EQUATIONS AND CONCEPTS

The **displacement of a particle** is defined as the difference between the final and initial position vectors. *For the displacement illustrated in the figure, the magnitude of $\Delta \vec{r}$ is less than the actual path length from the initial to the final position.*

$$\Delta \vec{r} \equiv \vec{r}_f - \vec{r}_i \tag{4.1}$$

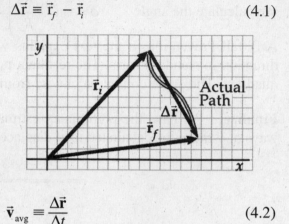

The **average velocity** of a particle which undergoes a displacement $\Delta \vec{r}$ in a time interval Δt equals the ratio $\Delta \vec{r}/t$. *The average velocity depends on the displacement vector and not on the length of path traveled.*

$$\vec{v}_{avg} \equiv \frac{\Delta \vec{r}}{\Delta t} \tag{4.2}$$

The **average speed** of a particle during any time interval is the ratio of the total distance of travel (length of path) to the total time. *If, during a time Δt a particle returns to the starting point, the displacement and average velocity over the interval will each be zero; however the distance traveled and the average speed will **not** be zero.*

$$\text{average speed} = \frac{\text{total distance}}{\text{total time}}$$

The **instantaneous velocity** of a particle equals the limit of the average velocity as $\Delta t \to 0$. *The magnitude of the instantaneous velocity is called the speed.*

$$\vec{v} \equiv \lim_{\Delta t \to 0} \frac{\Delta \vec{r}}{\Delta t} = \frac{d\vec{r}}{dt} \tag{4.3}$$

The **average acceleration** of a particle which undergoes a change in velocity $\Delta \vec{v}$ in a time interval Δt equals the ratio $\Delta \vec{v}/\Delta t$.

$$\vec{a}_{avg} \equiv \frac{\Delta \vec{v}}{\Delta t} = \frac{\vec{v}_f - \vec{v}_i}{t_f - t_i} \tag{4.4}$$

The **instantaneous acceleration** is defined as the limit of the average acceleration as $\Delta t \to 0$. *A particle experiences acceleration when the velocity vector undergoes a change in magnitude, direction, or both.*

$$\vec{a} \equiv \lim_{\Delta t \to 0} \frac{\Delta \vec{v}}{\Delta t} = \frac{d\vec{v}}{dt} \qquad (4.5)$$

The **position vector** for a particle in two-dimensional motion can be stated using unit vectors. The position as a function of time is given by Equation 4.9 below.

$$\vec{r} = x\hat{\mathbf{i}} + y\hat{\mathbf{j}} \qquad (4.6)$$

Motion in two dimensions with constant acceleration is described by equations for velocity and position which are vector versions of the one-dimensional kinematic equations.

$$\vec{v}_f = \vec{v}_i + \vec{a}t \qquad (4.8)$$

$$\vec{r}_f = \vec{r}_i + \vec{v}_i t + \tfrac{1}{2}\vec{a}t^2 \qquad (4.9)$$

A **projectile** moves with constant velocity along the horizontal direction and with constant acceleration (g) along the downward vertical direction. *The path of a projectile is a parabola.*

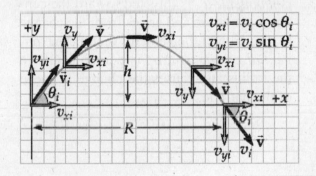

Equations describing motion of a projectile, which is launched from the origin at $t_i = 0$ with a positive y-component of velocity and returns to the initial horizontal level, are summarized below.

Position coordinates:

$$x_f = v_{xi}\,t = (v_i \cos \theta_i)t$$

$$y_f = (v_i \sin \theta_i)t - \tfrac{1}{2}gt^2$$

Velocity components:

$$v_x = v_i \cos \theta_i = \text{constant}$$

$$v_y = v_i \sin \theta_i - gt$$

Maximum height:

$$h = \frac{v_i^2 \sin^2 \theta_i}{2g} \qquad (4.12)$$

Range: The range is maximum when $\theta_i = 45°$

$$R = \frac{v_i^2 \sin 2\theta_i}{g} \qquad (4.13)$$

The **centripetal acceleration** vector for a particle in *uniform circular motion* has a constant magnitude given by Equation 4.14. The centripetal acceleration vector is always directed toward the center of the circular path, and, therefore, is constantly changing in direction.

$$a_c = \frac{v^2}{r} \tag{4.14}$$

The **period** (T) of a particle moving with constant speed in a circle of radius r is defined as the time interval required to complete one revolution. The frequency of the motion, f, equals $1/T$, and is stated in units of hertz (Hz).

$$T = \frac{2\pi r}{v} \tag{4.15}$$

The **total acceleration** of a particle moving on a curved path is the vector sum of the radial and the tangential components of acceleration (see figure below). The *radial component*, a_r, is directed toward the center of curvature and arises from the change in direction of the velocity vector. The *tangential component*, a_t, is perpendicular to the radius and causes the change in speed of the particle.

$$\vec{a} = \vec{a}_r + \vec{a}_t \tag{4.16}$$

$$a_t = \frac{d|v|}{dt} \tag{4.17}$$

$$a_r = -a_c = -\frac{v^2}{r} \tag{4.18}$$

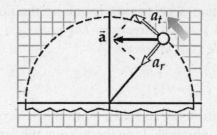

Galilean transformation equations relate observations made by an observer in one frame of reference (A) to observations of corresponding quantities made by an observer in a reference frame (B) moving with velocity $\vec{v}_{BA}$ with respect to the first. *In Equations 4.19 and 4.20, the first subscript represents the quantity to be measured and the second subscript represents the location of the observer.*

$$\vec{r}_{PA} = \vec{r}_{PB} + \vec{v}_{BA}t \tag{4.19}$$

$$\vec{u}_{PA} = \vec{u}_{PB} + \vec{v}_{BA} \tag{4.20}$$

Both observers measure the same value of acceleration when they are moving with constant velocity with respect to each other.

SUGGESTIONS, SKILLS, AND STRATEGIES

- A projectile moving under the influence of gravity has a parabolic trajectory and the equation of the path has the general form,

  $$y = Ax - Bx^2 \qquad \text{where} \qquad A = \tan \theta_i \qquad \text{and}$$

  $$B = \frac{g}{2v_i^2 \cos^2 \theta_i}$$

  This expression for y assumes that the particle leaves the origin at $t = 0$, with a velocity $\vec{v}_i$ which makes an angle θ_i with the horizontal. A sketch of y versus x for this situation is shown at the right.

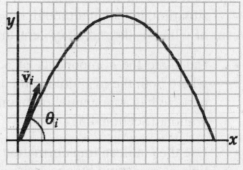

- Given the values for x and y at any time $t > 0$, an expression for the position vector $\vec{r}$ at that time can be written, using unit-vector notation, in the form $\vec{r} = x\hat{\mathbf{i}} + y\hat{\mathbf{j}}$. If v_x and v_y are known at any time, the velocity vector $\vec{v}$ can be written in the form $\vec{v} = v_x\hat{\mathbf{i}} + v_y\hat{\mathbf{j}}$. From this, the speed $v = \sqrt{v_x^2 + v_y^2}$ and the angle $\theta = \tan^{-1}(v_y/v_x)$ that the vector makes with the x axis can be determined.

- Remember, at maximum height $v_y = 0$; and at the point of impact (when $x = R$), $y = 0$ (for a projectile that lands at the same level from which it was launched).

PROBLEM-SOLVING STRATEGY: PROJECTILE MOTION

The following approach should be used in solving projectile motion problems:

- Select a coordinate system.

- Resolve the initial velocity vector into x and y components.

- Treat the horizontal motion and the vertical motion independently.

- Follow the techniques for solving problems with constant velocity to analyze the horizontal motion of the projectile.

- Follow the techniques for solving problems with constant acceleration to analyze the vertical motion of the projectile.

REVIEW CHECKLIST

You should be able to:

- Describe the displacement, velocity, and acceleration of a particle moving in the xy plane. (Sections 4.1 and 4.2)

- Find the velocity components and the coordinates of a projectile at any time t if the initial speed v_i and initial angle θ_i are known at $t = 0$. Also calculate the horizontal range R and maximum height h if v_i and θ_i are known. (Section 4.3)

- Calculate the acceleration of a particle moving in a circle with constant speed. In this situation, note that although $|\vec{v}|$ = constant, the *direction of $\vec{v}$ varies in time*, the result of which is the radial, or centripetal acceleration. (Section 4.4)

- Calculate the components of acceleration for a particle moving on a curved path, where both the magnitude and direction of $\vec{v}$ are changing with time. In this case, the particle has a tangential component of acceleration and a radial component of acceleration. (Section 4.5)

- Use the Galilean transformation equations to relate the position and velocity of a particle as measured by observers in relative motion. (Section 4.6)

ANSWER TO AN OBJECTIVE QUESTION

11. A sailor drops a wrench from the top of a sailboat's vertical mast while the boat is moving rapidly and steadily straight forward. Where will the wrench hit the deck? (a) ahead of the base of the mast (b) at the base of the mast (c) behind the base of the mast (d) on the windward side of the base of the mast (e) None of the choices (a) through (d) is true.

Answer (b) Here is one argument for this answer: When the sailor opens his fingers to release it, the wrench becomes a projectile. Its vertical motion is affected by gravitation, and is independent of its horizontal motion. The wrench continues moving horizontally with its speed prior to release, and this is the same as the constant forward speed of the boat. Thus the wrench falls down at a constant small distance from the moving mast.

Galileo suggested the idea for this question and gave another argument for answer (b): The reference frame of the boat is just as good as the Earth reference frame for describing the motion of the wrench. In the boat frame the wrench starts from rest and moves in free fall. It travels straight down next to the stationary mast, and lands at the base of the mast.

ANSWERS TO SELECTED CONCEPTUAL QUESTIONS

1. If you know the position vectors of a particle at two points along its path and also know the time interval during which it moved from one point to the other, can you determine the particle's instantaneous velocity? Its average velocity? Explain.

Answer Its instantaneous velocity cannot be determined at any point from this information. However, the average velocity over the time interval can be determined from its definition and the given information. The given information fits directly into the definition of the single value of the average velocity over that particular time interval. The infinitely many values of the instantaneous velocity, on the other hand, might show any sort of variability over the interval.

□ □ □ □

3.   A spacecraft drifts through space at a constant velocity. Suddenly, a gas leak in the side of the spacecraft gives it a constant acceleration in a direction perpendicular to the initial velocity. The orientation of the spacecraft does not change, so the acceleration remains perpendicular to the original direction of the velocity. What is the shape of the path followed by the spacecraft in this situation?

Answer   The spacecraft will follow a parabolic path, just like a projectile thrown off a cliff with a horizontal velocity. For the projectile, gravity provides an acceleration which is always perpendicular to the initial velocity, resulting in a parabolic path. For the spacecraft, the initial velocity plays the role of the horizontal velocity of the projectile. The leaking gas provides an acceleration that plays the role of gravity for the projectile. If the orientation of the spacecraft were to change in response to the gas leak (which is by far the more likely result), then the acceleration would change direction and the motion could become quite complicated.

5.   A projectile is launched at some angle to the horizontal with some initial speed v_i, and air resistance is negligible. (a) Is the projectile a freely falling body? (b) What is its acceleration in the vertical direction? (c) What is its acceleration in the horizontal direction?

Answer   (a) Yes. The projectile is a freely falling body, because the only force acting on it is the gravitational force exerted by the planet. (b) The vertical acceleration is the local gravitational acceleration, g downward. (c) The horizontal acceleration is zero.

7.   Explain whether or not the following particles have an acceleration: (a) a particle moving in a straight line with constant speed and (b) a particle moving around a curve with constant speed.

Answer   (a) The acceleration is zero, since the magnitude and direction of $\vec{v}$ remain constant. (b) The particle has a nonzero acceleration since the direction of $\vec{v}$ changes.

SOLUTIONS TO SELECTED END-OF-CHAPTER PROBLEMS

1.   A motorist drives south at 20.0 m/s for 3.00 min, then turns west and travels at 25.0 m/s for 2.00 min, and finally travels northwest at 30.0 m/s for 1.00 min. For this 6.00-min trip, find (a) the total vector displacement, (b) the average speed, and (c) the average velocity. Let the positive x axis point east.

Solution

Conceptualize: The motorist ends up somewhere southwest of her starting point. This specification will appear more precisely in the answers for total displacement and for average velocity, but the answer for speed will have no direction.

Categorize: We must use the method of vector addition and the definitions of average velocity and of average speed.

Analyze:

(a)   For each segment of the motion we model the car as a particle under constant velocity. Her displacements are

$$\Delta \vec{r} = (20.0 \text{ m/s})(180 \text{ s}) \textbf{ south} + (25.0 \text{ m/s})(120 \text{ s}) \textbf{ west} + (30.0 \text{ m/s})(60.0 \text{ s}) \textbf{ northwest}$$

Choosing $\hat{\textbf{i}} =$ east and $\hat{\textbf{j}} =$ north, we have

$$\Delta \vec{r} = (3.60 \text{ km})(-\hat{\textbf{j}}) + (3.00 \text{ km})(-\hat{\textbf{i}}) + (1.80 \text{ km})\cos 45.0°(-\hat{\textbf{i}}) + (1.80 \text{ km})\sin 45.0°(\hat{\textbf{j}})$$

$$\Delta \vec{r} = (3.00 + 1.27) \text{ km}(-\hat{\textbf{i}}) + (1.27 - 3.60) \text{ km}(\hat{\textbf{j}}) = (-4.27 \,\hat{\textbf{i}} - 2.33 \,\hat{\textbf{j}}) \text{ km}$$ ∎

The answer can also be written as

$$\Delta \vec{r} = \sqrt{(-4.27 \text{ km})^2 + (-2.33 \text{ km})^2} \text{ at } \tan^{-1}\left(\frac{2.33}{4.27}\right) = 28.6° \text{ south of west}$$

or $\Delta \vec{r} = 4.87$ km at $209°$ from the east.   ∎

(b)   The total distance or path-length traveled is $(3.60 + 3.00 + 1.80)$ km $= 8.40$ km

so average speed $= \left(\dfrac{8.40 \text{ km}}{6.00 \text{ min}}\right)\left(\dfrac{1.00 \text{ min}}{60.0 \text{ s}}\right)\left(\dfrac{1000 \text{ m}}{\text{km}}\right) = 23.3 \text{ m/s}$ ∎

(c) $\vec{v}_{\text{avg}} = \dfrac{\Delta \vec{r}}{t} = \dfrac{4.87 \text{ km}}{360 \text{ s}} = 13.5$ m/s at $209°$ ∎

or

$$\vec{v}_{\text{avg}} = \frac{\Delta \vec{r}}{t} = \frac{(-4.27 \textbf{ east} - 2.33 \textbf{ north})\text{km}}{360 \text{ s}} = (11.9 \textbf{ west} + 6.46 \textbf{ south})\text{m/s}$$

Finalize: The average velocity is necessarily in the same direction as the total displacement. The total distance and the average speed are scalars, with no direction. Distance must be greater than the magnitude of displacement for any motion that changes direction, and similarly average speed must be greater than average velocity.

---

7.   A fish swimming in a horizontal plane has velocity $\vec{v}_i = (4.00\hat{\textbf{i}} + 1.00\hat{\textbf{j}})$ m/s at a point in the ocean where the position relative to a certain rock is $\vec{r}_i = (10.0\hat{\textbf{i}} - 4.00\hat{\textbf{j}})$ m. After the fish swims with constant acceleration for 20.0 s, its velocity is $\vec{v} = (20.0\hat{\textbf{i}} - 5.00\hat{\textbf{j}})$ m/s. (a) What are the components of the acceleration of the fish? (b) What is the direction of the acceleration with respect to unit vector $\hat{\textbf{i}}$? (c) If the fish maintains constant acceleration, where is it at $t = 25.0$ s, and in what direction is it moving?

Solution

Conceptualize: The fish is speeding up and changing direction. We choose to write separate equations for the x and y components of its motion.

Categorize: Model the fish as a particle under constant acceleration. We use our old standard equations for constant-acceleration straightline motion, with x and y subscripts to make them apply to parts of the whole motion.

Analyze: At $t = 0$, $\vec{v}_i = (4.00\hat{i} + 1.00\hat{j})$ m/s and $\vec{r}_i = (10.00\hat{i} - 4.00\hat{j})$ m

At the first "final" point we consider, 20 s later, $\vec{v}_f = (20.0\hat{i} - 5.00\hat{j})$ m/s

(a) $a_x = \dfrac{\Delta v_x}{\Delta t} = \dfrac{20.0 \text{ m/s} - 4.00 \text{ m/s}}{20.0 \text{ s}} = 0.800$ m/s^2 ∎

$a_y = \dfrac{\Delta v_y}{\Delta t} = \dfrac{-5.00 \text{ m/s} - 1.00 \text{ m/s}}{20.0 \text{ s}} = -0.300$ m/s^2 ∎

(b) $\theta = \tan^{-1}\left(\dfrac{a_y}{a_x}\right) = \tan^{-1}\left(\dfrac{-0.300 \text{ m/s}^2}{0.800 \text{ m/s}^2}\right) = -20.6°$ or $339°$ from the $+x$ axis ∎

(c) With a new "final" point at $t = 25.0$ s, its coordinates are

$x_f = x_i + v_{xi}t + \frac{1}{2}a_x t^2$ and $y_f = y_i + v_{yi}t + \frac{1}{2}a_y t^2$

Evaluating,

$x_f = 10.0 \text{ m} + (4.00 \text{ m/s})(25.0 \text{ s}) + \frac{1}{2}(0.800 \text{ m/s}^2)(25.0 \text{ s})^2 = 360$ m

$y_f = -4.00 \text{ m} + (1.00 \text{ m/s})(25.0 \text{ s}) + \frac{1}{2}(-0.300 \text{ m/s}^2)(25.0 \text{ s})^2 = -72.8$ m

Its velocity components are

$v_{xf} = v_{xi} + a_y t = (4.00 \text{ m/s}) + (0.800 \text{ m/s}^2)(25.0 \text{ s}) = 24.0$ m/s

and $v_{yf} = v_{yi} + a_y t = (1.00 \text{ m/s}) - (0.300 \text{ m/s}^2)(25.0 \text{ s}) = -6.50$ m/s

Its location can be stated as the position vector $\vec{r}_f = (360\,\hat{i} - 72.8\,\hat{j})$ m ∎

or with the same meaning by giving its coordinates (360 m, −72.8 m).

Its direction of motion is the direction of its vector velocity,

$\theta = \tan^{-1}(v_y/v_x) = \tan^{-1}\left(\dfrac{-6.50 \text{ m/s}}{24.0 \text{ m/s}}\right) = -15.2° = 345°$ from the $+x$ axis. ∎

Finalize: The fish is not a projectile, but any motion with constant acceleration in a different direction from the velocity at one instant is motion along a parabola. The formal numerical solution is complete, but it would be good to have a diagram showing the rock, the x and y axes, the fish's initial position, the location 20 s later, the location at 25 s, the velocities tangent to the parabolic trajectory, and the constant vector acceleration. Sketch it!

9.   In a local bar, a customer slides an empty beer mug down the counter for a refill. The height of the counter is 1.22 m. The mug slides off the counter and strikes the floor 1.40 m from the base of the counter. (a) With what velocity did the mug leave the counter? (b) What was the direction of the mug's velocity just before it hit the floor?

Solution

Conceptualize: Based on our everyday experiences and the description of the problem, a reasonable speed of the mug would be a few m/s and it will hit the floor at some angle between 0° and 90°, probably about 45°. The mug is much more dense than the surrounding air, so its motion is free fall. It is a projectile.

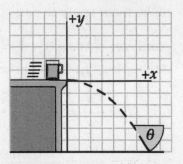

Categorize: We are looking for two different velocities, but we are only given two distances. Our approach will be to separate the vertical and horizontal motions. By using the height that the mug falls, we can find the time interval of the fall. Once we know the time, we can find the horizontal and vertical components of the velocity. For convenience, we will set the origin to be the point where the mug leaves the counter.

Analyze:

Vertical motion: $y_f = -1.22$ m $v_{yi} = 0$ $v_y = ?$ $a_y = -9.80$ m/s²

Horizontal motion: $x_f = 1.40$ m $v_x = ? = $ constant $a_x = 0$

(a)   To find the time interval of fall, we use the equation for motion

with constant acceleration $y_f = y_i + v_{yi}t + \frac{1}{2}a_y t^2$

Substituting, -1.22 m $= 0 + 0 - \frac{1}{2}(9.80$ m/s²$)t^2$

so $t = [2(1.22$ m$)$s²$/9.80$ m$]^{0.5} = 0.499$ s

Then $v_x = \dfrac{x_f}{t} = \dfrac{1.40 \text{ m}}{0.499 \text{ s}} = 2.81$ m/s   ∎

(b)   The mug hits the floor with a vertical velocity of $v_{yf} = v_{yi} + a_y t$ and an impact angle below the horizontal of $\theta = \tan^{-1}(v_{yf}/v_x)$

Evaluating v_{yf}, $v_{yf} = 0 - (9.80$ m/s²$)(0.499$ s$) = -4.89$ m/s

Thus, $\theta = \tan^{-1}\left(\dfrac{-4.89 \text{ m/s}}{2.81 \text{ m/s}}\right) = -60.2° = 60.2°$ below the horizontal.   ∎

Finalize: This was a multi-step problem that required several physics equations to solve; our answers agree with our initial expectations. Since the problem did not ask for the time, we could have eliminated this variable by substitution. We would have substituted the algebraic expression $t = \sqrt{2y/g}$ into two other equations. We chose instead to find a numerical value for the time interval as an intermediate step.

From the given information we could have found the diagonal distance from the edge of the counter to the impact point, and the elevation angle of a sightline between the two points. This would have been useless. The mug does not travel along this straight line. Its different accelerations in the horizontal and vertical directions require us to treat the horizontal and vertical motions separately. The original speed was an unknown, but the original *vertical* velocity counts as a known quantity. Its value must be zero because we assume the countertop is horizontal. With the original velocity horizontal, is the impact velocity vertical? No. Why not? Because the horizontal motion does not slow down. At the impact point the mug is still moving horizontally with its countertop speed.

11. A projectile is fired in such a way that its horizontal range is equal to three times its maximum height. What is the angle of projection?

Solution

Conceptualize: The angle must be between 0 and 90°. We guess it is around halfway between them.

Categorize: We could use the general equations for constant acceleration motion, applied to $a_y = -g$ and $a_x = 0$. The text derives specialized equations that give just the range and maximum height of a projectile fired over level ground, and we choose instead to use these limited-applicability equations.

Analyze: We want to find θ_i such that $R = 3h$. We can use the equations for the range and height of a projectile's trajectory,

$$R = \left(v_i^2 \sin 2\theta_i\right)/g \qquad \text{and} \qquad h = \left(v_i^2 \sin^2 \theta_i\right)/2g$$

We combine different requirements by mathematical substitution into $R = 3h$, thus:

$$\frac{v_i^2 \sin 2\theta_i}{g} = \frac{3v_i^2 \sin^2 \theta_i}{2g} \qquad \text{or} \qquad \frac{2}{3} = \frac{\sin^2 \theta_i}{\sin 2\theta_i}$$

But $\quad \sin 2\theta_i = 2 \sin \theta_i \cos \theta_i \quad$ so $\quad \dfrac{\sin^2 \theta_i}{\sin 2\theta_i} = \dfrac{\sin^2 \theta_i}{2\sin \theta_i \cos \theta_i} = \dfrac{\tan \theta_i}{2}$

Substituting and solving for θ_i gives $\qquad \theta_i = \tan^{-1}\left(\dfrac{4}{3}\right) = 53.1°$ ∎

Finalize: Such trigonometric relations as $\sin 2\theta = 2 \sin \theta \cos \theta$ and $\tan \theta = \sin \theta/\cos \theta$ are tabulated in the back of the book. With the range fully three times larger than the maximum height for a projection angle of 53°, the range must be an even greater multiple of maximum height at the maximum range angle of 45°. Apples plummet and intercontinental missiles take off straight up, but artillery shells have surprisingly flat trajectories.

17. A placekicker must kick a football from a point 36.0 m (about 40 yards) from the goal. Half the crowd hopes the ball will clear the crossbar, which is 3.05 m high. When kicked, the ball leaves the ground with a speed of 20.0 m/s at an angle of 53.0° to the horizontal. (a) By how much does the ball clear or fall short of clearing the crossbar? (b) Does the ball approach the crossbar while still rising or while falling?

Solution

Conceptualize: One might guess that the athlete wanted to kick the ball at a lower angle like 45°, so perhaps it will not go over the crossbar. Surely it will be on the way down from its high flight.

Categorize: Model the football as a projectile, moving with constant horizontal velocity and with constant vertical acceleration. We need a plan to get the necessary information to answer the yes-or-no questions. We will find the height of the ball when its horizontal displacement component is 36 m, to see whether it is more than 3.05 m. After that, we will find the time interval for which the ball rises and compare it with the time interval for it to travel $\Delta x = 36$ m, to see which interval is longer.

Analyze:

(a) To find the height of the football when it crosses above the goal line, we use the equations $y_f = y_i + v_{yi}t + \frac{1}{2}a_y t^2 = 0 + v_i \sin \theta_i t - \frac{1}{2}gt^2$ and $x_f = x_i + v_{xi}t = 0 + v_i \cos \theta_i t$

From the second of these we substitute $t = \dfrac{x_f}{v_i \cos \theta_i}$ into the first, to obtain the

"trajectory equation" $y_f = x_f \tan \theta_i - \dfrac{gx_f^2}{2v_i^2 \cos^2 \theta_i}$

We can use it directly to compute the altitude of the ball

with $\qquad\qquad\qquad\qquad x_f = 36.0 \text{ m}, \quad v_i = 20.0 \text{ m/s}, \quad \text{and} \quad \theta_i = 53.0°$

Thus: $\qquad\qquad\qquad\qquad y_f = (36.0 \text{ m})(\tan 53.0°) - \dfrac{(9.80 \text{ m/s}^2)(36.0 \text{ m})^2}{2(20.0 \text{ m/s})^2 \cos^2 53.0°}$

$$y_f = 47.774 - 43.834 = 3.939 \text{ m}$$

The ball clears the bar by $3.939 \text{ m} - 3.050 \text{ m} = 0.889 \text{ m}$ ∎

(b) The time interval the ball takes to reach the maximum height ($v_y = 0$) is

$t_1 = \dfrac{\left(v_i \sin \theta_i\right) - v_y}{g} = \dfrac{(20.0 \text{ m/s})(\sin 53.0°) - 0}{9.8 \text{ m/s}^2} = 1.63 \text{ s}$

The time interval to travel 36.0 m horizontally is

$$t_2 = \dfrac{x_f}{v_{xi}} = \dfrac{36.0 \text{ m}}{(20.0 \text{ m/s})(\cos 53.0°)} = 2.99 \text{ s}$$

Since $t_2 > t_1$, the ball clears the goal on its way down. ∎

Finalize: That 2.99 s is a long time for the faithful fans, and the 889 mm is a close call. In part (b) we could equally well have evaluated the vertical velocity of the ball at 2.99 s, to see that it is negative.

Notice that in part (b) we found the time interval for the ball to travel 36 m. We could have done this step first of all. Then we could have used this time to find the elevation of the ball relative to the bar from the standard equation $\Delta y = v_{yi} t + (\frac{1}{2})a_y t^2$. You can think of the trajectory equation as more complicated, but it just carries out precisely this pair of steps symbolically, instead of requiring a numerical substitution.

The last digit in our answer 0.889 m is uncertain. You might say it is entirely unknown, because the next digit is not quoted for the 3.05-m height of the crossbar. We choose to use consistently the standard announced in the Preface: When a problem quotes data to three significant digits, we quote answers to three significant digits.

---

27. The athlete shown in Figure P4.27 of the textbook rotates a 1.00-kg discus along a circular path of radius 1.06 m. The maximum speed of the discus is 20.0 m/s. Determine the magnitude of the maximum radial acceleration of the discus.

Solution

Conceptualize: The maximum radial acceleration occurs when maximum tangential speed is attained. Visualize the discus as keeping this speed constant for a while, so that its whole acceleration is its radially-inward acceleration.

Categorize: Model the discus as a particle in uniform circular motion. We evaluate its centripetal acceleration from the standard equation proved in the text.

Analyze: $a_c = \dfrac{v^2}{r} = \dfrac{(20.0 \text{ m/s})^2}{1.06 \text{ m}} = 377 \text{ m/s}^2$ ∎

Finalize: The athlete must keep a firm hold on the discus to give it so large an acceleration. We can call it "change-in-direction acceleration" to mean the same thing as radial acceleration or centripetal acceleration.

---

31. A train slows down as it rounds a sharp horizontal turn, going from 90.0 km/h to 50.0 km/h in the 15.0 s that it takes to round the bend. The radius of the curve is 150 m. Compute the acceleration at the moment the train speed reaches 50.0 km/h. Assume the train continues to slow down at this time at the same rate.

Solution

Conceptualize: If the train is taking this turn at a safe speed, then its acceleration should be significantly less than g, perhaps a few m/s^2. Otherwise it might jump the tracks! It should be directed toward the center of the curve and backwards, since the train is slowing.

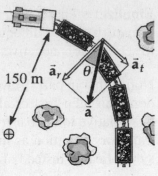

Categorize: Since the train is changing both its speed and direction, the acceleration vector will be the vector sum of the tangential and radial acceleration components. The tangential acceleration can be found from the changing speed and elapsed time, while the radial acceleration can be found from the radius of curvature and the train's speed.

Analyze: First, let's convert the speed units from km/h to m/s:

$$v_i = 90.0 \text{ km/h} = (90.0 \text{ km/h})\, (10^3 \text{ m/km})\, (1 \text{ h}/3\,600 \text{ s}) = 25.0 \text{ m/s}$$

$$v_f = 50.0 \text{ km/h} = (50.0 \text{ km/h})(10^3 \text{ m/km})\, (1 \text{ h}/3\,600 \text{ s}) = 13.9 \text{ m/s}$$

The tangential acceleration and radial acceleration are respectively

$$a_t = \frac{\Delta v}{\Delta t} = \frac{13.9 \text{ m/s} - 25.0 \text{ m/s}}{15.0 \text{ s}} = -0.741 \text{ m/s}^2 \quad \text{(backward)}$$

and

$$a_r = \frac{v^2}{r} = \frac{(13.9 \text{ m/s})^2}{150 \text{ m}} = 1.29 \text{ m/s} \quad \text{(inward)}$$

So the magnitude of the (whole) acceleration is

$$a = \sqrt{a_r^2 + a_t^2} = \sqrt{\left(1.29 \text{ m/s}^2\right)^2 + \left(-0.741 \text{ m/s}^2\right)^2} = 1.48 \text{ m/s}^2 \qquad \blacksquare$$

and its direction is inward and behind the radius, as represented in the diagram, at the angle

$$\theta = \tan^{-1}\left(\frac{|a_t|}{a_r}\right) = \tan^{-1}\left(\frac{0.741 \text{ m/s}^2}{1.29 \text{ m/s}^2}\right) = 29.9° \qquad \blacksquare$$

Finalize: The acceleration is clearly less than g, and it appears that most of the acceleration comes from the radial component, so it makes sense that the acceleration vector should point mostly inward toward the center of the curve and slightly backward due to the negative tangential acceleration.

---

32. Figure P4.32 represents the total acceleration of a particle moving clockwise in a circle of radius 2.50 m at a certain instant of time. For that instant, find (a) the radial acceleration of the particle, (b) the speed of the particle, and (c) its tangential acceleration.

Solution

Conceptualize: Visualize one instant in the history of a rubber stopper moving in a circle and speeding up as it does so.

Categorize: From the given magnitude and direction of the acceleration we can find both the centripetal and the tangential components. From the centripetal acceleration and radius we can find the speed in part (b).

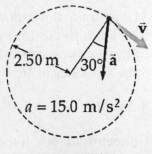

Figure P4.32

Analyze:

(a) The acceleration has an inward radial component:

$$a_c = a \cos 30.0° = (15.0 \text{ m/s}^2) \cos 30.0° = 13.0 \text{ m/s}^2 \text{ inward}$$ ∎

(b) The speed at the instant shown can be found by using

$$a_c = \frac{v^2}{r} \quad \text{or} \quad v = \sqrt{a_c r} = \sqrt{(13.0 \text{ m/s}^2)(2.50 \text{ m})} = 5.70 \text{ m/s}$$ ∎

(c) The acceleration also has a tangential component:

$$a_t = a \sin 30.0° = (15.0 \text{ m/s}^2) \sin 30.0° = 7.50 \text{ m/s}^2$$ ∎

Finalize: It is possible that the magnitude of the tangential acceleration is constant, but its direction must change to stay tangent to the circle. The speed of the particle must be changing, and then the radial acceleration must be increasing in magnitude and also swinging around in direction. The whole problem is about one instant of time.

———————

40. A river has a steady speed of 0.500 m/s. A student swims upstream a distance of 1.00 km and swims back to the starting point. (a) If the student can swim at a speed of 1.20 m/s in still water, how long does the trip take? (b) How much time is required in still water for the same length swim? (c) Intuitively, why does the swim take longer when there is a current?

Solution The idiom "How long does it take?" means "What time interval does it occupy?"

Conceptualize: If we think about the time interval for a trip as a function of the river's speed, we realize that if the stream is flowing at the same rate or faster than the student can swim, he will never reach the 1.00-km mark even after an infinite amount of time. Since the student can swim 1.20 km in 1 000 s, we should expect that the trip will definitely take longer than in still water, maybe about 2 000 s (about 30 minutes).

Categorize: The total time interval in the river is the longer time spent swimming upstream (against the current) plus the shorter time swimming downstream (with the current). For each part, we will use the basic equation $t = d/v$, where v is the speed of the student relative to the shore.

Analyze: (a)

$$t_{up} = \frac{d}{v_{student} - v_{stream}} = \frac{1\,000 \text{ m}}{1.20 \text{ m/s} - 0.500 \text{ m/s}} = 1\,430 \text{ s}$$

$$t_{dn} = \frac{d}{v_{student} + v_{stream}} = \frac{1\,000 \text{ m}}{1.20 \text{ m/s} + 0.500 \text{ m/s}} = 588 \text{ s}$$

Total time interval in the river is then
$$t_{river} = t_{up} + t_{dn} = 2.02 \times 10^3 \text{ s}$$ ∎

(b) In still water, $\quad t_{still} = \dfrac{d}{v} = \dfrac{2\,000 \text{ m}}{1.20 \text{ m/s}} = 1.67 \times 10^3 \text{ s} \quad$ or $\quad t_{still} = 0.827\, t_{river}$ ∎

(c) When there is a current, the student swims equal distances moving slowly and moving rapidly relative to the shore. But what counts for his average speed is the time he spends with each different speed. In the situation considered here, he spends more than twice as much time with the lower speed, so his average speed is lower than it would be without a current. This depressed average speed makes the round-trip time longer when the current is flowing. ∎

Finalize: As we predicted, it does take the student longer to swim up and back in the moving stream than in still water (21% longer in this case), and the amount of time agrees with our estimate.

———————————

43. A science student is riding on a flatcar of a train traveling along a straight horizontal track at a constant speed of 10.0 m/s. The student throws a ball into the air along a path that he judges to make an initial angle of 60.0° with the horizontal and to be in line with the track. The student's professor, who is standing on the ground nearby, observes the ball to rise vertically. How high does she see the ball rise?

Solution

Conceptualize: The student must throw the ball backward relative to the moving train. Then the velocity component toward the caboose that he gives the ball can cancel out with the 10 m/s of forward train velocity.

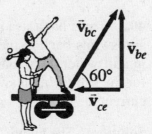

Categorize: We first do vector analysis to find the velocity with which the ball is thrown. Then we consider its vertical motion to find maximum height, on which the student and teacher will agree.

Analyze: As shown in the diagram, $\vec{v}_{be} = \vec{v}_{bc} + \vec{v}_{ce}$

with $\quad \vec{v}_{bc}$ = velocity of the ball relative to the car

$\qquad \vec{v}_{be}$ = velocity of the ball relative to the Earth

and $\quad \vec{v}_{ce}$ = velocity of the car relative to the Earth = 10.0 m/s

From the figure, we have $v_{ce} = v_{bc} \cos 60.0°$

So $\quad v_{bc} = \dfrac{10.0 \text{ m/s}}{\cos 60.0°} = 20.0 \text{ m/s}$

Again from the figure,

$$v_{be} = v_{bc} \sin 60.0° = (20.0 \text{ m/s})(0.866) = 17.3 \text{ m/s}$$

This is the initial velocity of the ball relative to the Earth. Now we can calculate the maximum height that the ball rises. We treat it as a particle under constant acceleration between the moment after release and the apex of its flight. From $v_{yf}^2 = v_{yi}^2 + 2a(y_f - y_i)$

$$y_f = y_i + \dfrac{v_{yf}^2 - v_{yi}^2}{2a} = 0 + \dfrac{0 - (17.3 \text{ m/s})^2}{2(-9.80 \text{ m/s}^2)} = 15.3 \text{ m} \qquad ∎$$

Finalize: Some students start out thinking that there is "a formula" to solve each particular problem. Here the calculation is fairly simple in each step, but you need skill and practice to string the steps together. The vector diagram of adding train-speed-relative-to-ground plus ball-speed-relative-to-train to get ball-speed-relative-to-ground deserves careful attention.

---

53. *Why is the following situation impossible?* Manny Ramírez hits a home run so that the baseball just clears the top row of bleachers, 24.0 m high, located 130 m from home plate. The ball is hit at 41.7 m/s at an angle of 35.0° to the horizontal, and air resistance is negligible.

Solution

Conceptualize: The professional baseball player can definitely hit the ball to give it this high velocity. It is possible, perhaps barely possible, that air resistance is negligible, perhaps because a wind is blowing along with the ball in the direction it is hit, or perhaps because a small osmium ball has been substituted for a regulation baseball. Therefore, there must be something wrong with the numbers quoted about the trajectory.

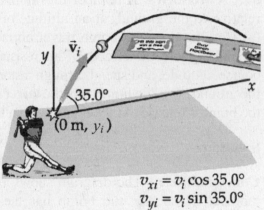

$$v_{xi} = v_i \cos 35.0°$$
$$v_{yi} = v_i \sin 35.0°$$

Categorize: Given the initial velocity, we can calculate the height change of the ball as it moves 130 m horizontally. So this is what we do, expecting the answer to be inconsistent with grazing the top of the bleachers. We assume the ball field is horizontal. We think of the ball as a particle in free fall (moving with constant acceleration) between the point just after it leaves the bat until it crosses above the cheap seats.

Analyze: The horizontal velocity of the ball is (41.7 m/s) cos 35.0° = 34.2 m/s

Its time in flight follows from $x = x_i + v_x t$

$$t = \frac{x - x_i}{v_x} = \frac{130 \text{ m} - 0}{34.2 \text{ m/s}} = 3.81 \text{ s}$$

The ball's final elevation is

$$y = y_i + v_{yi}t + \tfrac{1}{2}a_y t^2$$

$$= y_i + (41.7 \text{ m/s}) \sin 35.0° (3.81 \text{ s}) + \tfrac{1}{2}(-9.80 \text{ m/s}^2)(3.81 \text{ s})^2$$

$$y = y_i + 91.0 \text{ m} - 71.0 \text{ m} = y_i + 20.0 \text{ m}$$

For the final y coordinate to be 24.0 m at $x = 130$ m, the initial height would have to be

$$y_i = y - 20.0 \text{ m} = 24.0 \text{ m} - 20.0 \text{ m} = 3.94 \text{ m}.$$

A tennis player might hit a ball 2.5 m above the ground, but a baseball player could not hit a ball at 3.94 m elevation. ∎

Finalize: A reasonable value for the initial height is 1 m. Could the final height be 24 m if the free-fall acceleration were weaker than we have assumed? If we use $g = 9.78$ m/s^2, appropriate to an equatorial region, we have

$$24.0 \text{ m} = y_i + 91.0 \text{ m} + \tfrac{1}{2}(-9.78 \text{ m/s}^2)(3.81 \text{ s})^2$$

$$= y_i + 20.2 \text{ m}$$

$$y_i = 3.80 \text{ m, still impossibly large.}$$

---

59. A World War II bomber flies horizontally over level terrain with a speed of 275 m/s relative to the ground, at an altitude of 3.00 km. The bombardier releases one bomb. (a) How far does the bomb travel horizontally between its release and its impact on the ground? Ignore the effects of air resistance. (b) The pilot maintains the plane's original course, altitude, and speed through a storm of flak. Where is the plane when the bomb hits the ground? (c) The bomb hits the target seen in the telescopic bombsight at the moment of the bomb's release. At what angle from the vertical was the bombsight set?

Solution

Conceptualize: The diagram shows a lot. At the moment of release, the bomb has the horizontal velocity of the plane. Its motion is part of the downward half of a parabola, just as in the previously solved problem 9, about a mug sliding off the end of a bar.

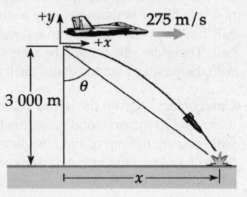

Categorize: We model the bomb as a particle with constant acceleration, equal to the downward free-fall acceleration, from the moment after release until the moment before impact. After we find its range it will be a right-triangle problem to find the bombsight angle.

Analyze:

(a) We take the origin at the point under the plane at bomb release. In its horizontal flight, the bomb has $v_{yi} = 0$ and $v_{xi} = 275$ m/s. We represent the height of the plane as y.

Therefore we have $y_f = y_i - \tfrac{1}{2}gt^2$, becoming $0 = y - \tfrac{1}{2}gt^2$ and $x_f = v_{xi}t$

We eliminate $t = \sqrt{(2y/g)}$ between these two equations to find

$$x_f = v_{xi}\sqrt{\frac{2y}{g}} = (275 \text{ m/s})\sqrt{\frac{(2)(3\,000 \text{ m})}{9.80 \text{ m/s}^2}} = 6\,800 \text{ m} \qquad \blacksquare$$

(b) The pilot may be trained to maintain constant velocity while bombs are being released, and to climb for safety afterward. If he cannot hear the bombardier call "Bombs away!"

he may bravely fly steady on for a long time. The plane and the bomb have the same constant **horizontal** velocity. Therefore, the plane will be 3 000 m above the bomb at impact, and a horizontal distance of 6 800 m from the point of release. ■

(c) Let θ represent the angle between the vertical and the direct line of sight to the target at the moment of release.

$$\theta = \tan^{-1}\left(\frac{x}{y}\right) = \tan^{-1}\left(\frac{6\ 800}{3\ 000}\right) = 66.2°$$ ■

Finalize: Compare this solution to that for the beer mug in problem 9. The different "stories" and very different numbers are consistent with the same physical model of projectile motion. An individual motion equation involves only the x component or only the y component of the motion, but the bombsight angle is computed from both of the perpendicular displacement components together.

---

61. A hawk is flying horizontally at 10.0 m/s in a straight line, 200 m above the ground. A mouse it has been carrying struggles free from its talons. The hawk continues on its path at the same speed for 2.00 s before attempting to retrieve its prey. To accomplish the retrieval, it dives in a straight line at constant speed and recaptures the mouse 3.00 m above the ground. (a) Assuming no air resistance acts on the mouse, find the diving speed of the hawk. (b) What angle did the hawk make with the horizontal during its descent? (c) For what time interval did the mouse experience free fall?

Solution

Conceptualize: We should first recognize the hawk cannot instantaneously change from slow horizontal motion to rapid downward motion. The hawk cannot move with infinite acceleration, but we assume that the time interval required for the hawk to accelerate is short compared to two seconds. Based on our everyday experiences, a reasonable diving speed for the hawk might be about 100 mi/h (-50 m/s) at some angle close to 90° and should last only a few seconds.

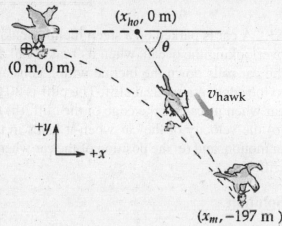

Categorize: We know the distance that the mouse and hawk move down, but to find the diving speed of the hawk, we must know the time interval of descent, so we will solve part (c) first. If the hawk and mouse both maintain their original horizontal velocity of 10 m/s (as the mouse should without air resistance), then the hawk only needs to think about diving straight down, but to a ground-based observer, the path will appear to be a straight line angled less than 90° below horizontal.

Analyze: (c) The mouse falls a total vertical distance $y = 200 \text{ m} - 3.00 \text{ m} = 197 \text{ m}$

The time interval of fall is found from $y = v_{yi}t - \frac{1}{2}gt^2$: $t = \sqrt{\dfrac{2(197 \text{ m})}{9.80 \text{ m/s}^2}} = 6.34 \text{ s}$ ∎

(a)   To find the diving speed of the hawk, we must first calculate the total distance covered from the vertical and horizontal components. We already know the vertical distance y; we just need the horizontal distance during the same time interval (minus the 2.00-s late start).

$x = v_{xi}(t - 2.00 \text{ s}) = (10.0 \text{ m/s})(6.34 \text{ s} - 2.00 \text{ s}) = 43.4 \text{ m}$

The total distance is $d = \sqrt{x^2 + y^2} = \sqrt{(43.4 \text{ m})^2 + (197 \text{ m})^2} = 202 \text{ m}$

So the hawk's diving speed is $v_{\text{hawk}} = \dfrac{d}{t - 2.00 \text{ s}} = \dfrac{202 \text{ m}}{4.34 \text{ s}} = 46.5 \text{ m/s}$ ∎

(b)   at an angle below the horizontal of $\theta = \tan^{-1}\left(\dfrac{y}{x}\right) = \tan^{-1}\left(\dfrac{197 \text{ m}}{43.4 \text{ m}}\right) = 77.6°$ ∎

Finalize: The answers appear to be consistent with our predictions, even though it is not possible for the hawk to reach its diving speed instantaneously. We typically make simplifying assumptions to solve complex physics problems, and sometimes these assumptions are not physically possible. After the idealized problem is understood, we can attempt to analyze the more complex, real-world problem. For this problem, if we considered the realistic effects of air resistance and the maximum diving acceleration attainable by the hawk, we might find that the hawk could not catch the mouse before it hit the ground.

---

63.   A car is parked on a steep incline, making an angle of 37.0° below the horizontal and overlooking the ocean, when its brakes fail and it begins to roll. Starting from rest at $t = 0$, the car rolls down the incline with a constant acceleration of 4.00 m/s², traveling 50.0 m to the edge of a vertical cliff. The cliff is 30.0 m above the ocean. Find (a) the speed of the car when it reaches the edge of the cliff, (b) the time interval elapsed when it arrives there, (c) the velocity of the car when it lands in the ocean, (d) the total time interval the car is in motion, and (e) the position of the car when it lands in the ocean, relative to the base of the cliff.

Solution

Conceptualize: The car has one acceleration while it is on the slope and a different acceleration when it is falling, so ...

Categorize: ... we must take the motion apart into two different sections. Our standard equations only describe a chunk of motion during which acceleration stays constant. We imagine the acceleration to change instantaneously at the brink of the cliff, but the velocity and the position must be the same just before point B and just after point B.

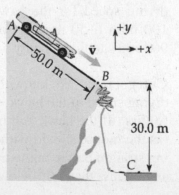

Analyze: From point A to point B (along the incline), the car can be modeled as a particle under constant acceleration in one dimension, starting from rest ($v_i = 0$). Therefore, taking s to be the position along the incline,

(a) $v_B^2 = v_i^2 + 2a(s_f - s_i) = 0 + 2a(s_B - s_A)$

   $v_B = \sqrt{2(4.00 \text{ m/s}^2)(50.0 \text{ m})} = 20.0 \text{ m/s}$ ∎

(b) We can find the elapsed time interval from $v_B = v_i + at$:

   $t_{AB} = \dfrac{v_B - v_i}{a} = \dfrac{20.0 \text{ m/s} - 0}{4.00 \text{ m/s}^2} = 5.00 \text{ s}$ ∎

(c) After the car passes the brink of the cliff, it becomes a projectile. At the edge of the cliff, the components of velocity v_B are:

   $v_{yB} = (-20.0 \text{ m/s}) \sin 37.0° = -12.0 \text{ m/s}$

   $v_{xB} = (20.0 \text{ m/s}) \cos 37.0° = 16.0 \text{ m/s}$

   There is no further horizontal acceleration, so

   $v_{xC} = v_{xB} = 16.0 \text{ m/s}$

   However, the downward (negative) vertical velocity is affected by free fall:

   $v_{yC} = \pm\sqrt{2a_y(\Delta y) + v_{yB}^2} = \pm\sqrt{2(-9.80 \text{ m/s}^2)(-30.0 \text{ m}) + (-12.0 \text{ m/s})^2}$

   $\phantom{v_{yC}} = -27.1 \text{ m/s}$

   We assemble the answer $\bar{\mathbf{v}}_C = (16.0\hat{\mathbf{i}} - 27.1\hat{\mathbf{j}}) \text{ m/s}$ ∎

(d) From point B to C, the time is $t_{BC} = \dfrac{v_{yC} - v_{yB}}{a_y} = \dfrac{(-27.1 \text{ m/s}) - (-12.0 \text{ m/s})}{(-9.80 \text{ m/s}^2)} = 1.53 \text{ s}$

   The total elapsed time interval is $t_{AC} = t_{AB} + t_{BC} = 6.53 \text{ s}$ ∎

(e) The horizontal distance covered is $\Delta x = v_{xB}\, t_{BC} = (16.0 \text{ m/s})(1.53 \text{ s}) = 24.5 \text{ m}$ ∎

Finalize: Part (a) had to be done separately from the others. Adding up the total time interval in motion in part (c) is not driven by an equation from back in the chapter, but just by thinking about the process. We could have calculated the time of fall before finding the final velocity, solving (c) before (b). Then we would have had to solve the quadratic equation $y_f = y_i + v_{yi}t + \frac{1}{2}a_y t^2$ for time.

---

67. A skier leaves the ramp of a ski jump with a velocity of $v = 10.0$ m/s at $\theta = 15.0°$ above the horizontal, as shown in Figure P4.67. The slope where she will land is inclined downward at $\phi = 50.0°$ and air resistance is negligible. Find (a) the distance from the end of the ramp to where the jumper lands and (b) her velocity components just before the landing. (c) Explain how you think the results might be affected if air resistance were included.

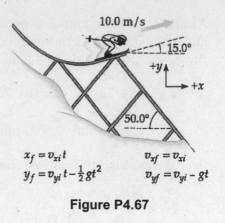

Figure P4.67

Solution

Conceptualize: The last question asks us to go beyond modeling the skier as moving in free fall, but we use that model in parts (a) and (b). The takeoff speed is not terribly fast, but the landing slope is so steep that we expect a distance on the order of a hundred meters, and a high landing speed.

Categorize: We will solve simultaneous equations to find the intersection between the skier's parabola and the hill's straight line.

Analyze: Set point "0" where the skier takes off, and point "2" where the skier lands. Define the coordinate system with x horizontal and y vertically up. The initial velocity is $\vec{v}_i = 10.0$ m/s at $15.0°$

with components $v_{xi} = v_i \cos \theta_i = (10.0 \text{ m/s}) \cos 15.0° = 9.66$ m/s

and $\quad v_{yi} = v_i \sin \theta_i = (10.0 \text{ m/s}) \sin 15.0° = 2.59$ m/s

(a) The skier travels horizontally $x_2 - x_0 = v_{xi} t = (9.66 \text{ m/s})t$ **[1]**

and vertically $y_2 - y_0 = v_{yi} t - \frac{1}{2}gt^2 = (2.59 \text{ m/s})t - (4.90 \text{ m/s}^2)t^2$ **[2]**

The skier hits the slope when $\dfrac{y_2 - y_0}{x_2 - x_0} = \tan(-50.0°) = -1.19$ **[3]**

To solve simultaneously, we substitute equations [1] and [2] into [3]

$$\frac{(2.59 \text{ m/s})t - \left(4.90 \text{ m/s}^2\right)t^2}{(9.66 \text{ m/s})t} = -1.19$$

The solution $t = 0$ refers to the takeoff point. We ignore it to have

$$-4.90t + 14.1 = 0 \quad \text{and} \quad t = 2.88 \text{ s}$$

Now solving equation [1], $\quad x_2 - x_0 = (9.66 \text{ m/s})t = 27.8$ m

From the diagram the downslope distance is

$$d = \frac{x_2 - x_0}{\cos 50.0°} = \frac{27.8 \text{ m}}{\cos 50.0°} = 43.2 \text{ m} \qquad \blacksquare$$

(b) The final horizontal velocity is $v_{xf} = v_{xi} = 9.66$ m/s $\qquad \blacksquare$

The vertical component we find from

$$v_{yf} = v_{yi} - gt = 2.59 \text{ m/s} - \left(9.80 \text{ m/s}^2\right)t$$

When $t = 2.88$ s, $\quad v_{yf} = -25.6$ m/s $\qquad \blacksquare$

(c)   The "drag" force of air resistance would necessarily decrease both components of the ski jumper's impact velocity. On the other hand, a "lift" force of the air could extend her time of flight and increase the distance of her jump. If the jumper has the profile of an airplane wing, she can deflect downward the air through which she passes, to make the air deflect her upward.   ∎

Finalize: The jump distance is indeed large, about half the length of a football field, and the impact speed is also large.

5

The Laws of Motion

EQUATIONS AND CONCEPTS

A **quantitative measurement of mass** (the term used to measure inertia) can be made by comparing the accelerations that a given force will produce on different bodies. If a given force acting on a body of mass m_1 produces an acceleration $\vec{a}_1$ and the same force acting on a body of mass m_2 produces an acceleration $\vec{a}_2$, the ratio of the two masses is defined as the inverse ratio of the magnitudes of the accelerations. *Mass is an inherent property of an object and is independent of the surroundings and the method of measurement.*

$$\frac{m_1}{m_2} \equiv \frac{a_2}{a_1} \qquad (5.1)$$

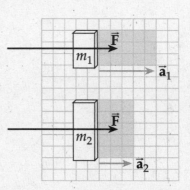

The **particle under a net force model** is used to analyze the motion of an object with acceleration. Newton's second law states that the acceleration is proportional to the net force acting on the object and is inversely proportional to the mass of the object. *The net force (or resultant force) is the vector sum of all the external forces acting on the object.*

$$\sum \vec{F} = m\vec{a} \qquad (5.2)$$

Three **component equations** are the equivalent of the vector equation expressing Newton's second law. *The orientation of the coordinate system can often be chosen so that the object has a nonzero acceleration along only one direction.*

$$\sum F_x = ma_x \qquad \sum F_y = ma_y \qquad \sum F_z = ma_z \qquad (5.3)$$

The **SI unit of force** is the newton (N), defined as the force that, when acting on a 1-kg mass, produces an acceleration of 1 m/s^2. *Calculations with Equation 5.2 must be made using a consistent set of units for the quantities force, mass, and acceleration.*

$$1 \text{ N} \equiv 1 \text{ kg} \cdot \text{m/s}^2 \qquad (5.4)$$

$$1 \text{ lb} \equiv 1 \text{ slug} \cdot \text{ft/s}^2 \qquad (5.5)$$

$$1 \text{ lb} \equiv 4.448 \text{ N}$$

The **magnitude of the gravitational force** exerted on an object is called the weight of the object. *Weight is not an inherent property of a body; it depends on the local value of g and varies with location.*

$$F_g = mg$$ (5.6)

$$\text{weight} = mg$$

Newton's third law states that the action force, $\vec{F}_{12}$, exerted by object 1 on object 2 is equal in magnitude and opposite in direction to the reaction force, $\vec{F}_{21}$, exerted by object 2 on object 1. *Remember, the two forces in an action reaction pair always act on two different objects—they cannot add to give a net force of zero.*

$$\vec{F}_{12} = -\vec{F}_{21}$$ (5.7)

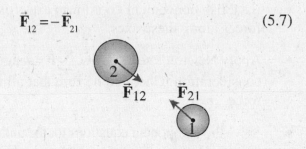

An **object in equilibrium** (modeled as a particle in equilibrium) has zero acceleration; the net force acting on the object is zero.

$$\sum \vec{F} = 0$$ (5.8)

The **force of static friction** between two surfaces in contact, but not in motion relative to each other, cannot be greater than $\mu_s n$, where n is the normal (perpendicular) force between the two surfaces. The coefficient of static friction, μ_s, is a dimensionless constant which depends on the nature of the pair of surfaces. *The equality sign holds when the two surfaces are on the verge of slipping (impending motion).*

$$f_s \leq \mu_s n$$ (5.9)

The **force of kinetic friction** applies when two surfaces are in relative motion. *The friction force is parallel to the surface on which an object is in contact and is directed opposite the direction of actual or impending motion.*

$$f_k = \mu_k n$$ (5.10)

SUGGESTIONS, SKILLS, AND STRATEGIES

The following procedure is recommended when dealing with problems involving the application of Newton's second law, including cases when an object is in static equilibrium:

*   Draw a simple, neat diagram showing all forces acting on the object of interest (a force diagram).

- Isolate the object of interest whose motion is being analyzed. Model the object as a particle and draw a free-body diagram for the object; that is, a diagram showing **all external forces acting on the object**. For systems containing more than one object, draw **separate diagrams for each object**. *Do not include forces that the object exerts on its surroundings.*

- Establish convenient coordinate axes for each object and find the components of the forces along these axes.

- Apply Newton's second law, $\sum \vec{F} = m\vec{a}$, in the x and y directions for each object under consideration. Check to be sure that all terms in the resulting equation have units of force.

- Solve the component equations for the unknowns. Remember that you must have as many independent equations as you have unknowns in order to obtain a complete solution. Often in solving such problems, one must also use the equations of kinematics (motion with constant acceleration) to find all the unknowns.

- Make sure that your final results are consistent with the free-body diagram and check the predictions of your solution in the case of extreme values of the variables.

REVIEW CHECKLIST

- You should be able to: State in your own words Newton's laws of motion, recall physical examples of each law, and identify the action–reaction force pairs in a multiple-body interaction problem as specified by Newton's third law. (Sections 5.2, 5.4, and 5.6)

- Apply Newton's laws of motion to various mechanical systems using the recommended procedure outlined in Suggestions, Skills, and Strategies and discussed in Section 5.7 of your textbook. Identify all external forces acting on the system, model each object as a particle, and draw separate free-body diagrams showing all external forces acting on each object. Write Newton's second law, $\sum \vec{F} = m\vec{a}$, in **component** form. (Section 5.7)

- Express the normal force in terms of other forces acting on an object and the acceleration of the object. Write out the equation which relates the coefficient of friction, force of friction, and normal force between an object and surface on which it rests or moves. (Section 5.8)

- Apply the equations of kinematics (which involve the quantities displacement, velocity, time, and acceleration) as described in Chapter 2 along with those methods and equations of Chapter 5 (involving mass, force, and acceleration) to the solutions of problems using Newton's second law. (Section 5.7)

- Solve several linear equations simultaneously for the unknown quantities. Recall that you must have as many independent equations as you have unknowns.

ANSWER TO AN OBJECTIVE QUESTION

11. If an object is in equilibrium, which of the following statements is *not* true? (a) The speed of the object remains constant. (b) The acceleration of the object is zero. (c) The net force acting on the object is zero. (d) The object must be at rest. (e) There are at least two forces acting on the object.

Answer Each of the statements can be true and three statements are necessarily true, but both statements (d) and (e) are not necessarily true. Statement (b) gives our definition of equilibrium, and (c) follows from Newton's second law. Statement (a) must be true because the speed must be constant when the velocity is constant. Statement (d) need not be true: the object might be at rest or it might be moving at constant velocity. And (e) need not be true: the object might be a meteoroid in intergalactic space, with no force exerted on it.

ANSWERS TO SELECTED CONCEPTUAL QUESTIONS

3. In the motion picture *It Happened One Night* (Columbia Pictures, 1934), Clark Gable is standing inside a stationary bus in front of Claudette Colbert, who is seated. The bus suddenly starts moving forward and Clark falls into Claudette's lap. Why did this happen?

Answer When the bus starts moving, the mass of Claudette is accelerated by the force of the back of the seat on her body. Clark is standing, however, and the only force on him is the friction exerted on his shoes by the floor of the bus. Thus, when the bus starts moving, his feet start accelerating forward, but the rest of his body experiences almost no accelerating force (only that due to his being attached to his accelerating feet!). As a consequence, his body tends to stay almost at rest, according to Newton's first law, relative to the ground. Relative to Claudette, however, he is moving toward her and falls into her lap. (Both performers won Academy Awards.)

□   □   □   □

9. A rubber ball is dropped onto the floor. What force causes the ball to bounce?

Answer When the ball hits the floor, it is compressed. As the ball returns to its original shape, it exerts a force on the floor. As described by Newton's third law, the floor exerts a force on the ball and thrusts it back into the air.

□   □   □   □

13. A weightlifter stands on a bathroom scale. He pumps a barbell up and down. What happens to the reading on the scale as he does so? **What If?** What if he is strong enough to actually *throw* the barbell upward? How does the reading on the scale vary now?

Answer If the barbell is not moving, the reading on the bathroom scale is the combined weight of the weightlifter and the barbell. At the beginning of the lift of the barbell, the barbell accelerates upward. By Newton's third law, the barbell pushes downward on the hands of the weightlifter with more force than its weight, in order to accelerate. As a result, he is pushed with more force into the scale, increasing its reading. Near the top of the lift, the weightlifter reduces the upward force, so that the acceleration of the barbell is downward, causing it to come to rest. While the barbell is coming to rest, it pushes with less force on the weightlifter's hands, so the reading on the scale is below the combined stationary weight. If the barbell is held at rest for an interval at the top of the lift, the scale reading is simply the combined weight. As it begins to be brought down, the reading decreases, as the force of the weightlifter on the barbell is reduced. The reading increases as the barbell is slowed down at the bottom.

If we now consider the throwing of the barbell, the variations in scale reading will be larger, since more force must be applied to throw the barbell upward rather than just lift it. Once the barbell leaves the weightlifter's hands, the reading will suddenly drop to just the weight of the weightlifter, and will rise suddenly when the barbell is caught.

□   □   □   □

17. Identify action–reaction pairs in the following situations: (a) a man takes a step (b) a snowball hits a girl in the back (c) a baseball player catches a ball (d) a gust of wind strikes a window.

Answer   There is no physical distinction between an action and a reaction. It is clearer to describe the pair of forces as together constituting the interaction between the two objects. (a) As a man takes a step, his foot exerts a backward friction force on the Earth; the Earth exerts a forward friction force on his foot. (b) In the second case, the snowball exerts force on the girl's back in the direction of its original motion. The girl's back exerts force on the snowball to stop its motion. (c) In the third case, the glove exerts force on the ball; the ball exerts force in the opposite direction on the glove. (d) In the fourth situation the air molecules exert force on the window; and the window exerts a force with an equal magnitude on the air molecules.

□   □   □   □

SOLUTIONS TO SELECTED END-OF-CHAPTER PROBLEMS

1.   A 3.00-kg object undergoes an acceleration given by $\vec{a} = (2.00\hat{i} + 5.00\hat{j})$ m/s². Find (a) the resultant force acting on the object and (b) the magnitude of the resultant force.

Solution

Conceptualize: The force will have a magnitude of several newtons and its direction will be in the first quadrant.

Categorize: We use Newton's second law to find the force as a vector and then the Pythagorean theorem to find its magnitude.

Analyze: (a) The total vector force is

$$\sum \vec{F} = m\vec{a} = (3.00 \text{ kg})(2.00\hat{i} + 5.00\hat{j}) \text{ m/s}^2 = (6.00\hat{i} + 15.0\hat{j})\text{N} \qquad \blacksquare$$

(b) Its magnitude is $|\vec{F}| = \sqrt{(F_x)^2 + (F_y)^2} = \sqrt{(6.00 \text{ N})^2 + (15.0 \text{ N})^2} = 16.2 \text{ N}$ $\qquad \blacksquare$

Finalize: Use of unit-vector notation makes it absolutely straightforward to find the total force from the acceleration.

---

3. A toy rocket engine is securely fastened to a large puck that can glide with negligible friction over a horizontal surface, taken as the *xy* plane. The 4.00-kg puck has a velocity of $3.00\hat{i}$ m/s at one instant. Eight seconds later, its velocity is $(8.00\hat{i} + 10.0\hat{j})$ m/s. Assuming the rocket engine exerts a constant horizontal force, find (a) the components of the force and (b) its magnitude.

Solution

Conceptualize: The apparatus described is good for demonstrating how a rocket works. The puck is gaining velocity directed into the first quadrant, so that both components of the net force on it are positive and several newtons in size, as is the magnitude of the force.

Categorize: We solve a motion problem to find the acceleration of the puck. The only horizontal force is the force exerted by the rocket motor, which we find from Newton's second law. Then the Pythagorean theorem gives the magnitude of the force.

Analyze: We use the particle under constant acceleration and particle under net force models. We first calculate the acceleration of the puck:

$$\vec{a} = \frac{\Delta \vec{v}}{\Delta t} = \frac{(8.00\hat{i} + 10.0\hat{j})\text{m/s} - 3.00\hat{i} \text{ m/s}}{8.00 \text{ s}} = 0.625\hat{i} \text{ m/s}^2 + 1.25\hat{j} \text{ m/s}^2$$

In $\sum \vec{F} = m\vec{a}$, the only horizontal force is the thrust $\vec{F}$ of the rocket:

(a) $\vec{F} = (4.00 \text{ kg})(0.625\hat{i} \text{ m/s}^2 + 1.25\hat{j} \text{ m/s}^2) = 2.50\hat{i} \text{ N} + 5.00\hat{j} \text{ N}$ $\qquad \blacksquare$

(b) Its magnitude is $|\vec{F}| = \sqrt{(2.50 \text{ N})^2 + (5.00 \text{ N})^2} = 5.59 \text{ N}$ $\qquad \blacksquare$

Finalize: Compare this problem with problem 1. Acceleration is the one quantity connecting a kinematic description of motion and Newton's account of what forces do. Use of unit-vector notation makes it absolutely straightforward to find the total force from the acceleration.

---

7. An electron of mass 9.11×10^{-31} kg has an initial speed of 3.00×10^5 m/s. It travels in a straight line, and its speed increases to 7.00×10^5 m/s in a distance of 5.00 cm. Assuming its acceleration is constant, (a) determine the magnitude of the force exerted on the electron and (b) compare this force with the weight of the electron, which we ignored.

Solution

Conceptualize: Visualize the electron as part of a beam in a vacuum tube, speeding up in response to an electric force. Only a very small force is required to accelerate an electron because of its small mass, but this force is much greater than the weight of the electron if the gravitational force can be neglected.

Categorize: Since this is simply a linear acceleration problem, we can use Newton's second law to find the force as long as the electron does not approach relativistic speeds (as long as its speed is much less than 3×10^8 m/s), which is certainly the case for this problem. We know the initial and final velocities, and the distance involved, so from these we can find the acceleration needed to determine the force.

Analyze: (a) From $v_f^2 = v_i^2 + 2ax$ and $\Sigma F = ma$

we can solve for the acceleration and then the force: $a = \dfrac{v_f^2 - v_i^2}{2x}$

Substituting to eliminate a, $\sum F = \dfrac{m\left(v_f^2 - v_i^2\right)}{2x}$

Substituting the given information,

$$\sum F = \frac{\left(9.11 \times 10^{-31}\ \text{kg}\right)\left[\left(7.00 \times 10^5\ \text{m/s}\right)^2 - \left(3.00 \times 10^5\ \text{m/s}\right)^2\right]}{2(0.050\,0\ \text{m})}$$

$$\sum F = 3.64 \times 10^{-18}\ \text{N}$$ ■

(b)   The Earth exerts on the electron the force called weight,

$$F_g = mg = \left(9.11 \times 10^{-31}\ \text{kg}\right)\left(9.80\ \text{m/s}^2\right) = 8.93 \times 10^{-30}\ \text{N}$$ ■

The ratio of the electrical force to the weight is $F/F_g = 4.08 \times 10^{11}$ ■

Finalize: The force that causes the electron to accelerate is indeed a small fraction of a newton, but it is much greater than the gravitational force. In general, it is quite reasonable to ignore the weight of the electron in problems about electric forces.

---

13. Two forces $\vec{F}_1$ and $\vec{F}_2$ act on a 5.00-kg object. Taking $F_1 = 20.0$ N and $F_2 = 15.0$ N, find the accelerations of the object for the configurations of forces shown in parts (a) and (b) of Figure P5.13.

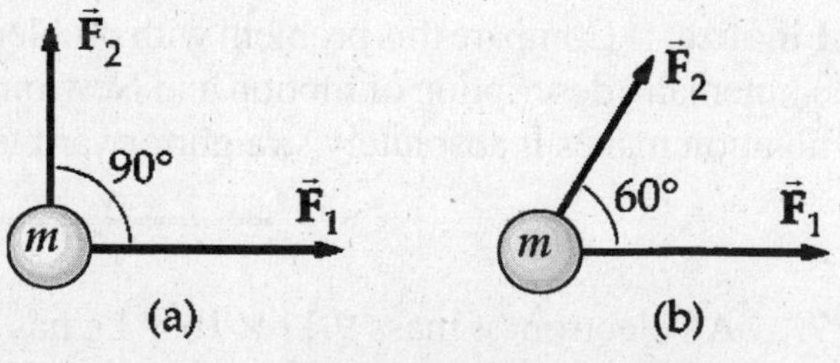

Solution

Conceptualize: We are reviewing that forces are vectors. The acceleration will be

Figure P5.13

somewhat greater in magnitude in situation (b). In both cases the acceleration will be a few meters per second in every second, directed into the first quadrant.

Categorize: We must add the forces as vectors. Then Newton's second law tells us the acceleration.

Analyze: We use the particle under a net force model. $m = 5.00$ kg

(a) $\sum \vec{F} = \vec{F}_1 + \vec{F}_2 = (20.0\hat{i} + 15.0\hat{j})$ N

Newton's second law gives

$$\vec{a} = \frac{\sum \vec{F}}{m} = (4.00\hat{i} + 3.00\hat{j}) \ \text{m/s}^2 = 5.00 \ \text{m/s}^2 \text{at } 36.9°$$ ∎

(b) $\sum \vec{F} = \vec{F}_1 + \vec{F}_2 = \left[20.0\hat{i} + (15.0\cos 60°\hat{i} + 15.0\sin 60°\hat{j}) \right] = (27.5\hat{i} + 13.0\hat{j})\text{N}$

$$\vec{a} = \frac{\sum \vec{F}}{m} = (5.50\hat{i} + 2.60\hat{j})\text{m/s}^2 = 6.08 \ \text{m/s}^2 \text{ at } 25.3°$$ ∎

Finalize: The problem did not explicitly ask for the magnitude and direction of the acceleration, so we could have stopped with the answer stated in unit-vector form. We see that the acceleration is indeed larger in magnitude in part (b) than in part (a) and in a direction closer to the x axis.

---

25. A bag of cement whose weight is F_g hangs in equilibrium from three wires as shown in Figure P5.25. Two of the wires make angles θ_1 and θ_2 with the horizontal. Assuming the system is in equilibrium, show that the tension in the left-hand wire is

$$T_1 = \frac{F_g \cos\theta_2}{\sin(\theta_1 + \theta_2)}$$

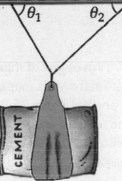

Solution

Conceptualize: We follow the same steps as we would in the numerical problem 24 in the text. F_g, θ_1, and θ_2 count as known quantities. For the bag of cement to be in equilibrium, the tension T_3 in the vertical wire must be equal to F_g.

Figure P5.25

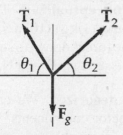

Categorize: From Newton's second law for an object with zero acceleration we can write down x and y component equations. The magnitudes of T_1 and T_2 are unknown, but we can take components of these two tensions just as if they were known forces. Then we can eliminate T_2 by substitution and solve for T_1.

Analyze: We use the particle in equilibrium model. Draw a free-body diagram for the knot where the three ropes are joined. Choose the x axis to be horizontal and apply Newton's second law in component form.

$$\sum F_x = 0: \qquad T_2 \cos \theta_2 - T_1 \cos \theta_1 = 0 \qquad\qquad\qquad \text{[1]}$$

$$\sum F_y = 0: \qquad T_2 \sin \theta_2 + T_1 \sin \theta_1 - F_g = 0 \qquad\qquad \text{[2]}$$

Solve equation [1] for $\quad T_2 = \dfrac{T_1 \cos \theta_1}{\cos \theta_2}$

Substitute this expression for T_2 into equation [2]:

$$\left(\frac{T_1 \cos \theta_1}{\cos \theta_2} \right) \sin \theta_2 + T_1 \sin \theta_1 = F_g$$

Solve for $\quad T_1 = \dfrac{F_g \cos \theta_2}{\cos \theta_1 \, \sin \theta_2 + \sin \theta_1 \, \cos \theta_2}$

Use the trigonometric identity found in Appendix B.4,

$$\sin(\theta_1 + \theta_2) = \cos \theta_1 \, \sin \theta_2 + \sin \theta_1 \, \cos \theta_2$$

to find $\quad T_1 = \dfrac{F_g \cos \theta_2}{\sin(\theta_1 + \theta_2)}$ ∎

Finalize: The equation indicates that the tension is directly proportional to the weight of the bag. As $\sin(\theta_1 + \theta_2)$ approaches zero (as the angle between the two upper ropes approaches 180°) the tension goes to infinity. Making the right-hand rope horizontal maximizes the tension in the left-hand rope, according to the proportionality of T_1 to $\cos \theta_2$. If the right-hand rope is vertical, the tension in the left-hand rope is zero. All this information is contained in the symbolic answer, and not in the numerical answer to problem 24.

---

27. An object of mass $m = 1.00$ kg is observed to have an acceleration $\vec{\mathbf{a}}$ with a magnitude of 10.0 m/s² in a direction 60.0° east of north. Figure P5.27 shows a view of the object from above. The force $\vec{\mathbf{F}}_2$ acting on the object has a magnitude of 5.00 N and is directed north. Determine the magnitude and direction of the one other horizontal force $\vec{\mathbf{F}}_1$ acting on the object.

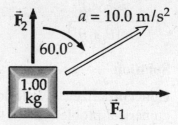

Figure P5.27

Solution

Conceptualize: The net force acting on the mass is $\Sigma F = ma = (1\ \text{kg})(10\ \text{m/s}^2) = 10$ N. If we sketch a vector diagram of the forces drawn to scale, subtracting according to $\vec{\mathbf{F}}_1 = \Sigma \vec{\mathbf{F}} - \vec{\mathbf{F}}_2$, we see that $F_1 \approx 9$ N, to the east.

Categorize: We can find a more precise result by examining the forces in terms of vector components. For convenience, we choose directions east and north along $\hat{\mathbf{i}}$ and $\hat{\mathbf{j}}$, respectively.

Analyze: The acceleration is

$$\vec{a} = \left[(10.0 \cos 30.0°)\hat{i} + (10.0 \sin 30.0°)\hat{j}\right] \text{m/s}^2 = (8.66\hat{i} + 5.00\hat{j}) \text{ m/s}^2$$

From Newton's second law,

$$\Sigma\vec{F} = m\vec{a} = (1.00 \text{ kg})(8.66\hat{i} \text{ m/s}^2 + 5.00\hat{j} \text{ m/s}^2) = (8.66\hat{i} + 5.00\hat{j})\text{N} \quad \text{and} \quad \Sigma\vec{F} = \vec{F}_1 + \vec{F}_2$$

so the force we want is

$$\vec{F}_1 = \Sigma\vec{F} - \vec{F}_2 = (8.66\hat{i} + 5.00\hat{j} - 5.00\hat{j})\text{ N} = 8.66\hat{i} \text{ N} = 8.66 \text{ N east} \qquad \blacksquare$$

Finalize: Our calculated answer agrees with the prediction from the force diagram. It is merely a coincidence that force $\vec{F}_1$ has zero for its northward component.

---

33. In the system shown in Figure P5.33, a horizontal force $\vec{F}_x$ acts on an object of mass $m_2 = 8.00$ kg. The horizontal surface is frictionless. Consider the acceleration of the sliding object as a function of F_x. (a) For what values of F_x does the object of mass $m_1 = 2.00$ kg accelerate upward? (b) For what values of F_x is the tension in the cord zero? (c) Plot the acceleration of the m_2 object versus F_x. Include values of F_x from −100 N to +100 N.

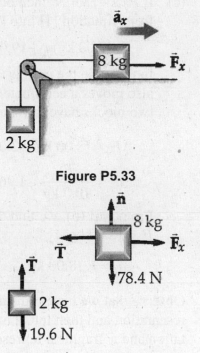

Figure P5.33

Solution

Conceptualize: The applied horizontal force can be either positive or negative—either right or left. As the force increases through sufficiently large positive values, the rightward acceleration will increase linearly. When F_x is negative, the acceleration will be negative, and maybe the string will go slack. There can be especially interesting behavior for small positive values of the force, when it is counterbalancing some but not all of the weight of the two kilograms.

Categorize: We use the particle under a net force model for each of the two objects separately. Then we combine the equations.

Analyze: The blocks' weights are:

$$F_{g1} = m_1 g = (8.00 \text{ kg})(9.80 \text{ m/s}^2) = 78.4 \text{ N}$$

and $\quad F_{g2} = m_2 g = (2.00 \text{ kg})(9.80 \text{ m/s}^2) = 19.6 \text{ N}$

Let T be the tension in the connecting cord and draw a force diagram for each block.

(a) For the 2-kg mass, with the y axis directed upwards,

$$\Sigma F_y = ma_y \quad \text{yields} \quad T - 19.6 \text{ N} = (2.00 \text{ kg})a_y \qquad [1]$$

Thus, we find that $a_y > 0$ when $T > 19.6$ N

For acceleration of the system of two blocks $F_x > T$, so $F_x > 19.6$ N whenever the 2-kg mass accelerates upward.   ■

(b) Looking at the force diagram for the 8.00-kg mass, and taking the $+x$ direction to be directed to the right, we can apply Newton's law in the horizontal direction:

From $\Sigma F_x = ma_x$ we have $-T + F_x = (8.00 \text{ kg})a_x$ [2]

If $T = 0$, the cord goes slack; the 2-kg object is in free fall. The 8-kg object can have an acceleration to the left of larger magnitude:

$a_x \leq -9.80$ m/s^2 with $F_x \leq -78.4$ N   ■

(c) If $F_x \geq -78.4$ N, then both equations [1] and [2] apply. Substituting the value for T from equation [1] into [2],

$-(2.00 \text{ kg})a_y - 19.6 \text{ N} + F_x = (8.00 \text{ kg})a_x$

In this case the cord is taut. Whenever one block moves a centimeter the other block also moves a centimeter. The two blocks have the same speed at every instant. The two blocks have the same acceleration: $a_x = a_y$

$F_x = (8.00 \text{ kg} + 2.00 \text{ kg})a_x + 19.6$ N

$a_x = \dfrac{F_x}{10.0 \text{ kg}} - 1.96 \text{ m/s}^2 \qquad (F_x \geq -78.4 \text{ N})$ [3]

From part (b), we find that if $F_x \leq -78.4$ N, then $T = 0$ and equation [2] becomes $F_x = (8.00 \text{ kg})a_x$:

$a_x = F_x / 8.00 \text{ kg} \qquad (F_x \leq -78.4 \text{ N})$ [4]

Observe that we have translated the pictorial representation into a simplified pictorial representation and then into a mathematical representation. We proceed to a tabular representation and a graphical representation of equations [3] and [4]: Here is the graph:

F_x	a_x
−100 N	−12.5 m/s^2
−50.0 N	−6.96 m/s^2
0	−1.96 m/s^2
50.0 N	3.04 m/s^2
100 N	8.04 m/s^2
150 N	13.04 m/s^2

Note that slope changes at $F_x = -78.4$ N

Finalize: When F_x has a value like −80 N, the cord is slack, the 2-kg block is in free fall, and the block on the table has a larger-magnitude acceleration to the left. When F_x has a value like −50 N or 0 or +10 N, the blocks move together in the negative direction with some smaller-magnitude acceleration. When F_x is + 19.6 N, it just counterbalances the hanging weight. The blocks can move together with zero acceleration. When F_x has a larger positive value, the blocks accelerate together in the positive direction. Their acceleration is a linear function of the force as predicted.

---

35. In Example 5.8, we investigated the apparent weight of a fish in an elevator. Now consider a 72.0-kg man standing on a spring scale in an elevator. Starting from rest, the elevator ascends, attaining its maximum speed of 1.20 m/s in 0.800 s. It travels with this constant speed for the next 5.00 s. The elevator then undergoes a uniform acceleration in the negative y direction for 1.50 s and comes to rest. What does the spring scale register (a) before the elevator starts to move, (b) during the first 0.800 s, (c) while the elevator is traveling at constant speed, and (d) during the time interval it is slowing down?

Solution

Conceptualize: Based on sensations experienced riding in an elevator, we expect that the man should feel slightly heavier when the elevator first starts to ascend, lighter when it comes to a stop, and his normal weight when the elevator is not accelerating. His apparent weight is registered by the spring scale beneath his feet, so the scale force measures the force he feels through his legs, as described by Newton's third law.

Categorize: We draw a force diagram and apply Newton's second law for each part of the elevator trip to find the scale force. The acceleration can be found from the change in speed divided by the elapsed time.

Analyze: Consider the force diagram of the man shown as two arrows. The force F is the upward force exerted on the man by the scale, and his weight is

$$F_g = mg = (72.0 \text{ kg})(9.80 \text{ m/s}^2) = 706 \text{ N}$$

With $+y$ defined to be upwards, Newton's 2nd law gives

$$\Sigma F_y = +F_s - F_g = ma$$

Thus, we calculate the upward scale force to be

$$F_s = 706 \text{ N} + (72.0 \text{ kg})a \qquad\qquad [1]$$

where a is the acceleration the man experiences as the elevator changes speed.

(a) Before the elevator starts moving, the elevator's acceleration is zero ($a = 0$). Therefore, equation [1] gives the force exerted by the scale on the man as 706 N upward, and the man exerts a downward force of 706 N on the scale. ∎

(b) During the first 0.800 s of motion, the man accelerates at a rate of

$$a_x = \frac{\Delta v}{\Delta t} = \frac{1.20 \text{ m/s} - 0}{0.800 \text{ s}} = 1.50 \text{ m/s}^2$$

Substituting a into equation [1] then gives

$$F = 706 \text{ N} + (72.0 \text{ kg})(1.50 \text{ m/s}^2) = 814 \text{ N}$$ ∎

(c) While the elevator is traveling upward at constant speed, the acceleration is zero and equation [1] again gives a scale force $F = 706$ N.   ∎

(d) During the last 1.50 s, the elevator first has an upward velocity of 1.20 m/s, and then comes to rest with an acceleration of

$$a = \frac{\Delta v}{\Delta t} = \frac{0 - 1.20 \text{ m/s}}{1.50 \text{ s}} = -0.800 \text{ m/s}^2$$

Thus, the force of the man on the scale is

$$F = 706 \text{ N} + (72.0 \text{ kg})(-0.800 \text{ m/s}^2) = 648 \text{ N}$$ ∎

Finalize: The calculated scale forces are consistent with our predictions. In part (b), the force of the scale is no larger than the downward gravitational force, so why does the man keep moving upward? Johannes Kepler invented the term inertia, taken from a Latin word for laziness, to sound like an explanation for this motion in the absence of a net force. When the total force acting on an object is zero, the object clings to the status quo in its motion. Such an object moves with no acceleration, just like an object feeling no forces at all and described by Newton's first law. This problem could be extended to a couple of extreme cases. If the acceleration of the elevator were +9.80 m/s^2, then the man would feel twice as heavy, and if $a = -9.80$ m/s^2 (free fall), then he would feel "weightless," even though his true weight ($F_g = mg$) would remain the same.

---

43. A 3.00-kg block starts from rest at the top of a 30.0° incline and slides a distance of 2.00 m down the incline in 1.50 s. Find (a) the magnitude of the acceleration of the block, (b) the coefficient of kinetic friction between block and plane, (c) the friction force acting on the block, and (d) the speed of the block after it has slid 2.00 m.

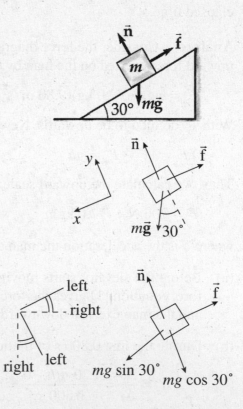

Solution

Conceptualize: The acceleration will be a couple of m/s^2 down the incline, the friction force several newtons, the coefficient of friction a fraction around 0.5, and the final speed a couple of m/s.

Categorize: Every successful physics student (this means you) learns to solve inclined-plane problems.

Hint one: Set one axis in the direction of motion. In this case, take the x axis downward along the incline, so that $a_y = 0$ becomes a known quantity.

Hint two: Recognize that the 30.0° angle between the *x* axis and horizontal implies a 30.0° angle between the weight vector and the *y* axis. This relationship is described by the theorem written down in Appendix B.3, "Angles are equal if their sides are perpendicular, right side to right side and left side to left side." Either you learned this theorem in geometry class, or you learn it now, since it is a theorem used often in physics.

Hint three: The 30.0° angle lies between $m\vec{g}$ and the *y* axis, so split the weight vector into its *x* and *y* components:

$$mg_x = +mg \sin 30.0° \qquad mg_y = -mg \cos 30.0°$$

Note with care that the sine gives the *x* component and the cosine gives the *y* component.

Analyze: We use the particle under constant acceleration and particle under net force models.

(a) At constant acceleration, $x_f = v_i t + \frac{1}{2}at^2$

Solving, $a = \dfrac{2(x_f - v_i t)}{t^2} = \dfrac{2(2.00 \text{ m} - 0)}{(1.50 \text{ s})^2} = 1.78 \text{ m/s}^2$ ∎

From the acceleration, we can calculate the friction force, answer (c), next.

(c) Take the positive *x* axis down parallel to the incline, in the direction of the acceleration. We apply the second law.

$$\Sigma F_x = mg \sin \theta - f = ma$$

Solving, $f = m(g \sin \theta - a)$

Substituting, $f = (3.00 \text{ kg})[(9.80 \text{ m/s}^2)\sin 30.0° - 1.78 \text{ m/s}^2] = 9.37 \text{ N}$ ∎

(b) Applying Newton's law in the *y* direction (perpendicular to the incline), we have no burrowing-in or taking-off motion. Then the *y* component of acceleration is zero:

$$\Sigma F_y = n - mg \cos \theta = 0$$

Thus $n = mg \cos \theta$

Because $f = \mu n$

we have $\mu = \dfrac{f}{mg \cos\theta} = \dfrac{9.37\text{N}}{(3.00 \text{ kg})(9.80 \text{ m/s}^2)\cos 30.0°} = 0.368$ ∎

(d) $v_f = v_i + at$, so $v_f = 0 + (1.78 \text{ m/s}^2)(1.50 \text{ s}) = 2.67 \text{ m/s}$ ∎

Finalize: You can think of this problem as having many steps, but then get used to recognizing the sense that each step makes. Note that 2 m/1.5 s does not give a useful answer. The average speed is not the final speed that part (d) asks for. Only two objects exert forces on the block, namely the incline and the Earth. But the incline exerts both normal and friction forces, and we resolve the gravitational force into its down-incline and into-incline components.

47. Two blocks connected by a rope of negligible mass are being dragged by a horizontal force (Fig. P5.47). Suppose $F = 68.0\,\text{N}$, $m_1 = 12.0\,\text{kg}$, $m_2 = 18.0\,\text{kg}$, and the coefficient of kinetic friction between each block and the surface is 0.100. (a) Draw a free-body diagram for each block. Determine (b) the acceleration of the system, and (c) the tension T in the rope.

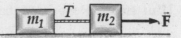

Figure P5.47

Solution

Conceptualize: If the coefficient of friction were zero, the acceleration would be $68\,\text{N}/30\,\text{kg} \approx 2\,\text{m/s}^2$. Thus we expect acceleration about $1\,\text{m/s}^2$. The tension will be less than one-half of 68 N.

Categorize: Because the cord has constant length, both blocks move the same number of centimeters in each second and so move with the same acceleration. To find just this acceleration, we could model the 30-kg system as a particle under net force. That method would not help to finding the tension, so we treat the two blocks as separate accelerating particles.

Analyze:

(a) The free-body diagrams for m_1 and m_2 are: The tension force exerted by block 1 on block 2 is the same size as the tension force exerted by object 2 on object 1. The tension in a light string is a constant along its length, and tells how strongly the string pulls on objects at both ends.

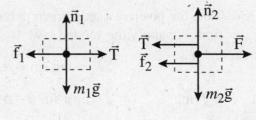

■

(b) We use the free-body diagrams to apply Newton's second law.

For m_1: $\Sigma F_x = T - f_1 = m_1 a$ or $T = m_1 a + f_1$ [1]

And also $\Sigma F_y = n_1 - m_1 g = 0$ or $n_1 = m_1 g$

Also, the definition of the coefficient of friction gives

$$f_1 = \mu n_1 = (0.100)(12.0\,\text{kg})(9.80\,\text{m/s}^2) = 11.8\,\text{N}$$

For m_2: $\Sigma F_x = F - T - f_2 = ma$ [2]

Also from the y component $n_2 - m_2 g = 0$ or $n_2 = m_2 g$

And again $f_2 = \mu n_2 = (0.100)(18.0\,\text{kg})(9.80\,\text{m/s}^2) = 17.6\,\text{N}$

Substituting T from equation [1] into [2], we get

$$F - m_1 a - f_1 - f_2 = m_2 a \quad \text{or} \quad F - f_1 - f_2 = m_2 a + m_1 a$$

Solving for a, $a = \dfrac{F - f_1 - f_2}{m_1 + m_2} = \dfrac{(68.0 - 11.8 - 17.6)\,\text{N}}{(12.0 + 18.0)\,\text{kg}} = 1.29\,\text{m/s}^2$ ■

(c)　From equation [1],　　$T = m_1 a + f_1 = (12.0 \text{ kg})(1.29 \text{ m/s}^2) + 11.8 \text{ N} = 27.2 \text{ N}$　　■

Finalize:　Our answers agree with our expectations. The solution contains a built-in check. When we combined equations [1] and [2] we got $+F - f_1 - f_2$ as the net forward force on the two objects together, and across the equality sign, $+m_2 a + m_1 a$ as the measure of its effect on the motion of the system. If any of these positive or negative signs were different, it would reveal a mistake.

57.　An inventive child named Nick wants to reach an apple in a tree without climbing the tree. Sitting in a chair connected to a rope that passes over a frictionless pulley (Fig. P5.57), Nick pulls on the loose end of the rope with such a force that the spring scale reads 250 N. Nick's true weight is 320 N, and the chair weighs 160 N. Nick's feet are not touching the ground. (a) Draw one pair of diagrams showing the forces for Nick and the chair considered as separate systems and another diagram for Nick and the chair considered as one system. (b) Show that the acceleration of the system is *upward* and find its magnitude. (c) Find the force Nick exerts on the chair.

Solution

Conceptualize:　It is not possible to lift yourself by tugging on your boot laces, but surprisingly easy to lift yourself with a bosun's chair having a rope going over a pulley. We expect an upward acceleration of a few m/s² and a force of child on chair of less than a hundred newtons.

Categorize:　Child and chair will be two particles under net forces.

Analyze:　(a)

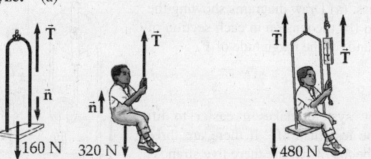

Figure P5.57

Note that the same-size force *n* acts up on Nick and down on chair, and cancels out in the system picture. The same-size force $T = 250$ N acts up on Nick and up on chair, and appears twice in the system diagram.

(b)　First consider Nick and the chair together as the system. Note that **two** ropes support the system, and $T = 250$ N in each rope.

Applying $\Sigma F = ma$,　　$2T - (160 \text{ N} + 320 \text{ N}) = ma$

where $\qquad m = \dfrac{480\ N}{9.80\ m/s^2} = 49.0\ kg$

Solving for a gives $\qquad a = \dfrac{(500 - 480)\ N}{49.0\ kg} = 0.408\ m/s^2$ ∎

(c) On Nick, we apply $\Sigma F = ma$: $\qquad n + T - 320\ N = ma$

where $\qquad m = \dfrac{320\ N}{9.80\ m/s^2} = 32.7\ kg$

The normal force is the one remaining unknown:

$$n = ma + 320\ N - T$$

Substituting, $\qquad n = (32.7\ kg)(0.408\ m/s^2) + 320\ N - 250\ N$

and $\qquad n = 83.3\ N$ ∎

Finalize: The normal force is much smaller than Nick's weight. His acceleration is quite low. If we had written $\Sigma F = ma$ equations separately for the two objects from their separate force diagrams, and then added the equations, we would have obtained

$$T - n - 160\ N + T + n - 320\ N = m_1 a + m_2 a$$

which becomes the equation we used, $2T - (160\ N + 320\ N) = m_{total}\,a$

---

61. An object of mass M is held in place by an applied force $\vec{F}$ and a pulley system as shown in Figure P5.61. The pulleys are massless and frictionless. (a) Draw diagrams showing the forces on each pulley. Find (b) the tension in each section of rope, T_1, T_2, T_3, T_4, and T_5 and (c) the magnitude of $\vec{F}$.

Solution

Conceptualize: A pulley system makes it easier to lift a load. We expect T_1 to be less than Mg. If there are three strands of rope, should T_1 be $Mg/3$? Or are there five strands? This will be made clear by . . .

Categorize: . . . the force diagrams of object M, the movable pulley, and the stationary pulley, that we draw using the particle-under-net-force model for each.

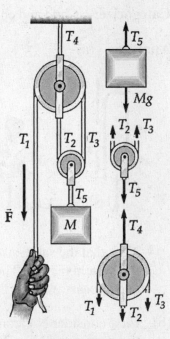

Figure P5.61

Analyze:

(a) Draw force diagrams. ∎

All forces are in the vertical direction. The lifting can be done at constant speed, with zero acceleration and total force zero on each object.

(b) For M, $\Sigma F = 0 = T_5 - Mg$

so $T_5 = Mg$

Assume frictionless pulleys. The tension is constant throughout a light, continuous rope. Therefore, $T_1 = T_2 = T_3$

For the bottom pulley $\Sigma F = 0 = T_2 + T_3 - T_5$

So $2T_2 = T_5$

and by substitution $T_1 = T_2 = T_3 = \frac{1}{2}Mg$ ■

For the top pulley, $\Sigma F = 0 = T_4 - T_1 - T_2 - T_3$

Solving, $T_4 = T_1 + T_2 + T_3 = \frac{3}{2}Mg$ ■

and as noted above, $T_5 = Mg$ ■

(c) The applied force is $F = T_1 = \frac{1}{2}Mg$ ■

Finalize: The moving pulley is supported by two strings, so the tension in each is half the weight of the load. The ceiling must exert a force greater than the weight of the load. To lift the load at 3 cm/s, the hand must pull the rope at 6 cm/s.

---

67. What horizontal force must be applied to a large block of mass M shown in Figure P5.67 so that the small blocks remain stationary relative to M? Assume all surfaces and the pulley are frictionless. Notice that the force exerted by the string accelerates m_2.

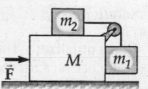

Figure P5.67

Solution

Conceptualize: What can keep m_1 from falling? Only tension in the cord connecting it with m_2. This tension pulls forward on m_2 to accelerate that mass. We might guess that the acceleration is proportional to both m_1 and g and inversely proportional to m_2, so perhaps $a = m_1 g / m_2$. If the entire system accelerates at this rate, then m_2 need not slide on M to achieve this acceleration. We should also expect the applied force to be proportional to the total mass of the system.

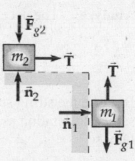

Categorize: We will use $\Sigma F = ma$ on each object, so we draw force diagrams for the $M + m_1 + m_2$ system, and also for blocks m_1 and m_2. Remembering that normal forces are always perpendicular to the contacting surface, and always **push** on a body, draw n_1 and n_2 as shown. Note that m_1 is in contact with the cart, and therefore feels a normal force exerted by the cart. Remembering that ropes always **pull** on bodies toward the center of the rope, draw the tension force $\vec{T}$. Finally, draw the gravitational force on each block, which always points downwards.

Analyze: For vertical forces on m_1 $T - m_1 g = 0$ or $T = m_1 g$

For m_2 $T = m_2 a$ or $a = T/m_2$

Substituting for T, we have $a = m_1 g/m_2$

For all three blocks $F = (M + m_1 + m_2)a$

Therefore $F = (M + m_1 + m_2)(m_1 g/m_2)$ ∎

Finalize: This problem did not have a numerical solution, but we were still able to reason about the algebraic form of the solution. This technique does not always work, especially for complex situations, but often we can think through a problem to see if an equation for the solution makes sense based on the physical principles we know.

---

69. A car accelerates down a hill (Fig. P5.69), going from rest to 30.0 m/s in 6.00 s. A toy inside the car hangs by a string from the car's ceiling. The ball in the figure represents the toy, of mass 0.100 kg. The acceleration is such that the string remains perpendicular to the ceiling. Determine (a) the angle θ and (b) the tension in the string.

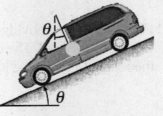

Figure P5.69

Solution

Conceptualize: The forces on the toy are just like the forces on a block on a frictionless incline. We may anticipate an equation like $a = g \sin\theta$ so $(30 \text{ m/s} - 0)/6 \text{ s} = 9.8 \text{ m/s}^2 \sin\theta$ and $\theta \approx 30°$. It is quite a steep hill. The tension will be a bit less than 1 N, a bit less than the weight of the toy.

Categorize: The car is a particle under constant acceleration and the toy is a particle under net force.

Analyze: The acceleration is obtained from $v_f = v_i + at$:

$$a = (v_f - v_i)/t = (30.0 \text{ m/s} - 0)/6.00 \text{ s}$$
$$a = 5.00 \text{ m/s}^2$$

Because the string stays perpendicular to the ceiling, we know that the toy moves with the same acceleration as the van, 5.00 m/s² parallel to the hill. We take the x axis in this direction, so

$$a_x = 5.00 \text{ m/s}^2 \quad \text{and} \quad a_y = 0$$

The only forces on the toy are the string tension in the y direction and the planet's gravitational force, as shown in the force diagram.

The size of the latter is $\quad mg = (0.100 \text{ kg})(9.80 \text{ m/s}^2) = 0.980 \text{ N}$

(a) Using $\quad \Sigma F_x = ma_x \quad$ gives $\quad (0.980 \text{ N})\sin\theta = (0.100 \text{ kg})(5.00 \text{ m/s}^2)$

   Then $\quad \sin\theta = 0.510 \quad$ and $\quad \theta = 30.7° \quad$ ■

(b) Using $\quad \Sigma F_y = ma_y \quad$ gives $\quad +T - (0.980 \text{ N})\cos\theta = 0$

   $\quad T = (0.980 \text{ N})\cos 30.7° = 0.843 \text{ N} \quad$ ■

Finalize: This problem is a good review about inclined planes. We take the x axis in the direction of motion along the plane. We identify the weight as being at angle θ to the y axis, and we compute its components. Make sure you know why no normal force acts on the toy: it is not in contact with any solid surface.

6

Circular Motion and Other Applications of Newton's Laws

EQUATIONS AND CONCEPTS

The **net force exerted on an object** of mass m, moving uniformly in a circular path of radius r is directed toward the center of the circle. This force causes the object's centripetal acceleration. *Equation 6.1 is a statement of Newton's second law along the radial direction. The F in the equation refers only to the radial components of the forces.*

$$\sum F = ma_c = m\frac{v^2}{r} \tag{6.1}$$

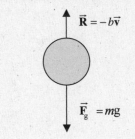

A **resistive force** $\vec{R}$ will be exerted on an object moving through a medium (gas or liquid). The form of Equation 6.2 assumes that the resistive force is proportional to the speed of the object. *The constant b has a value that depends on the properties of the medium and the dimensions and shape of the object.*

$$\vec{R} = -b\vec{v} \tag{6.2}$$

$\vec{R} = -b\vec{v}$

$\vec{F}_g = mg$

When **falling in a viscous medium**, the motion of an object is described by a differential equation. *Equation 6.5 gives the speed as a function of time, when the object is released from rest at $t = 0$.*

$$\frac{dv}{dt} = g - \frac{b}{m}v \tag{6.4}$$

$$v = \frac{mg}{b}\left(1 - e^{-bt/m}\right) = v_T\left(1 - e^{-t/\tau}\right) \tag{6.5}$$

The **time constant** τ is the time at which an object, released from rest, will achieve a speed equal to 63.2% of the terminal speed.

$$\tau = \frac{m}{b}$$

Terminal speed is achieved as the magnitude of the resistive force approaches the weight of the falling object.

88

SUGGESTIONS, SKILLS, AND STRATEGIES

Section 6.4 deals with the motion of an object through a viscous medium. The method of solution for Equation 6.4, when the resistive force $R = -bv$, is shown below:

$$\frac{dv}{dt} = g - \frac{b}{m}v \tag{6.4}$$

In order to solve this equation, it is convenient to change variables. If we let $y = g - (b/m)v$, it follows that $dy = -(b/m)dv$. With these substitutions, Equation 6.4 becomes

$$-\left(\frac{m}{b}\right)\frac{dy}{dt} = y \qquad \text{or} \qquad \frac{dy}{y} = -\left(\frac{b}{m}\right)dt$$

Integrating this expression (now that the variables are separated) gives

$$\int \frac{dy}{y} = -\frac{b}{m}\int dt \qquad \text{or} \qquad \ln y = -\frac{b}{m}t + \text{constant}$$

This is equivalent to

$$y = (\text{const})e^{-bt/m} = g - (b/m)v$$

Taking $v = 0$ and $t = 0$, we see that $\quad \text{const} = g$

so

$$v = \frac{mg}{b}\left(1 - e^{-bt/m}\right) = v_t\left(1 - e^{-t/\tau}\right) \tag{6.5}$$

where

$$\tau = m/b$$

REVIEW CHECKLIST

- You should be able to: Apply Newton's second law to uniform and nonuniform circular motion. (Sections 6.1 and 6.2)

- Identify situations in which an observer is in a noninertial frame of reference (an accelerating frame), and explain the apparent violation of Newton's laws of motion in terms of "fictitious" forces. (Section 6.3)

- Use the equations describing motion in a viscous medium to determine the speed of an object as a function of time and the value of the terminal speed. (Section 6.4)

ANSWER TO AN OBJECTIVE QUESTION

3. A child is practicing for a BMX race. His speed remains constant as he goes counterclockwise around a level track with two straight sections and two nearly semicircular sections as shown in the aerial view of Figure OQ6.3. (a) Rank the magnitudes of his acceleration at the points A, B, C, D, and E, from largest to smallest. If his acceleration is the same size at two points, display that fact in your ranking. If his acceleration is zero, display that fact. (b) What are the directions of his velocity at points A, B, and C? For each point choose one: north, south, east, west, or nonexistent. (c) What are the directions of his acceleration at points A, B, and C?

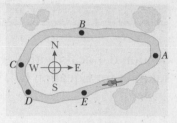

Figure OQ6.3

Answer (a) The child's speed is not changing, so his acceleration is zero at points B and E, on the straightaways. His acceleration, called centripetal acceleration, is largest where the radius of curvature of his path is smallest, at point A. Accelerations at C and D are equal in magnitude because the curve radius is constant for these points. Then the answer is $A > C = D > B = E = 0$.

(b) The direction of the child's velocity is the direction of the track, taken counterclockwise. So it is north at A, west at B, and south at C.

(c) The direction of the acceleration is toward the center of curvature of each curve. So it is west at A, nonexistent at B, and east at C.

ANSWERS TO SELECTED CONCEPTUAL QUESTIONS

4. A falling skydiver reaches terminal speed with her parachute closed. After the parachute is opened, what parameters change to decrease this terminal speed?

Answer From the expression for the force of air resistance and Newton's law, we derive the equation that describes the motion of the skydiver:

$$m\frac{dv_y}{dt} = mg - \frac{D\rho A}{2}v_y^2$$

where D is the coefficient of drag of the skydiver, and A is the area of his or her body projected onto a plane perpendicular to the motion. At terminal speed,

$$a_y = dv_y/dt = 0 \quad \text{and} \quad v_T = \sqrt{2mg/D\rho A}$$

When the parachute opens, the coefficient of drag D and the effective area A both increase, thus reducing the velocity of the skydiver.

Modern parachutes also add a third term, lift, to change the equation to

$$m\frac{dv_y}{dt} = mg - \frac{D\rho A}{2}v_y^2 - \frac{L\rho A}{2}v_x^2$$

where v_y is the vertical velocity, and v_x is the horizontal velocity. This lift is clearly seen for a "Paraplane," an ultralight airplane made from a fan, a chair, and a parachute.

□   □   □   □

7. It has been suggested that rotating cylinders about 20 km in length and 8 km in diameter be placed in space and used as colonies. The purpose of the rotation is to simulate gravity for the inhabitants. Explain this concept for producing an effective imitation of gravity.

Answer The colonists stand and walk around on the gently curved interior wall of the cylinder, with their heads pointing toward the axis of rotation. The normal force exerted on the inhabitants by the cylindrical wall causes their centripetal acceleration. If the rotation rate is adjusted to such a speed that this normal force is equal to their weight on Earth, then this artificial gravity would seem to the inhabitants to be the same as normal gravity.

□   □   □   □

9. Why does a pilot tend to black out when pulling out of a steep dive?

Answer When pulling out of a dive, blood leaves the pilot's head because the pilot's blood pressure is not great enough to compensate for both the gravitational force and the centripetal acceleration of the airplane. This loss of blood from the brain can cause the pilot to lose consciousness.

□   □   □   □

SOLUTIONS TO SELECTED END-OF-CHAPTER PROBLEMS

1. A light string can support a stationary hanging load of 25.0 kg before breaking. An object of mass $m = 3.00$ kg attached to the string rotates on a frictionless horizontal table in a circle of radius $r = 0.800$ m, and the other end of the string is held fixed. What range of speeds can the object have before the string breaks?

Solution

Conceptualize: The object will be able to move with any speed between zero and a maximum of several meters per second.

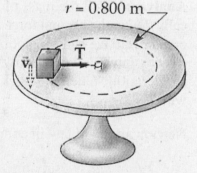

$r = 0.800$ m

Categorize: We use the models of a particle under a net force and a particle in uniform circular motion.

Analyze: The string will break if the tension T exceeds

$$T_{max} = mg = (25.0 \text{ kg})(9.80 \text{ m/s}^2) = 245 \text{ N}$$

As the 3.00-kg mass rotates in a horizontal circle, the tension provides the centripetal acceleration:

$$a = v^2/r$$

From $\sum F = ma,$ $T = mv^2/r$

Solving, $v^2 = \dfrac{rT}{m} = \dfrac{(0.800\ \text{m})T}{3.00\ \text{kg}} \leq \dfrac{0.800\ \text{m}}{3.00\ \text{kg}} T_{max}$

Substituting $T_{max} = 245$ N, we find $v^2 \leq 65.3$ m²/s²

And $0 < v \leq 8.08$ m/s   ∎

Finalize: We did not have to consider the gravitational or normal forces because they are perpendicular to the motion.

---

9.   A crate of eggs is located in the middle of the flat bed of a pickup truck as the truck negotiates a curve in the flat road. The curve may be regarded as an arc of a circle of radius 35.0 m. If the coefficient of static friction between crate and truck is 0.600, how fast can the truck be moving without the crate sliding?

Solution

Conceptualize: That's a tight curve, so the speed may be rather low, say about 10 m/s.

Categorize: It makes sense to assume that the truck can make the curve, since it can use the friction of rubber on concrete. The crate is a particle under net force moving with constant speed and with the greatest acceleration it can get from friction exerted by the truck bed.

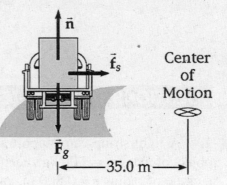

Analyze: Call the mass of the egg crate m. The forces on it are its weight $F_g = mg$ vertically down, the normal force n of the truck bed vertically up, and static friction f_s directed to oppose relative sliding motion of the crate over the truck bed. The friction force is directed radially inward. It is the only horizontal force on the crate, so it must provide the centripetal acceleration. When the truck has maximum speed, friction f_s will have its maximum value with $f_s = \mu_s n$.

Newton's second law in component form becomes

$\quad \sum F_y = ma_y$ giving $n - mg = 0$ or $n = mg$

$\quad \sum F_x = ma_x$ giving $f_s = ma_r$

From these three equations, $\mu_s n = \dfrac{mv^2}{r}$ and $\mu_s mg = \dfrac{mv^2}{r}$

The mass divides out, leaving for the maximum speed

$$v = \sqrt{\mu_s gr} = \sqrt{(0.600)(9.80 \text{ m/s}^2)(35.0 \text{ m})} = 14.3 \text{ m/s} \qquad \blacksquare$$

Finalize: We were roughly right about the speed. It is static friction that acts on the moving crate because the crate is not quite sliding relative to the truck bed. If the truck turns a bit more sharply, the crate will start to slide and a smaller force of kinetic friction will act on it. Then the crate will move on a larger-radius curve than the truck, and fall off the outer edge of the truck bed.

---

11.   A coin placed 30.0 cm from the center of a rotating horizontal turntable slips when its speed is 50.0 cm/s.   (a) What force causes the centripetal acceleration when the coin is stationary relative to the turntable?   (b) What is the coefficient of static friction between coin and turntable?

Solution

Conceptualize: If you shift the turntable to higher and higher rotation rates, letting it turn steadily to test each, the coin will stay put at all lower speeds up to some maximum. We expect the coefficient of friction to be some pure number between 0 and 1, or maybe 2 for sticky rubber.

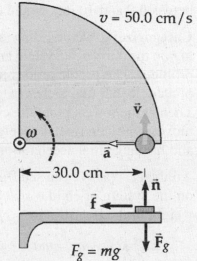

Categorize: The top view diagram shows the constant maximum speed and the centripetal acceleration. The side view shows the forces. We model the coin as a particle under a net force.

Analyze:

(a)   The force of static friction causes the centripetal acceleration. $\blacksquare$

(b)   The forces on the coin are the normal force, the weight, and the force of static friction. The only force in the radial direction is the friction force. Take the x axis pointing toward the center of the turntable.

Therefore, $\Sigma F_x = ma_x$ $f = m\dfrac{v^2}{r}$

The normal force balances the weight, so $\Sigma F_y = ma_y$ $n - mg = 0$ $n = mg$

The friction force follows the empirical rule $f \le \mu_s n = \mu_s mg$

The coin is ready to slip when the maximum friction force is exactly enough to cause the centripetal acceleration: $\mu_s mg = mv^2/r$

We substitute $r = 30$ cm, $v = 50$ cm/s, and $g = 980$ cm/s², to find

$$\mu_s = v^2/rg = \frac{(50.0 \text{ cm/s})^2}{(30.0 \text{ cm})(980 \text{ cm/s}^2)} = 0.085\,0$$ ■

Finalize: Note that the coin's mass divides out at the last step. A trick was choosing the *x* axis toward the center of the circle, in the direction of the coin's instantaneous acceleration. Compare this problem with the previous one in this manual. The force diagrams are the same and the derived equations are the same, even though the numbers and the objects are different.

---

14. A 40.0-kg child swings in a swing supported by two chains, each 3.00 m long. The tension in each chain at the lowest point is 350 N. Find (a) the child's speed at the lowest point and (b) the force exerted by the seat on the child at the lowest point. (Ignore the mass of the seat.)

Solution

Conceptualize: If the tension in each chain is 350 N at the lowest point, then the force of the seat on the child should be twice this force or 700 N. The child's speed is not so easy to determine, but somewhere between 0 and 10 m/s would be reasonable for the situation described.

Categorize: We first draw a force diagram that shows the forces acting on the child-seat system and apply Newton's second law to solve the problem. The problem does not mention that the child's path is an arc of a circle, but this is clearly true, if the top ends of the chains are fixed. Then at the lowest point the child's motion is changing in direction: He moves with centripetal acceleration even as his speed is not changing and his tangential acceleration is zero.

Analyze: (a) We can see from the diagram that the only forces acting on the system of child + seat are the tensions in the two chains and the weight of the boy:

$$\Sigma F = 2T - mg = ma$$

$$\Sigma F = F_{net} = 2(350 \text{ N}) - (40.0 \text{ kg})(9.80 \text{ m/s}^2) = 308 \text{ N (up)} = mv^2/r$$

Because $a = v^2/r$ is the centripetal acceleration,

$$v = \sqrt{\frac{F_{net}r}{m}} = \sqrt{\frac{(308 \text{ N})(3.00 \text{ m})}{40.0 \text{ kg}}} = 4.81 \text{ m/s}$$ ■

(b) The normal force exerted by the seat on the child accelerates the child in the same way that the total tension in the chains accelerates the child-seat system. Therefore the normal force is

$$\bar{n} = 2(350 \text{ N}) = 700 \text{ N (upward)}$$ ■

Finalize: Our answers agree with our predictions. It may seem strange that there is a net upward force on the boy and he is not here moving upward. We must remember that a net force causes an acceleration, but not necessarily immediate motion in the direction of the force. In this case, the acceleration describes a change in the direction of the sideways velocity. The boy feels about twice as heavy as normal, so his experience is equivalent to an acceleration of about 2 *g*'s.

---

17. An adventurous archeologist (*m* = 85.0 kg) tries to cross a river by swinging from a vine. The vine is 10.0 m long, and his speed at the bottom of the swing is 8.00 m/s. The archeologist doesn't know that the vine has a breaking strength of 1 000 N. Does he make it across the river without falling in?

Solution

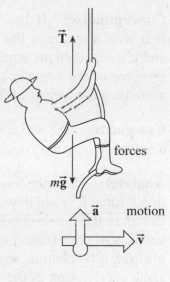

Conceptualize: The scientist does not move all the way around a circle, but he moves on an arc of a circle of radius 10 m. His direction of motion is changing, so he has change-in-direction acceleration, called centripetal acceleration. As he passes the bottom of the swing his speed is no longer increasing and not yet decreasing. At this moment he has no change-in-speed (tangential) acceleration but only acceleration $\dfrac{v^2}{r}$ upward.

Categorize: We can think of this as a straightforward particle-under-a-net-force problem. We will find the value of tension required under the assumption that the man passes the bottom of the swing. If it is less than 1 000 N, he will make it across the river.

Analyze: The forces acting on the archaeologist are the force of the Earth $m\vec{g}$ and the force from the vine, $\vec{T}$. At the lowest point in his motion, $\vec{T}$ is upward and $m\vec{g}$ is downward as in the force diagram. Thus, Newton's second law gives

$$\Sigma F_y = ma_y \qquad T - mg = \frac{mv^2}{r}$$

Solving for *T*, with $v = 8.00$ m/s, $r = 10.0$ m, and $m = 85.0$ kg, gives

$$T = m\left(g + \frac{v^2}{r}\right) = (85.0\ \text{kg})\left(9.80\ \text{m/s}^2 + \frac{(8.00\ \text{m/s})^2}{10.0\ \text{m}}\right) = 1.38 \times 10^3\ \text{N}$$

Since *T* **exceeds** the breaking strength of the vine (1 000 N), our hero **doesn't make it!** The vine breaks **before** he reaches the bottom of the swing. ∎

Finalize: If the vine does not break, the scientist's forward motion at the bottom can be thought of as an example of Newton's first law. He keeps moving because no backward or forward force acts on him. All of the forces exerted on him are perpendicular to his motion. The forces explain his acceleration according to Newton's second law, in contrast to explaining his velocity. The third law is also involved in saying that the force of the man on the vine is equal in magnitude to the force of the vine on him.

---

21. An object of mass $m = 0.500$ kg is suspended from the ceiling of an accelerating truck as shown in Figure P6.21. Taking $a = 3.00$ m/s², find (a) the angle θ that the string makes with the vertical and (b) the tension T in the string.

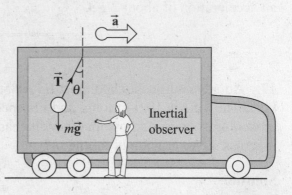

Figure P6.21

Solution

Conceptualize: If the horizontal acceleration were zero, then the angle would be 0°, and if $a = g$, then the angle would be 45°. Since the acceleration is 3.00 m/s², a reasonable estimate of the angle is about 20°. Similarly, the tension in the string should be slightly more than the weight of the object, which is about 5 N.

Categorize: We will apply Newton's second law to the hanging object as a particle under a net force.

Analyze: The only forces acting on the suspended object are the gravitational force $F_g = mg$ down and the force of tension T forward and upward, as shown in the free-body diagram. Note that the angle θ is between T and the y axis, so $T\cos\theta$ is the y component. We assume that the object is not swinging like a pendulum, so that θ is constant and the object moves with the same acceleration as the truck. Applying Newton's second law in the x and y directions,

$$\sum F_x = T\sin\theta = ma \qquad\qquad\qquad [1]$$

$$\sum F_y = T\cos\theta - mg = 0 \qquad \text{or} \qquad T\cos\theta = mg \qquad [2]$$

(a)   Dividing equation [1] by [2],

$$\tan\theta = \frac{a}{g} = \frac{3.00 \text{ m/s}^2}{9.80 \text{ m/s}^2} = 0.306$$

Solving for θ, $\theta = \tan^{-1}0.306 = 17.0°$ ■

(b)   From equation [1], $T = \dfrac{ma}{\sin\theta} = \dfrac{(0.500 \text{ kg})(3.00 \text{ m/s}^2)}{\sin(17.0°)} = 5.12$ N   ■

Finalize: Our answers agree with our original estimates. We used the same equations [1] and [2] as in Example 6.7. We did not need the idea of a fictitious force.

---

23. A person stands on a scale in an elevator. As the elevator starts, the scale has a constant reading of 591 N. As the elevator later stops, the scale reading is 391 N. Assuming the magnitude of the acceleration is the same during starting and stopping, determine (a) the weight of the person, (b) the person's mass, and (c) the acceleration of the elevator.

Solution

Conceptualize: The weight of the person should be about 491 N, her mass about 50 kg, and the magnitude of the acceleration about 100 N/50 kg = 2 m/s^2.

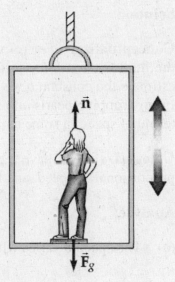

Categorize: Take the person as a particle under net force, moving with the acceleration of the elevator. We must write separate equations for the starting and stopping processes and then solve simultaneously.

Analyze: The scale reads the upward normal force exerted by the floor on the passenger. The maximum force occurs during upward acceleration (when starting an upward trip or ending a downward trip). The minimum normal force occurs with downward acceleration. For each respective situation,

$\sum F_y = ma_y$ becomes for starting $+591\ \text{N} - mg = +ma$

and for stopping $+391\ \text{N} - mg = -ma$

where a represents the magnitude of the acceleration.

(a) These two simultaneous equations can be added to eliminate a and solve for mg.

$$+591\ \text{N} - mg + 391\ \text{N} - mg = 0$$

or $982\ \text{N} - 2mg = 0$

$$F_g = mg = \frac{982\ \text{N}}{2} = 491\ \text{N} \qquad \blacksquare$$

(b) $$m = \frac{F_g}{g} = \frac{491\ \text{N}}{9.80\ \text{m/s}^2} = 50.1\ \text{kg} \qquad \blacksquare$$

(c) Substituting back gives $+591\ \text{N} - 491\ \text{N} = (50.1\ \text{kg})a$

$$a = \frac{100\ \text{N}}{50.1\ \text{kg}} = 2.00\ \text{m/s}^2 \qquad \blacksquare$$

Finalize: Our estimates were right on. We solved the problem in the frame of reference of an observer outside the elevator in an inertial frame. We did not need the idea of fictitious forces.

31. A small, spherical bead of mass 3.00 g is released from rest at $t=0$ from a point under the surface of a viscous liquid. The terminal speed is observed to be $v_T = 2.00$ cm/s. Find (a) the value of the constant b that appears in Equation 6.2, (b) the time t at which the bead reaches $0.632v_T$, and (c) the value of the resistive force when the bead reaches terminal speed.

Solution

Conceptualize: For (c) we can identify that at terminal speed the acceleration is zero, the fluid friction force just counterbalances the weight, and that this is about 30 mN. To estimate the constant b we can write 0.03 N $\approx b\,0.02$ m/s, so b is on the order of 1 N·s/m. From dropping pearls into shampoo we estimate that the bead attains a good fraction of its terminal speed in some tenths of a second.

Categorize: We have a particle under net force in the special case of a resistive force proportional to speed, and feeling also the gravitational force.

Analyze:

(a) The speed v varies with time according to Equation 6.5,

$$v = \frac{mg}{b}\left(1-e^{-bt/m}\right) = v_T\left(1-e^{-bt/m}\right)$$

where $v_T = mg/b$ is the terminal speed.

Hence,

$$b = \frac{mg}{v_T} = \frac{\left(3.00\times10^{-3}\text{ kg}\right)\left(9.80\text{ m/s}^2\right)}{2.00\times10^{-2}\text{ m/s}} = 1.47\text{ N·s/m} \qquad\blacksquare$$

(b) To find the time interval for v to reach $0.632v_T$, we substitute $v = 0.632v_T$ into Equation 6.5, giving

$$0.632v_T = v_T\left(1-e^{-bt/m}\right) \qquad\text{or}\qquad 0.368 = e^{-(1.47t/0.003\,00)}$$

Solve for t by taking the natural logarithm of each side of the equation:

$$\ln(0.368) = -\frac{1.47\,t}{3.00\times10^{-3}} \qquad\text{or}\qquad -1 = -\frac{1.47\,t}{3.00\times10^{-3}}$$

or $t = m/b = 2.04\times10^{-3}$ s $\blacksquare$

(c) At terminal speed, $R = v_T b = mg$

Therefore,

$$R = \left(3.00\times10^{-3}\text{ kg}\right)\left(9.80\text{ m/s}^2\right) = 2.94\times10^{-2}\text{ N} \qquad\blacksquare$$

Finalize: We were right about the estimated size of the proportionality constant b and the maximum resistive force, but the characteristic time for the bead to approach terminal speed is in milliseconds and not tenths of a second. The parameter m/b that we found in part (b) is a time interval, and is called the *time constant* of the exponential function's pattern of approach to the terminal speed.

35. A motorboat cuts its engine when its speed is 10.0 m/s and then coasts to rest. The equation describing the motion of the motorboat during this period is $v = v_i e^{-ct}$, where v is the speed at time t, v_i is the initial speed at $t = 0$, and c is a constant. At $t = 20.0$ s, the speed is 5.00 m/s. (a) Find the constant c. (b) What is the speed at $t = 40.0$ s? (c) Differentiate the expression for $v(t)$ and thus show that the acceleration of the boat is proportional to the speed at any time.

Solution

Conceptualize:　Twenty seconds can be called the *half-life* in the exponential decay process described by $v = v_i e^{-ct}$. So for c we estimate something like $1/(20 \text{ s}) = 0.05/\text{s}$. The speed after 40 s should be 2.5 m/s.

Categorize:　In part (c) we will show that we have a particle feeling just a net force of fluid friction proportional to its speed. Each part is solved from the given equation $v = v_i e^{-ct}$.

Analyze:

(a)　We must fit the equation $v = v_i e^{-ct}$ to the two data points:

At $t = 0$, $v = 10.0$ m/s　　so　　$v = v_i e^{-ct}$　becomes　$10.0 \text{ m/s} = v_i e^0 = (v_i)(1)$

which gives　　$v_i = 10.0$ m/s

At $t = 20.0$ s, $v = 5.00$ m/s so the equation becomes

$5.00 \text{ m/s} = (10.0 \text{ m/s})e^{-c(20.0 \text{ s})}$　　giving　　$0.500 = e^{-c(20.0 \text{ s})}$

or　　　　$\ln(0.500) = (-c)(20.0 \text{ s})$

so　　　　$c = \dfrac{-\ln(0.500)}{20.0 \text{ s}} = 0.034\,7 \text{ s}^{-1}$ ∎

(b)　At all times,　$v = (10.0 \text{ m/s})e^{-(0.034\,7)t}$

At $t = 40.0$ s　　$v = (10.0 \text{ m/s})e^{-(0.034\,7)(40.0)} = 2.50$ m/s ∎

(c)　The acceleration is the rate of change of the velocity:

$$a = \frac{dv}{dt} = \frac{d}{dt}v_i e^{-ct} = v_i \left(e^{-ct}\right)(-c) = -c\left(v_i e^{-ct}\right) = -cv = \left(-0.034\,7 \text{ s}^{-1}\right)v$$

Thus, the acceleration is a negative constant times the speed. ∎

Finalize:　Our estimates were good. It is interesting that the previous problem in this manual and the present problem can apply the same model of fluid friction proportional to speed to objects as different as a bead in shampoo and a motorboat in water. The time variation of the speed is on the order of ten thousand times slower in this problem.

---

49. Interpret the graph in Figure 6.16(b), which describes the results for falling coffee filters, discussed in Example 6.10. Proceed as follows: (a) Find the slope of the straight line including its units. (b) From Equation 6.6, $R = \dfrac{1}{2}D\rho Av^2$, identify the theoretical slope of a graph of resistive force versus squared speed. (c) Set the experimental and theoretical slopes equal to each other and proceed to calculate the drag coefficient of the filters. Model the cross-sectional area of the filters as that of a circle of radius 10.5 cm and take the density of air to be 1.20 kg/m^3. (d) Arbitrarily choose the eighth data point on the graph and find its vertical separation from the line of best fit. Express this scatter as a percentage. (e) In a short paragraph state what the graph demonstrates and compare it to the theoretical prediction. You will need to make reference to the quantities plotted on the axes, to the shape of the graph line, to the data points, and to the results of parts (c) and (d).

Solution

Conceptualize: On a graph, a straight line through the origin displays a proportionality. The graph shows experimental measurements as data points, a theoretical pattern as a best-fit line, and reveals uncertainty as scatter of the points from the line. The slope of the line can give information about things that were held constant during the various trials, for experimental control.

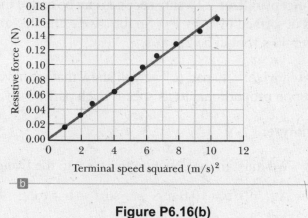

Figure P6.16(b)

Categorize: We use the graph itself and the equation $R = \dfrac{1}{2}D\rho Av^2$ that the graph can be thought of as testing.

Analyze: (a) The graph line is straight, so we may use any two points on it to find the slope. It is convenient to take the origin as one point, and we read (9.9 m^2/s^2, 0.16 N) as the coordinates of another point. Then the slope is

$$\text{slope} = \frac{0.16\ \text{N} - 0}{9.9\ \text{m}^2/\text{s}^2 - 0} = 0.016\ 2\ \text{kg/m} \qquad\blacksquare$$

(b) In $R = \dfrac{1}{2}D\rho Av^2$ we identify the vertical-axis variable as R and the horizontal-axis variable as v^2. Then the slope is $\dfrac{R}{v^2} = \dfrac{D\rho A}{2}$ $\qquad\blacksquare$

(c) We follow the directions in the problem statement:

$$\frac{D\rho A}{2} = 0.016\ 2\ \text{kg/m}$$

$$D = \frac{2(\text{slope})}{\rho A} = \frac{2(0.016\ 2\ \text{kg/m})}{(1.20\ \text{kg/m}^3)\pi(0.105\ \text{m})^2} = 0.778 \qquad\blacksquare$$

(d) From the data table in the text Example 6.10, the eighth data point is at force $mg = 8(1.64 \times 10^{-7}\ \text{kg})(9.80\ \text{m/s}^2) = 0.129\ \text{N}$ and horizontal coordinate $(2.80\ \text{m/s})^2$. The vertical coordinate of the line is here $(0.016\ 2\ \text{kg/m})(2.80\ \text{m/s})^2 = 0.127\ \text{N}$. The scatter

percentage is $\dfrac{0.129\ \text{N} - 0.127\ \text{N}}{0.127\ \text{N}} = 1.5\%.$ $\qquad\blacksquare$

(e) The interpretation of the graph can be stated like this:

For stacked coffee filters falling at terminal speed, a graph of air resistance force as a function of squared speed demonstrates that the force is proportional to the speed squared within the experimental uncertainty, estimated as 2%. This proportionality agrees with that described by the theoretical equation $R = \frac{1}{2}D\rho Av^2$. The value of the constant slope of the graph implies that the drag coefficient for coffee filters is $D = 0.78 \pm 2\%$.

Finalize: A graph shows a lot! Some students need reminders that the slope has units. The slope is found by reading coordinates from the line, not by pulling numbers from the data table. We chose two points far apart on the line to minimize uncertainty in our slope calculation. When it comes to data, avoid underestimating the uncertainty. It is inadequate to write that the graph shows a "linear function" or a "correlation" or a "relationship," or that "the greater the speed, the greater the force," or that R and v^2 are proportional "to each other." It is wrong to call the function a "direct proportionality" and very bad to call it an "indirect proportionality." The fraction between 0 and 1 is a reasonable value for the drag coefficient.

53. Because the Earth rotates about its axis, a point on the equator experiences a centripetal acceleration of 0.033 7 m/s², whereas a point at the poles experiences no centripetal acceleration. If a person at the equator has a mass of 75.0 kg, calculate (a) the gravitational force (true weight) on the person, and (b) the normal force (apparent weight) on the person. (c) Which force is greater? Assume the Earth is a uniform sphere and take $g = 9.800$ m/s², as measured directly at the poles.

Solution

Conceptualize: If the person were standing at a pole, the normal force would be equal in magnitude to the true weight of the person. At the equator, because the centripetal acceleration is a small fraction (~0.3%) of g, we should expect that a person would have an apparent weight that is just slightly less than her true weight, due to the rotation of the Earth.

Categorize: We will apply Newton's second law and the equation for centripetal acceleration to the person as a particle under net force.

Analyze: (a) The gravitational force exerted by the planet on the person is

$$mg = (75.0 \text{ kg})(9.80 \text{ m/s}^2) = 735 \text{ N} \quad \text{down} \qquad \blacksquare$$

Let n represent the force exerted on the person by a scale, which is an upward force whose size is her "apparent weight." The true weight is mg down. For the person at the equator, summing up forces on the object in the direction towards the Earth's center gives $\Sigma F = ma$

$$mg - n = ma_c$$

where $\qquad a_c = v^2/R_E = 0.033\ 7 \text{ m/s}^2$

is the centripetal acceleration directed toward the center of the Earth.

Thus, we can solve part (c) before part (b) by noting that $n = m(g - a_c) < mg$

(c) or $mg = n + ma_c > n$. The gravitational force is greater than the normal force. ∎

(b) If $m = 75.0$ kg   and $g = 9.800$ m/s², at the Equator we have

$$n = m(g - a_c) = (75.0 \text{ kg})(9.800 \text{ m/s}^2 - 0.033\ 7 \text{ m/s}^2) = 732 \text{ N}$$ ∎

Finalize: As we expected, the person does appear to weigh about 0.3% less at the equator than the poles. We might extend this problem to consider the effect of the Earth's bulge on a person's weight. Since the Earth is fatter at the equator than the poles, would you expect g itself to be less than 9.80 m/s² at the equator and slightly more at the poles?

─────────────

59. An amusement park ride consists of a large vertical cylinder that spins about its axis fast enough that any person inside is held up against the wall when the floor drops away (Fig. P6.59). The coefficient of static friction between person and wall is μ_s, and the radius of the cylinder is R. (a) Show that the maximum period of revolution necessary to keep the person from falling is $T = (4\pi^2 R \mu_s / g)^{1/2}$. (b) If the rate of revolution of the cylinder is made to be somewhat larger, what happens to the magnitude of each one of the forces acting on the person? What happens in the motion of the person? (c) If the rate of revolution of the cylinder is instead made to be somewhat smaller, what happens to the magnitude of each one of the forces acting on the person? How does the motion of the person change?

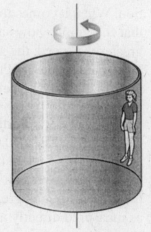

Solution

Figure P6.59

Conceptualize: A certain minimum rate of rotation is required so that the normal force will be large enough to make the friction force large enough to support the person's weight.

Categorize: Take the person as a particle under net force and as a particle in uniform circular motion.

Analyze:

(a) The wall's normal force pushes inward: $\Sigma F_{inward} = ma_{inward}$
becomes

$$n = \frac{mv^2}{R} = \frac{m}{R}\left(\frac{2\pi R}{T}\right)^2 = \frac{4\pi^2 Rm}{T^2}$$

The friction and weight balance: $\Sigma F_{upward} = ma_{upward}$ becomes $+f - mg = 0$ so
with the person just ready to start sliding down, $f_s = \mu_s n = mg$

Substituting, $\mu_s n = \mu_s \dfrac{4\pi^2 Rm}{T^2} = mg$

Solving, $T^2 = \dfrac{4\pi^2 R\mu_s}{g}$ gives $T = \sqrt{\dfrac{4\pi^2 R\mu_s}{g}}$ ∎

(b) The gravitational and friction forces remain constant. The normal force increases. The person remains in motion with the wall. The less-than sign applies in $f_s \leq \mu_s n$.

(c) The gravitational force remains constant. The normal and friction forces decrease. The person slides downward relative to the wall into the pit.

Finalize: Why is the normal force horizontally inward? Because the wall is vertical, and on the outside.

Why is there no upward normal force? Because there is no floor.

Why is the friction force directed upward? The friction opposes the possible relative motion of the person sliding down the wall.

Why is it not kinetic friction? Because person and wall are moving together in parts (a) and (b), stationary with respect to each other.

Why is there no outward force on her? No other object pushes out on her.

She pushes out on the wall as the wall pushes inward on her.

---

63. A model airplane of mass 0.750 kg flies with a speed of 35.0 m/s in a horizontal circle at the end of a 60.0-m control wire. The forces exerted on the airplane are shown in Figure P6.63: the tension in the control wire, the gravitational force, and aerodynamic lift that acts at $\theta = 20.0°$ inward from the vertical. Compute the tension in the wire, assuming it makes a constant angle of 20.0° with the horizontal.

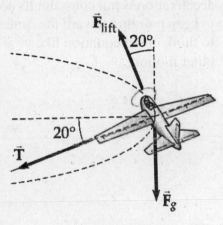

Figure P6.63

Solution

Conceptualize: A quick wrong calculation can be good for an estimate of the string tension:

$$mv^2/r = 0.75 \text{ kg}(35 \text{ m/s})^2/60 \text{ m} \approx 0.75 \text{ kg } (20 \text{ m/s}^2) = 15 \text{ N}$$

Categorize: The plane is a particle under net force in uniform circular motion. The force diagram keeps track of the several forces on it.

Analyze: The plane's acceleration is toward the center of the circle of motion, so it is horizontal.

The radius of the circle of motion is (60.0 m) cos 20.0° = 56.4 m and the acceleration is

$$a_c = \frac{v^2}{r} = \frac{(35 \text{ m/s})^2}{56.4 \text{ m}} = 21.7 \text{ m/s}^2$$

We can also calculate the weight of the airplane:

$$F_g = mg = (0.750 \text{ kg})(9.80 \text{ m/s}^2) = 7.35 \text{ N}$$

We define our axes for convenience. In this case, two of the forces—one of them our force of interest—are directed along the 20.0° lines. We define the x axis to be directed in the $(+\vec{T})$ direction, and the y axis to be directed in the direction of lift. With these definitions, the x component of the centripetal acceleration is

$$a_{cx} = a_c \cos(20°)$$

and $\Sigma F_x = ma_x$ yields $T + F_g \sin 20.0° = ma_{cx}$

Solving for T, $T = ma_{cx} - F_g \sin 20.0°$

Substituting, $T = (0.750 \text{ kg})(21.7 \text{ m/s}^2) \cos 20.0° - (7.35 \text{ N}) \sin 20.0°$

Computing, $T = (15.3 \text{ N}) - (2.51 \text{ N}) = 12.8 \text{ N}$ ∎

Finalize: The real string tension is a bit less than our estimate. The lift force has an inward component and provides a bit of the force required for keeping the plane on the circle. Would it be correct to call the plane a particle under constant acceleration? All the forces on it are constant in magnitude, but its acceleration is changing in direction so its acceleration is not constant. Its acceleration must change among north, south, east, and west to keep pointing toward the center of the circle of motion. Make sure you are never tempted to think up an equation like $v_f^2 = v_i^2 + 2a_c(x_f - x_i)$ for an object in circular motion or in any other motion.

7

Energy of a System

EQUATIONS AND CONCEPTS

The **work done on a system by a constant force** $\vec{F}$ is defined to be the product of the magnitude of the force, magnitude of the displacement of the point of application of the force, and the cosine of the angle between the force and the displacement vectors. *Work is a scalar quantity and represents a transfer of energy. The SI unit of work is the N·m. One N·m = 1 joule (J).*

$$W \equiv F\Delta r \cos\theta \qquad (7.1)$$

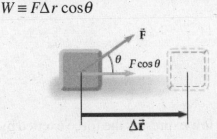

The **scalar product** (or **dot product**) of any two vectors $\vec{A}$ and $\vec{B}$ is defined to be a scalar quantity whose magnitude is equal to the product of the magnitudes of the two vectors and the cosine of the angle between the directions of the two vectors.

$$\vec{A} \cdot \vec{B} \equiv AB \cos\theta \qquad (7.2)$$

Work as a scalar product is shown in Equation 7.3. When using Equation 7.3, the force must be constant in magnitude and direction and $\vec{F} \cdot \Delta\vec{r}$ is read "$\vec{F}$ dot $\Delta\vec{r}$."

$$W = F\Delta r \cos\theta = \vec{F} \cdot \Delta\vec{r} \qquad (7.3)$$

Work done by a given force can be positive, negative, or zero depending on the value of θ. The figure at right shows four forces acting on an object on an incline. Illustrated below are cases for four different values of θ.

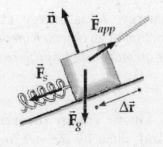

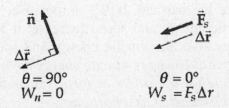

105

The **scalar product of two vectors** can be expressed in terms of their x, y, and z components.

$$\vec{A} \cdot \vec{B} = A_x B_x + A_y B_y + A_z B_z \qquad (7.6)$$

The **work done by a variable force** on a particle which has been displaced along the x axis from x_i to x_f is given by an integral expression. *Graphically, the work done equals the area under the F_x versus x curve.*

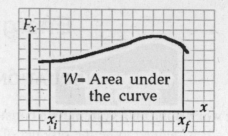

$$W = \int_{x_i}^{x_f} F_x \, dx \qquad (7.7)$$

Hooke's law expresses the force exerted by a spring which is stretched or compressed from the equilibrium position. *The negative sign signifies that the force exerted by the spring is always directed opposite the displacement from equilibrium.*

$$F_s = -kx \qquad (7.9)$$

The **work done *by* a spring** on an object which undergoes an arbitrary displacement from x_i to x_f is given by Equation 7.12.

$$W_s = \int_{x_i}^{x_f} (-kx)\,dx = \tfrac{1}{2}kx_i^2 - \tfrac{1}{2}kx_f^2 \qquad (7.12)$$

Work **done *on* a spring** by an applied force during a displacement from x_i to x_f is the negative of Equation 7.12.

$$W_{app} = \int_{x_i}^{x_f} (kx)\,dx = \tfrac{1}{2}kx_f^2 - \tfrac{1}{2}kx_i^2 \qquad (7.13)$$

Kinetic energy, K, is energy associated with the motion of an object. *Kinetic energy is a scalar quantity and has the same units as work.*

$$K \equiv \tfrac{1}{2}mv^2 \qquad (7.16)$$

The **work–kinetic energy theorem** states that the work done by the *net force* on a particle equals the change in kinetic energy of the particle. If W_{net} is positive, the kinetic energy and speed increase; if W_{net} is negative, the kinetic energy and speed decrease. *The work-kinetic energy theorem is valid for a particle or for a system that can be modeled as a particle.*

$$W_{ext} = K_f - K_i = \Delta K \qquad (7.17)$$

The **gravitational potential energy** associated with an object at any point in space is the product of the object's weight and the vertical coordinate relative to an arbitrary reference level.

$$U_g \equiv mgy \qquad (7.19)$$

An **elastic potential energy function** is associated with a mass-spring system which has been deformed a distance x from the equilibrium position. The force constant k has units of N/m and is characteristic of a particular spring. *The elastic potential energy of a deformed spring is always positive.*

$$U_s \equiv \tfrac{1}{2}kx^2 \qquad (7.22)$$

The **total mechanical energy** of a system is the sum of the kinetic and potential energies. *In Equation 7.24, U represents the total of all forms of potential energy (e.g. gravitational and elastic).*

$$E_{mech} \equiv K + U \qquad (7.24)$$

A **potential energy function** U can be defined for a conservative force such that the work done by a conservative force equals the decrease in the potential energy of the system. *The work done on an object by a conservative force depends only on the initial and final positions of the object and equals zero around a closed path. If more than one conservative force acts, then a potential energy function is associated with each force.*

$$W_{int} = U_i - U_f = -\Delta U \qquad (7.23)$$

$$\Delta U = U_f - U_i = -\int_{x_i}^{x_f} F_x\, dx \qquad (7.26)$$

When F_x is constant,

$$\Delta U = -F_x(\Delta x)$$

The **relationship between a conservative force and the potential energy function** is expressed by Equation 7.28.

$$F_x = -\frac{dU}{dx} \qquad (7.28)$$

Similar equations apply along the y and z directions.

$$F_y = -\frac{dU}{dy} \qquad F_z = -\frac{dU}{dz}$$

SUGGESTIONS, SKILLS, AND STRATEGIES

In order to apply the system model to problem solving, you must be able to identify the particular system boundary (an imaginary surface that separates the system of interest and the environment surrounding the system). A system may be a single object or particle, a collection of objects or particles, or a region of space, and may vary in size and shape.

The scalar (or dot) product is introduced as a new mathematical skill in this chapter.

$$\vec{A} \cdot \vec{B} = AB \cos\theta \quad \text{where} \quad \theta \text{ is the angle between } \vec{A} \text{ and } \vec{B}.$$

Also, $\quad \vec{A} \cdot \vec{B} = A_x B_x + A_y B_y + A_z B_z$

$\vec{A} \cdot \vec{B}$ is a scalar, and the order of the product can be interchanged.

That is, $\quad \vec{A} \cdot \vec{B} = \vec{B} \cdot \vec{A}$

Furthermore, $\vec{A} \cdot \vec{B}$ can be positive, negative, or zero depending on the value of θ. (That is, $\cos\theta$ varies from -1 to $+1$.) If vectors are expressed in unit-vector form, then the dot product is conveniently carried out using the multiplication table for unit vectors:

$$\hat{i} \cdot \hat{i} = \hat{j} \cdot \hat{j} = \hat{k} \cdot \hat{k} = 1; \qquad \hat{i} \cdot \hat{j} = \hat{i} \cdot \hat{k} = \hat{k} \cdot \hat{j} = 0$$

The definite integral is introduced as a method to calculate the work done by a varying force. In Section 7.4, it is shown that the incremental quantity of work done by a variable force F_x in displacing a particle a small distance Δx is given by

$$\Delta W \approx F_x \Delta x$$

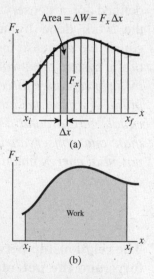

ΔW equals the area of the shaded rectangle in Figure (a) at right. The total work, W, done by F_x as the particle is displaced from x_i to x_f is approximately equal to the sum of the ΔW terms (i.e. the sum of the areas of the shaded rectangles).

As the width of the displacements (Δx) approach zero and the number of terms in the sum increases to infinity, the actual work done by F_x as the particle is displaced from x_i to x_f is equal to the definite integral stated in Equation 7.7:

$$W = \lim_{x \to \infty} \sum_{x_i}^{x_f} F_x \Delta x = \int_{x_i}^{x_f} F_x dx$$

This definite integral equals the area under the curve as illustrated in Figure (b) at right.

REVIEW CHECKLIST

You should be able to:

- Calculate the work done by a constant force using Equations 7.1 or 7.3. (Section 7.2)

- Calculate the scalar or dot product of any two vectors $\vec{A}$ and $\vec{B}$ using the definition $\vec{A} \cdot \vec{B} = AB \cos \theta$, or by writing $\vec{A}$ and $\vec{B}$ in unit-vector notation and using Equation 7.6. Also, find the value of the angle between two vectors. (Section 7.3)

- Calculate the work done by a variable force from the force vs. distance curve and by evaluating the integral, $\int F(x)dx$. (Section 7.4)

- Make calculations using the work-kinetic energy theorem. (Section 7.5)

- Calculate the potential energy function for a system when the conservative force acting on the system is given. (Section 7.6)

- Calculate the work done by conservative and nonconservative forces acting between two points. (Section 7.7)

- Calculate the force components as the negative derivative of the potential function. (Section 7.8)

- Plot the potential as a function of coordinate and determine points of equilibrium. (Section 7.9)

ANSWERS TO SELECTED OBJECTIVE QUESTIONS

5. Bullet 2 has twice the mass of bullet 1. Both are fired so that they have the same speed. If the kinetic energy of bullet 1 is K, is the kinetic energy of bullet 2 (a) $0.25K$ (b) $0.5K$ (c) $0.71K$ (d) K (e) $2K$?

Answer (e) The kinetic energy of the more massive bullet is twice that of the lower mass bullet.

□   □   □   □

8. As a simple pendulum swings back and forth, the forces acting on the suspended object are (a) the gravitational force, (b) the tension in the supporting cord, and (c) air resistance. (i) Which of these forces, if any, does no work on the pendulum at any time? (ii) Which of these forces does negative work on the pendulum at all times during its motion?

Answer (i) (b) The tension in the supporting cord does no work, because the motion of the pendulum is always perpendicular to the cord, and therefore to the force exerted by the cord. (ii) (c) The air resistance does negative work at all times when the pendulum bob is moving, because the air resistance is always acting in a direction opposite to the motion.

□   □   □   □

11. If the speed of a particle is doubled, what happens to its kinetic energy? (a) It becomes four times larger. (b) It becomes two times larger. (c) It becomes $\sqrt{2}$ times larger. (d) It is unchanged. (e) It becomes half as large.

Answer   (a) Kinetic energy, K, is proportional to the square of the velocity. Therefore if the speed is doubled, the kinetic energy will increase by a factor of four.

□   □   □   □

14. A certain spring that obeys Hooke's law is stretched by an external agent. The work done in stretching the spring by 10 cm is 4 J. How much additional work is required to stretch the spring an additional 10 cm? (a) 2 J   (b) 4 J   (c) 8 J   (d) 12 J   (e) 16 J.

Answer   (d) The force required to stretch the spring is proportional to the distance the spring is stretched, and since the work required is proportional to the force **and** to the distance, then W is proportional to x^2. This means if the extension of the spring is doubled, the work will increase by a factor of 4, so that to achieve a total extension of $x = 20$ cm from $x = 0$ requires net work $W = 16$ J, which means 12 J of additional work for the additional extension from $x = 10$ cm to $x = 20$ cm.

□   □   □   □

ANSWER TO A CONCEPTUAL QUESTION

5. Can kinetic energy be negative? Explain.

Answer   No. Kinetic energy $= mv^2/2$. Since v^2 is always positive, K is always positive.

□   □   □   □

SOLUTIONS TO SELECTED END-OF-CHAPTER PROBLEMS

1.   A block of mass $m = 2.50$ kg is pushed a distance $d = 2.20$ m along a frictionless horizontal table by a constant applied force of magnitude $F = 16.0$ N directed at an angle $\theta = 25.0°$ below the horizontal as shown in Figure P7.1. Determine the work done on the block by (a) the applied force, (b) the normal force exerted by the table, (c) the gravitational force, and (d) the net force on the block.

Solution

Conceptualize: Each answer should be between +100 J and −100 J, but watch out for which are positive, which are negative, and which are zero.

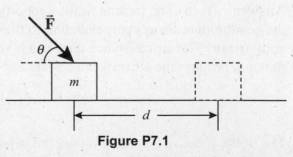

Figure P7.1

Categorize: We apply the definition of work by a constant force in the first three parts, but then in the fourth part we add up the answers. The total (net) work is the sum of the amounts of work done by the individual forces, and is the work done by the total (net) force. This identification is not represented by an equation in the chapter text, but is something you know by thinking about it, without relying on an equation in a list. Do not think that you can write down every idea in a "formula."

Analyze: The definition of work by a constant force is $W = F\Delta r \cos\theta$

(a) By the applied force, $W_{app} = (16.0 \text{ N})(2.20 \text{ m})\cos(25.0°) = 31.9$ J ∎

(b) By the normal force, $W_n = n\Delta r \cos\theta = n\Delta r \cos(90°) = 0$ ∎

(c) By force of gravitation, $W_g = F_g\Delta r \cos\theta = mg\Delta r \cos(90°) = 0$ ∎

(d) Net work done on the block: $W_{net} = W_{app} + W_n + W_g = 31.9$ J ∎

Finalize: The angle θ in the definition of work is always between force and (instantaneous) displacement. Draw the two vectors as starting from the same point to identify the angle most clearly. It need never be thought of as negative or as larger than 180°. Work measures the effectiveness of a force in changing the speed of an object. The forces perpendicular to the motion do zero work. None of the four forces here is pushing backward to tend to slow down the object, so none does negative work.

---

6. Spiderman, whose mass is 80.0 kg, is dangling on the free end of a 12.0-m-long rope, the other end of which is fixed to a tree limb above. By repeatedly bending at the waist, he is able to get the rope in motion, eventually getting it to swing enough that he can reach a ledge when the rope makes a 60.0° angle with the vertical. How much work was done by the gravitational force on Spiderman in this maneuver?

Solution

Conceptualize: The force of the Earth on the man is fairly large and tends to slow his upward motion. So it should do hundreds of joules of negative work on him.

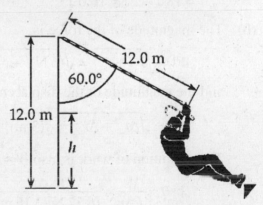

Categorize: As he pumps up his motion, the webslinger swings through several or many cycles, each going farther than the last. Gravity does positive work on him in each downswing and negative work in each upswing. We can simply use the definition of work by a constant force between the original starting point and the final endpoint.

Analyze: The work done is $W = \vec{F} \cdot \Delta\vec{r}$, where the gravitational force is

$$\vec{F}_g = -mg\hat{j} = (80.0 \text{ kg})(-9.80 \text{ m/s}^2\hat{j}) = -784\hat{j} \text{ N}$$

The hero travels through net displacement

$$\Delta \vec{\mathbf{r}} = L \sin 60°\hat{\mathbf{i}} + L(1 - \cos 60°)\hat{\mathbf{j}} = (12.0 \text{ m}) \sin 60°\hat{\mathbf{i}} + (12.0 \text{ m})(1 - \cos 60°)\hat{\mathbf{j}}$$

Thus the work is

$$W = \vec{\mathbf{F}} \cdot \Delta \vec{\mathbf{r}} = (-784\hat{\mathbf{j}} \text{ N}) \cdot (10.39\hat{\mathbf{i}} \text{ m} + 6.00\hat{\mathbf{j}} \text{ m}) = -4.70 \times 10^3 \text{ J}$$ ∎

Finalize: It is thousands of joules of negative work. A particle cannot do work on itself, but in the man-Earth system, Spiderman is not a structureless particle, so there is no contradiction in identifying his original and final kinetic energies as both zero. His muscles do thousands of joules of positive work. To identify the net displacement you need to study the diagram with care.

---

9.   A force $\vec{\mathbf{F}} = (6\hat{\mathbf{i}} - 2\hat{\mathbf{j}})$ N acts on a particle that undergoes a displacement $\Delta \vec{\mathbf{r}} = (3\hat{\mathbf{i}} + \hat{\mathbf{j}})$ m. Find (a) the work done by the force on the particle and (b) the angle between $\vec{\mathbf{F}}$ and $\Delta \vec{\mathbf{r}}$.

Solution

Conceptualize: The work will be on the order of ten joules and the angle less than 45°. Think about a sketch of the two vectors starting from the same point…

Categorize: …but we can get away without drawing a diagram. The definition of the dot product makes this an easy calculation.

Analyze:

(a)   We use the mathematical representation of the definition of work. $W = \vec{\mathbf{F}} \cdot \Delta \vec{\mathbf{r}} =$
$(6\hat{\mathbf{i}} - 2\hat{\mathbf{j}}) \cdot (3\hat{\mathbf{i}} + 1\hat{\mathbf{j}}) = (6 \text{ N})(3 \text{ m}) + (-2 \text{ N})(1 \text{ m}) + 0 + 0$

$$= 18 \text{ J} - 2 \text{ J} = 16.0 \text{ J}$$ ∎

(b)   The magnitude of the force is

$$|\vec{\mathbf{F}}| = \sqrt{F_x^2 + F_y^2} = \sqrt{(6 \text{ N})^2 + (-2 \text{ N})^2} = 6.32 \text{ N}$$

and the magnitude of the displacement is

$$|\Delta \vec{\mathbf{r}}| = \sqrt{\Delta r_x^2 + \Delta r_y^2} = \sqrt{(3 \text{ m})^2 + (1 \text{ m})^2} = 3.16 \text{ m}$$

The definition of work is also $W = F\Delta r \cos \theta$ so we have

$$\cos\theta = \frac{W}{F\Delta r} = \frac{16.0 \text{ J}}{(6.32 \text{ N})(3.16 \text{ m})} = 0.800$$

and $\theta = \cos^{-1}(0.800) = 36.9°$ ∎

Finalize: Disastrous mistakes you might make include writing the work as $(18\hat{\mathbf{i}} - 2\hat{\mathbf{j}})$J and not recognizing that there is real and useful knowledge in having more than one way to calculate work.

---

15. A particle is subject to a force F_x that varies with position as shown in Figure P7.15. Find the work done by the force on the particle as it moves (a) from $x = 0$ to $x = 5.00$ m, (b) from $x = 5.00$ m to $x = 10.0$ m, and (c) from $x = 10.0$ m to $x = 15.0$ m. (d) What is the total work done by the force over the distance $x = 0$ to $x = 15.0$ m?

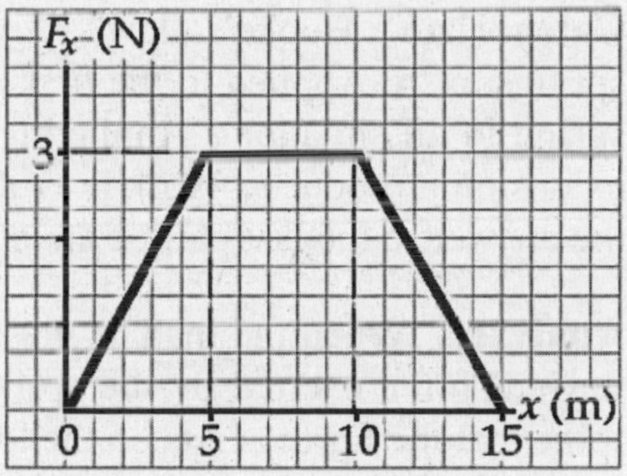

Figure P7.15

Solution

Conceptualize: A few newtons, always pulling forward, over several meters does work between 10 J and 100 J.

Categorize: We use the graphical representation of the definition of work. W equals the area under the force-displacement curve.

Analyze: This definition is still written $W = \int F_x\, dx$ but it is computed geometrically by identifying triangles and rectangles on the graph.

(a) For the region $0 \leq x \leq 5.00$ m $W = \dfrac{(3.00\ \text{N})(5.00\ \text{m})}{2} = 7.50$ J   ∎

(b) For the region 5.00 m $\leq x \leq 10.0$ m $W = (3.00\ \text{N})(5.00\ \text{m}) = 15.0$ J   ∎

(c) For the region 10.0 m $\leq x \leq 15.0$ m $W = \dfrac{(3.00\ \text{N})(5.00\ \text{m})}{2} = 7.50$ J   ∎

(d) For the region $0 \leq x \leq 15.0$ m $W = (7.50\ \text{J} + 7.50\ \text{J} + 15.0\ \text{J}) = 30.0$ J   ∎

Finalize: Make sure you know how to find the area of a triangle. The particle on which this force acts will have good reason to speed up. Without knowing its mass or original speed we know its change in kinetic energy is 30 J if this force is the only force doing work on it.

---

17. When a 4.00-kg object is hung vertically on a certain light spring that obeys Hooke's law, the spring stretches 2.50 cm. If the 4.00-kg object is removed, (a) how far will the spring stretch if a 1.50-kg block is hung on it? (b) How much work must an external agent do to stretch the same spring 4.00 cm from its unstretched position?

Solution

Conceptualize: Hooke's law is a rule of direct proportionality. The second load is less than half of the original load. Then the second extension will be correspondingly less than half of the first extension; we estimate 1 cm. The third load, producing the 4-cm extension, is larger than either of the first two. The force producing it would be about the weight of 7 kg, or about 70 N. The work that this load does in stretching the spring, though, will be much less than $(70\ \text{N})(0.04\ \text{m}) = 2.8$ J, because it is a varying force and not a constant force that stretches the spring.

Categorize: Hooke's "law" $F_{spring} = -kx$ is an empirical description of the forces some springs exert. Applied to the first loading situation, this law will tell us the k value of this spring in all situations. Applied to the second load, the same law will tell us the smaller extension it produces. Work is a different physical quantity, and a different rule will give us the answer to the last part of the problem.

Analyze: When the load of mass $M = 4$ kg is hanging on the spring in equilibrium, the upward force exerted by the spring on the load is equal in magnitude to the downward force that the Earth exerts on the load, given by $w = Mg$. Then we can write Hooke's law as $Mg = +kx$. The spring constant, force constant, stiffness constant, or Hooke's-law constant of the spring is given by

$$k = \frac{F}{x} = \frac{Mg}{x} = \frac{(4.00)\,(9.80)\,\text{N}}{2.50 \times 10^{-2}\,\text{m}} = 1.57 \times 10^3\,\text{N/m}$$

(a) For the 1.50-kg mass the extension is

$$x = \frac{mg}{k} = \frac{(1.50\,\text{kg})\,\left(9.80\,\text{m/s}^2\right)}{1.57 \times 10^3\,\text{N/m}} = 0.009\,38\,\text{m} = 0.938\,\text{cm}\qquad \blacksquare$$

(b) The chapter text proves, from the definition of work by a variable force and Hooke's law, that the work an outside agent does to distort a spring from a configuration with extension x_i to extension x_f is

$$W_{app} = \frac{1}{2}kx_f^2 - \frac{1}{2}kx_i^2$$

Then we do not need to calculate the force exerted by the third load. The work it does in stretching the spring to $x = 4$ cm, from $x = 0$ at its natural length is

$$W_{app} = \frac{1}{2}\left(1.57 \times 10^3\,\text{N/m}\right)\left(4.00 \times 10^{-2}\,\text{m}\right)^2 - 0 = 1.25\text{J}\qquad \blacksquare$$

Finalize: Our answer to (a) agrees with our estimate, and our answer to (b) is in the range we identified as reasonable. Notice that the length of the spring is not given or asked for, either in the undistorted or a stretched configuration. The x in Hooke's law always represents a *change* in length from the natural length. It can be a compression distance as well as an extension. From $\frac{1}{2}kx^2$, positive work is done either to stretch or to compress a spring. In contrast, the spring exerts oppositely directed, negative and positive, forces $-kx$ when it is stretched or compressed. Further, x need not be horizontal. In this problem it is vertical in the first loading situation and in that of part (a); x could be in any direction in part (b). There are lots of things you might be confused by: the spring exerts equal-magnitude forces on the objects attached to both of its ends. Hooke's law describes the strength of the force at the "fixed" end, as well as at the end where we visualize measuring the distortion distance. While Hooke's law does not describe all springs, notably failing for large distortions, the law describes things that you would not think of as springs. The small sag of a sturdy bench when you sit on it, or of a floor when you stand on it, accurately follows the rule Hooke first wrote as "ut tensio sic vis," or "as with the distortion, so with the force."

21. A light spring with spring constant k_1 is hung from an elevated support. From its lower end a second light spring is hung, which has spring constant k_2. An object of mass m is hung at rest from the lower end of the second spring. (a) Find the total extension distance of the pair of springs. (b) Find the effective spring constant of the pair of springs as a system.

Solution

Conceptualize: If the pair of suspended springs were surrounded by a cardboard box, with just the hook at the bottom of the second spring revealed, we could exert different forces on the hook, observe the extensions they produce, and prove that the pair of springs constitute a single device described by Hooke's law.

Categorize: In terms of the symbols stated in the problem, we will find the extension distances of the two springs separately. That will lead us to the total extension of the pair. If our answer to part (a) shows that the total extension is proportional to the force causing it, we will have a mathematical proof that Hooke's law describes the pair, and the effective spring constant of the system will be identifiable in the expression for total extension.

Analyze: (a) The force mg is the tension in each of the springs. The bottom of the upper (first) spring moves down by distance $x_1 = |F|/k_1 = mg/k_1$. The top of the second spring moves down by this distance, and the second spring also stretches by $x_2 = mg/k_2$. The bottom of the lower spring then moves down by distance

$$x_{\text{total}} = x_1 + x_2 = mg/k_1 + mg/k_2 = mg(1/k_1 + 1/k_2)$$ ■

(b) From the last equation we have $mg = \dfrac{x_1 + x_2}{\frac{1}{k_1} + \frac{1}{k_2}}$.

This is of the form $|F| = \left(\dfrac{1}{1/k_1 + 1/k_2}\right)(x_1 + x_2)$

The downward displacement is opposite in direction to the upward force the springs exert on the load, so we may write $F = -k_{eff}\, x_{\text{total}}$ with the effective spring constant for the pair of springs given by

$$k_{eff} = \dfrac{1}{1/k_1 + 1/k_2}$$ ■

Finalize: The pair of springs are described as *in series*. We have proved that Hooke's law describes the linked pair. It would be misleading to call the effective spring constant a "total spring constant." In fact, the value of k_{eff} is always less than the smaller of k_1 and k_2. For example, if $k_1 = 8$ N/m and $k_2 = 12$ N/m, then $1/(1/k_1 + 1/k_2) = 4.8$ N/m. It is easier to stretch the pair of springs in series by a centimeter than it is to stretch either spring by itself.

---

35. A 2 100-kg pile driver is used to drive a steel I-beam into the ground. The pile driver falls 5.00 m before coming into contact with the top of the beam, and it drives the beam 12.0 cm farther into the ground before coming to rest. Using energy considerations, calculate the average force the beam exerts on the pile driver while the pile driver is brought to rest.

Solution

Conceptualize: A pile driver works like a hammerhead, falling vertically between guides. Anyone who has hit his or her thumb with a hammer knows that the resulting force is greater than just the weight of the hammer, so we should also expect the force of the pile driver to be significantly greater than its weight:

$F \gg mg \sim 20$ kN. The force **on** the pile driver will be directed upwards.

Categorize: The initial and final kinetic energies of the driver are zero. The average force stopping the driver can be found from the work done by the gravitational force that starts its motion.

Analyze: Choose the initial point when the driver is elevated, before its release, and the final point when it comes to rest again 5.12 m below. Two forces do work on the pile driver: the gravitational force (weight) and the normal force exerted by the beam on the pile driver.

$$K_f - K_i = W_{ext} \qquad \text{so that} \qquad 0 - 0 = mgd_w \cos(0) + nd_n \cos(180°)$$

Where $d_w = 5.12$ m, $d_n = 0.120$ m, and $m = 2\,100$ kg

We identified the angles by noting that the weight vector is in the direction of motion and the beam exerts a force on the pile driver opposite the direction of motion.

$$\text{Solving, } n = \frac{-mgd_w \cos 0°}{d_n \cos 180°} = \frac{-(2\,100 \text{ kg})\,(9.80 \text{ m/s}^2)\,(5.12 \text{ m})\,(1)}{0.120 \text{ m}\,(-1)}$$

$$\text{Computing,} \qquad n = \frac{1.05 \times 10^5 \text{ J}}{0.120 \text{ m}} = 878 \text{ kN upward} \qquad ■$$

Finalize: The normal force is larger than 20 kN as we expected. It is actually about 43 times greater than the weight of the pile driver, which is why this machine is so effective.

As an additional calculation, let us show that the work done by the gravitational force on an object can in general be represented by *mgh*, where *h* is the vertical height that the object falls. Then we can apply this result to the problem above.

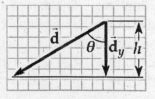

From the diagram, where $\vec{\mathbf{d}}$ is the displacement of the object, and *h* is the height through which the object falls, $h = \left|\vec{\mathbf{d}}_y\right| = d\cos\theta$

Since $\quad F_g = mg, \qquad mgh = F_g d \cos\theta = \vec{\mathbf{F}}_g \cdot \vec{\mathbf{d}}$

In this problem, $\quad mgh = n(d_n)$ so $n = mgh/d_n$

and again $\quad n = (2\,100 \text{ kg})(9.80 \text{ m/s}^2)(5.12 \text{ m})/(0.120 \text{ m}) = 878 \text{ kN}$ $\qquad ■$

43. A 4.00-kg particle moves from the origin to position ⓒ, having coordinates $x = 5.00$ m and $y = 5.00$ m (Fig. P7.43). One force on the particle is the gravitational force acting in the negative y direction. Using Equation 7.3, calculate the work done by the gravitational force on the particle as it goes from O to ⓒ along (a) the purple path OⒶⒸ, (b) the red path OⒷⒸ, and (c) the direct blue path OⒸ. (d) Your results should all be identical. Why?

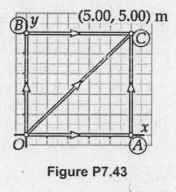

Figure P7.43

Solution

Conceptualize: Visualize carrying a gallon milk container from the ground floor up a staircase to the third floor of a dormitory, and then along the corridor to your room. The work you do will be several hundred joules, and the work done by the planet's gravitation in the lifting process will be negative several hundred joules.

Categorize: We apply the definition of work by a constant force, add up amounts of work over sequential displacements, and keep track of results.

Analyze: The gravitational force is downward

$$F_g = mg = (4.00 \text{ kg})(9.80 \text{ m/s}^2) = 39.2 \text{ N}$$

(a) The work on the path by way of point Ⓐ is $\quad W_{OAC} = W_{OA} + W_{AC}$

By the definition, $\quad W_{OAC} = F_g d_{OA} \cos(90°) + F_g d_{AC} \cos(180°)$

$$W_{OAC} = (39.2 \text{ N})(5.00 \text{ m})(0) + (39.2 \text{ N})(5.00 \text{ m})(-1) = -196 \text{ J} \quad \blacksquare$$

(b) We do the same steps for a trip by way of point Ⓑ: $\quad W_{OBC} = W_{OB} + W_{BC}$

$$W_{OBC} = (39.2 \text{ N})(5.00 \text{ m}) \cos 180° + (39.2 \text{ N})(5.00 \text{ m}) \cos 90° = -196 \text{ J} \quad \blacksquare$$

(c) And for the straight path

$$W_{OC} = F_g d_{OC} \cos 135° = (39.2 \text{ N})(5\sqrt{2}\text{m}) \left(-\frac{\sqrt{2}}{2} \right) = -196 \text{ J} \quad \blacksquare$$

(d) The results should all be the same since the gravitational force is conservative, and the work done by a conservative force is independent of the path. $\quad \blacksquare$

Finalize: The results are all the same, and are, as estimated, negative some hundreds of joules. If the object fell back to its starting point, the Earth would do +196 J of work on it in the downward trip. If it started from rest and fell freely, it would have 196 J of kinetic energy just before landing and we could find its speed from $(1/2)mv^2 = 196$ J. The answer would agree with a calculation based on a particle with constant acceleration.

45. A force acting on a particle moving in the xy plane is given by $\vec{F} = (2y\hat{i} + x^2\hat{j})$, where $\vec{F}$ is in newtons and x and y are in meters. The particle moves from the origin to a final position having coordinates $x = 5.00$ m and $y = 5.00$ m as shown in Figure P7.45. Calculate the work done by $\vec{F}$ on the particle as it moves along (a) the purple path $O\text{Ⓐ}\text{Ⓒ}$, (b) the red path $O\text{Ⓑ}\text{Ⓒ}$, and (c) the direct blue path $O\text{Ⓒ}$. (d) Is $\vec{F}$ conservative or nonconservative? (e) Explain your answer to part (d).

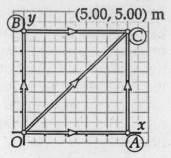

Figure P7.45

Solution

Conceptualize: In contrast to problem 43, the force is variable, so it may be nonconservative. The force is directed upward and to the right, so it will do positive work of some tens of joules on the particle.

Categorize: To account for the work done by a variable force, we must integrate the force over the path.

Analyze: In the following integrals, remember that

$$\hat{i}\cdot\hat{i} = \hat{j}\cdot\hat{j} = 1 \qquad \text{and} \qquad \hat{i}\cdot\hat{j} = 0$$

(a) The work is $W = \int_{\text{starting point}}^{\text{endpoint}} \vec{F}\cdot d\vec{r}$. We compute it for each step of the motion. For the displacement from O to Ⓐ,

$$W_{OA} = \int_0^{5.00}\left(2y\hat{i}+x^2\hat{j}\right)\cdot\left(\hat{i}dx\right) = \int_0^{5.00} 2y\,dx = 2y\int_0^{5.00} dx = 2yx\Big]_{x=0,\,y=0}^{x=5.00,\,y=0} = 0$$

Next, $\quad W_{AC} = \int_0^{5.00}\left(2y\hat{i}+x^2\hat{j}\right)\cdot\left(\hat{j}dy\right) = \int_0^{5.00} x^2\,dy = x^2\int_0^{5.00} dy = x^2 y\Big]_{x=5.00,\,y=0}^{x=5.00,\,y=5.00} = 125\text{ J}$

So the work along the path by way of Ⓐ is $\quad W_{AC} = 0 + 125\text{ J} = 125\text{ J}$ ∎

We repeat the same steps for the rest of the calculations:

(b) $W_{OB} = \int_0^{5.00}\left(2y\hat{i}+x^2\hat{j}\right)\cdot\left(\hat{j}dy\right) = \int_0^{5.00} x^2\,dy = x^2\int_0^{5.00} dy = x^2 y\Big]_{x=0...}^{x=0...} = 0$

$W_{BC} = \int_0^{5.00}\left(2y\hat{i}+x^2\hat{j}\right)\cdot\left(\hat{i}dx\right) = \int_0^{5.00} 2y\,dx = 2y\int_0^{5.00} dx = 2(5.00)x\Big]_{x=0}^{x=5.00} = 50.0\text{ J}$

$W_{OBC} = 0 + 50.0\text{ J} = 50.0\text{ J}$ ∎

(c) $W_{OC} = \int\left(2y\hat{i}+x^2\hat{j}\right)\cdot\left(\hat{i}dx+\hat{j}dy\right) = \int_{x=0,\,y=0}^{x=5.00,\,y=5.00}\left(2y\,dx+x^2\,dy\right)$

Since $x = y$ along $O\text{Ⓒ}$, $dx = dy$ and

$$W_{OC} = \int_0^{5.00}\left(2x+x^2\right)dx = 2\frac{x^2}{2}+\frac{x^3}{3}\Big]_{x=0}^{5} = 25+\frac{125}{3}-0 = 66.7\text{ J}$$ ∎

(d) 125 J, 50 J, and 66.7 J are all different amounts of work. $\vec{F}$ is nonconservative. ∎

(e) The work done on the particle depends on the path followed by the particle. ∎

Finalize: You have been learning it in calculus class so that you can do it here! The integrals of $x\,dx$ and $x^2\,dx$ may seem the most familiar, but the integral of dx is encountered remarkably often. To do the integral of $y\,dx$, you must know how y depends on x. Study with care how the integrals are set up, and how you read what $d\vec{r}$ is from the path.

---

47. The potential energy of a system of two particles separated by a distance r is given by $U(r) = A/r$, where A is a constant. Find the radial force $\bar{F}_r$ that each particle exerts on the other.

Solution

Conceptualize: The force will be a function of the separation distance r. If A is positive, the potential energy is high when the particles are close together, so the force will describe the particles repelling each other.

Categorize: We use the relation of force to potential energy as…

Analyze: …the force is the negative derivative of the potential energy with respect to distance:

$$F_r = -\frac{dU}{dr} = -\frac{d}{dr}\left(Ar^{-1}\right) = -A(-1)\,r^{-2} = \frac{A}{r^2}$$ ∎

Finalize: With A positive, the positive force means radially away from the other object. This describes an inverse-square-law force of repulsion, as between two negative point electric charges. The force gets one ninth as large if the distance gets three times larger, and the potential energy gets one third as large.

---

49. A single conservative force acts on a 5.00-kg particle within a system due to its interaction with the rest of the system. The equation $F_x = 2x + 4$ describes the force, where F_x is in newtons and x is in meters. As the particle moves along the x axis from $x = 1.00$ m to $x = 5.00$ m, calculate (a) the work done by this force on the particle, (b) the change in the potential energy of the system, and (c) the kinetic energy the particle has at $x = 5.00$ m if its speed is 3.00 m/s at $x = 1.00$ m.

Solution

Conceptualize: The force grows linearly in strength from $2(1) + 4 = 6$ N to 14 N. We could draw a graph and find the area of a trapezoid to get about $\left(\dfrac{6+14}{2}\right)$ N$(5-1)$ m $= 40$ J. You should make sure you know where the + sign and the − sign in this expression are coming from. But…

Categorize: …it will be straightforward to calculate the work from its definition as an integral. Then the definition of potential energy and the work-kinetic energy theorem are applied in parts (b) and (c).

Analyze:

(a) For a particle moving along the x axis, the definition of work by a variable force is

$$W_F = \int_{x_i}^{x_f} F_x \, dx$$

Here $F_x = (2x + 4)$ N, $x_i = 1.00$ m, and $x_f = 5.00$ m

So $W_F = \int_{1.00\,m}^{5.00\,m} (2x + 4) dx$ N $\cdot$ m $= x^2 + 4x]_{1.00m}^{5.00m}$ N $\cdot$ m $= \left(5^2 + 20 - 1 - 4 \right)$ J $= 40.0$ J ∎

(b) The change in potential energy of the system equals the negative of the internal work done by the conservative force.

$$\Delta U = -W_F = -40.0 \text{ J}$$ ∎

(c) The work-kinetic energy theorem gives $K_f - K_i = W_{net}$

Then $K_f = K_i + W = \frac{1}{2}mv_i^2 + W = \frac{1}{2}(5.00 \text{ kg})(3.00 \text{ m/s})^2 + 40.0 \text{ J} = 62.5 \text{ J}$ ∎

Finalize: It would be natural to tie off the problem by finding the final speed of the particle from $\frac{1}{2}m\,v_f^2 = K_f$ $v_f = \sqrt{\left(2\,K_f / m \right)}$

$$v_f = (2[62.5 \text{ m}^2/\text{s}^2]/5)^{1/2} = 5.00 \text{ m/s}$$

To preview the next chapter, we could also write the equation describing this situation $\Delta K = W_{net}$ as $\Delta K + \Delta U = 0$ or $K_i + U_i = K_f + U_f$. The situation can be thought of as 40 J of original potential energy turning into 40 J of extra kinetic energy as the total energy of the system is constant. We say the total system energy is conserved.

---

67.   A light spring has unstressed length 15.5 cm. It is described by Hooke's law with spring constant 4.30 N/m. One end of the horizontal spring is held on a fixed vertical axle, and the other end is attached to a puck of mass m that can move without friction over a horizontal surface. The puck is set into motion in a circle with a period of 1.30 s. (a) Find the extension of the spring x as it depends on m. Evaluate x for (b) $m = 0.070\ 0$ kg, (c) $m = 0.140$ kg, (d) $m = 0.180$ kg, and (e) $m = 0.190$ kg. (f) Describe the pattern of variation of x as it depends on m.

Solution

Conceptualize: The puck speed is constant at any particular setting, but to make its velocity change in direction the spring must exert force on the puck, and must exert more force on a more massive puck. The spring can do this by stretching more, but there is some feedback: The more the spring extends, the larger the radius of the puck's motion and the higher its speed. We expect the extension to increase as the mass goes up, but it is hard to guess the pattern. From the square in mv^2/r, could the extension be proportional to the square of the mass?

Categorize: We will model the puck as a particle in uniform circular motion with $\Sigma F = mv^2/r$ and $v = 2\pi r/T$, model the spring as described by Hooke's law $|F| = kx$, and assemble an expression for x by combining simultaneous equations by substitution.

Analyze: (a) With the spring horizontal, the radius of the puck's motion is

$$r = 0.155 \text{ m} + x.$$

The spring force causes the centripetal acceleration according to $kx = mv^2/r$ so we have

$$(4.3 \text{ N/m})x = mv^2/r = m(2\pi r/T)^2/r = 4\pi^2 \, mr/(1.3 \text{ s})^2$$

which becomes

$$(4.3 \text{ kg/s}^2)x = (23.4/\text{s}^2)m(0.155 \text{ m} + x) = 3.62 \, m \text{ m/s}^2 + 23.4 \, mx/\text{s}^2$$

or $4.300 \text{ kg } x - 23.36 \, m \, x = 3.6208 \, m$ m

We solve for x to obtain $x = \dfrac{3.62m}{(4.30 \text{ kg} - 23.4m)}$ meters

which we write as $x = \dfrac{3.62m}{4.30 - 23.4m}$, where x is in m and m is in kg. ∎

(b) We expect just to substitute from now on. When the mass is 0.070 0 kg the

extension is $x = \dfrac{3.62 \text{ m}(0.070 \ 0 \text{ kg})}{(4.30 \text{ kg} - 23.4[0.070 \ 0 \text{ kg}])} = 0.095 \ 1 \text{ m}$ ∎

This is a nice reasonable extension, with the spring 25.1 cm long.

(c) We double the puck mass and find

$$x = \dfrac{3.620 \ 8 \text{ m}(0.140 \text{ kg})}{(4.30 \text{ kg} - 23.360[0.140 \text{ kg}])} = 0.492 \text{ m}$$ ∎

The extension is more than twice as big.

(d) Just 40 grams more and $x = \dfrac{3.62 \text{ m}(0.180 \text{ kg})}{(4.30 \text{ kg} - 23.4[0.180 \text{ kg}])} = 6.85 \text{ m}$ ∎

we have to move to the gymnasium floor.

(e) When the denominator of the fraction goes to zero the extension becomes infinite. This happens for $4.30 \text{ kg} - 23.4 \, m = 0$; that is, for $m = 0.184$ kg. For any larger mass the spring cannot constrain the extension. The situation for 0.190 kg is impossible. ∎

(f) The extension starts from zero for $m = 0$ and is proportional to m when m is only a few grams. Then it grows faster and faster, diverging to infinity for $m = 0.184$ kg. ∎

Finalize: The answer to part (f) is already a good summary. A reasonable-sounding problem does not always have one answer. It might have two or more, or, as in part (e) here, none at all.

8

Conservation of Energy

EQUATIONS AND CONCEPTS

The **principle of conservation of energy** is described mathematically by Equation 8.1. E represents the total energy of a system and T is the quantity of energy transferred across the system boundary by any transfer mechanism.

$$\Delta E_{\text{system}} = \sum_{} T \qquad (8.1)$$

The **total mechanical energy** of a system is the sum of the kinetic and potential energies. *In Equation 8.7, U represents the total of all types of potential energy.*

$$E_{\text{mech}} = K + U \qquad (8.7)$$

Energy conservation in an isolated system:

When no nonconservative forces are acting:

The **mechanical energy** of the system is conserved.

$$\Delta K + \Delta U = 0 \qquad (8.6)$$
or
$$K_f + U_f = K_i + U_i \qquad (8.10)$$

When nonconservative forces are acting:

The **total energy** of the system is conserved.

$$\Delta E_{\text{system}} = 0 \qquad (8.9)$$

The **internal energy of a system** will change as a result of transfer of energy across the system boundary by friction forces. *The increase in internal energy is equal to the decrease in kinetic energy.*

$$\Delta E_{\text{int}} = f_k d \qquad (8.15)$$

When friction and other forces are present, the mechanical energy of the system will not be conserved. In Equation 8.16, ΔU is the change in all forms of potential energy. *Note also that Equation 8.16 is equivalent to Equation 8.10 when the friction force is zero.*

$$\Delta E_{\text{mech}} = \Delta K + \Delta U = -f_k d \qquad (8.16)$$
(isolated system)

$$\Delta E_{\text{mech}} = -f_k d + \sum W_{\text{other forces}} \qquad (8.17)$$
(nonisolated system)

The **general expression for power** defines power as any type of energy transfer. The rate at which energy crosses the boundary of a system by any transfer mechanism is dE/dt.

$$P \equiv \frac{dE}{dt} \qquad (8.18)$$

The **average power** supplied by a force is the ratio of the work done by that force to the time interval over which it acts.

$$P_{avg} = \frac{W}{\Delta t}$$

The **instantaneous power** is equal to the limit of the average power as the time interval approaches zero.

$$P = \frac{dW}{dt} = \vec{\mathbf{F}} \cdot \vec{\mathbf{v}} \qquad (8.19)$$

The **SI unit of power** is J/s, which is called a watt (W).

$1\ W = 1\ J/s$

$1\ kW \cdot h = 3.60 \times 10^6\ J$

The **unit of power in the US customary system** is the horsepower (hp).

$1\ hp = 746\ W$

SUGGESTIONS, SKILLS, AND STRATEGIES

CONSERVATION OF ENERGY

Take the following steps in applying the principle of conservation of energy:

- Identify the system of interest, which may consist of more than one object.

- Select a reference position for the zero point of gravitational potential energy.

- Determine whether or not nonconservative forces are present.

- **If only conservative forces are present** (mechanical energy is conserved), you can write the total initial energy at some point as the sum of the kinetic and potential energies at that point ($K_i + U_i$). Then, write an expression for the total energy ($K_f + U_f$) at the final point of interest. Since mechanical energy is conserved, you can use Equation 8.10, $K_f + U_f = K_i + U_i$.

- **If nonconservative forces such as friction are present** (and thus mechanical energy is not conserved), first write expressions for the total initial and total final energies. In this case, the difference between the two total energies is equal to the decrease in mechanical energy due to the presence of nonconservative force(s); and Equation 8.16 applies: $\Delta E_{mech} = \Delta K + \Delta U = -fd$.

REVIEW CHECKLIST

You should be able to:

• Identify a mechanical system of interest, determine whether or not nonconservative forces are present and apply Equation 8.10 or Equation 8.16 as appropriate. (Sections 8.1, 8.2, 8.3, and 8.4)

• Make calculations of average and instantaneous power (Section 5)

ANSWER TO AN OBJECTIVE QUESTION

6. A ball of clay falls freely to the hard floor. It does not bounce noticeably, and it very quickly comes to rest. What then has happened to the energy the ball had while it was falling? (a) It has been used up in producing the downward motion. (b) It has been transformed back into potential energy. (c) It has been transferred into the ball by heat. (d) It is in the ball and floor (and walls) as energy of invisible molecular motion. (e) Most of it went into sound.

Answer (d) is the only correct answer. Energy is never 'used up' when it changes into another form. The ball of clay on the floor does not have gravitational potential energy, elastic potential energy, or any other kind of potential energy. The ball has risen in temperature, but not because of receiving energy transferred by heat from a stove. The ball would fall and stop in very nearly the same way if it were in a vacuum, radiating no sound energy at all. Nearly all of the kinetic energy that the ball has at impact is degraded into kinetic energy of rapidly damped vibrations in the clay and the floor, turning into kinetic energy of random molecular vibration. The energy of invisible molecular motion is named *internal energy*. Even the tiny fraction of a joule of energy that went off with the sound turns completely into internal energy when the sound is absorbed.

ANSWERS TO SELECTED CONCEPTUAL QUESTIONS

3. One person drops a ball from the top of a building while another person at the bottom observes its motion. Will these two people agree (a) on the value of the gravitational potential energy of the ball–Earth system? (b) On the change in potential energy? (c) On the kinetic energy of the ball at some point in its motion?

Answer (a) The two will not necessarily agree on the potential energy, since this depends on the origin—which may be chosen differently for the two observers. (b) However, the two must agree on the value of the change in potential energy, which is independent of the choice of the reference frames. (c) The two will also agree on the kinetic energy of the ball, assuming both observers are at rest with respect to each other, and hence measure the same v.

□   □   □   □

4. You ride a bicycle. In what sense is your bicycle solar-powered?

Answer The energy to move the bicycle comes from your body. The source of that energy is the food that you ate at some previous time. The energy in the food, assuming that we focus on vegetables, came from the growth of the plant, for which photosynthesis is a major factor. The light for the photosynthesis comes from the Sun. The argument for meats has a couple of extra steps, but also goes through the process of photosynthesis in the plants eaten by animals. Thus, the source of the energy to ride the bicycle is the Sun, and your bicycle is solar-powered!

□   □   □   □

5. A bowling ball is suspended from the ceiling of a lecture hall by a strong cord. The ball is drawn away from its equilibrium position and released from rest at the tip of the demonstrator's nose as shown in Figure CQ8.5. The demonstrator remains stationary. (a) Explain why the ball does not strike her on its return swing. (b) Would this demonstrator be safe if the ball were given a push from its starting position at her nose?

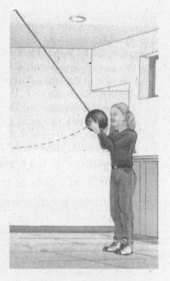

Answer (a) The total energy of the system (the bowling ball and the Earth) must be conserved. Since the system initially has a potential energy mgh, and the ball has no kinetic energy, it cannot have any kinetic energy when returning to its initial position. Of course, air resistance will cause the ball to return to a point slightly below its initial position. (b) On the other hand, if the ball is given a forward push anywhere along its path, the demonstrator's nose will be in big trouble.

Figure CQ8.5

□   □   □   □

SOLUTIONS TO SELECTED END-OF-CHAPTER PROBLEMS

3. A block of mass 0.250 kg is placed on top of a light vertical spring of force constant 5 000 N/m and pushed downward so that the spring is compressed by 0.100 m. After the block is released from rest, it travels upward and then leaves the spring. To what maximum height above the point of release does it rise?

Solution

Conceptualize: The spring exerts on the block a maximum force of (5 000 N/m)(0.1 m) = 500 N. This is much larger than the 2.5-N weight of the block, so we may expect the spring cannon to fire the block a couple of meters into the air.

Categorize: Thinking about the spring force in the Conceptualize step is good for estimation, but does not fit into a problem solution at our mathematical level. The block moves with changing acceleration while the spring pushes on it, so none of our theory about a particle under constant acceleration works. This is a conservation of energy problem.

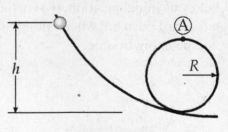

Analyze: In both the initial and final states, the block is not moving. We choose the zero configuration of gravitational energy to be the block's starting point. Therefore, the initial and final energies of the block-spring-Earth system are:

initial $\qquad E_i = K_i + U_i = 0 + (U_g + U_s)_i = 0 + (0 + \frac{1}{2}kx^2)$

final $\qquad E_f = K_f + U_f = 0 + (U_g + U_s)_f = 0 + (mgh + 0)$

Then we have simply $\qquad E_i = E_f \qquad$ or $\qquad mgh = \frac{1}{2}kx^2$

Solving, $\qquad h = \dfrac{kx^2}{2mg} = \dfrac{(5000 \text{ N/m})(0.100 \text{ m})^2}{2(0.250 \text{ kg})(9.80 \text{ m/s}^2)} = 10.2 \text{ m}$ ∎

Finalize: This spring cannon can bombard squirrels in treetops. The initial and final pictures help you to explain why no kinetic energy term appears in the energy equation and why there is no spring energy in the final picture. Get used to the idea that there can be one or several kinds of potential energy. We are free to choose $y = 0$ where we like for gravitational energy, but we must choose $x = 0$ at the unstressed configuration of the spring.

---

5. A bead slides without friction around a loop-the-loop (Fig. P8.5). The bead is released from a height $h = 3.50R$. (a) What is its speed at point Ⓐ? (b) How large is the normal force on the bead at point Ⓐ if its mass is 5.00 g?

Solution

Figure P8.5

Conceptualize: Since the bead is released above the top of the loop, it will start with enough potential energy to later reach point Ⓐ and still have excess kinetic energy. The energy of the bead at point Ⓐ will be proportional to h and g. If it is moving relatively slowly, the track will exert an upward force on the bead, but if it is whipping around fast, the normal force will push it toward the center of the loop.

Categorize: The speed at the top can be found from the conservation of energy for the bead-track-Earth system, and the normal force can be found from Newton's second law.

Analyze:

(a) We define the bottom of the loop as the zero level for the gravitational potential energy.

Since $v_i = 0, \qquad E_i = K_i + U_i = 0 + mgh = mg(3.50R)$

The total energy of the bead at point Ⓐ can be written as

$$E_A = K_A + U_A = \tfrac{1}{2}mv_A^2 + mg(2R)$$

Since mechanical energy is conserved, $\qquad E_i = E_A$

and we get $\qquad mg(3.50R) = \tfrac{1}{2}mv_A^2 + mg(2R)$

simplifying, $\qquad v_A^2 = 3.00\,gR$

solving, $\qquad v_A = \sqrt{3.00gR}$ ∎

(b) To find the normal force at the top, we construct a force diagram as shown, where we assume that n is downward, like mg. Newton's second law gives $\Sigma F = ma_c$, where a_c is the centripetal acceleration.

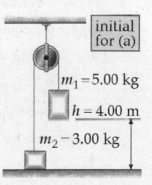

$$\Sigma F_y = ma_y$$

$$n + mg = \frac{mv_A^2}{R} = \frac{m(3.00gR)}{R} = 3.00mg$$

$$n = 3.00mg - mg = 2.00\ mg$$

$$n = 2.00\big(5.00 \times 10^{-3}\,\text{kg}\big)\big(9.80\ \text{m/s}^2\big) = 0.098\ 0\ \text{N downward}$$ ∎

Finalize: Our answer represents the speed at point Ⓐ as proportional to the square root of the product of g and R, but we must not think that simply increasing the diameter of the loop will increase the speed of the bead at the top. Instead, the speed will increase with increasing release height, which for this problem was defined in terms of the radius. The normal force may seem small, but it is twice the weight of the bead.

---

7. Two objects are connected by a light string passing over a light, frictionless pulley as shown in Figure P8.7. The object of mass m_1 = 5.00 kg is released from rest at a height $h = 4.00$ m above the table. Using the isolated system model, (a) determine the speed of the object of mass $m_2 = 3.00$ kg just as the 5.00-kg object hits the table and (b) find the maximum height above the table to which the 3.00-kg object rises.

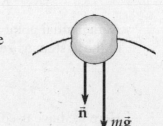

Figure P8.7

Solution

Conceptualize: We expect a speed of a few meters per second.

Categorize: The configuration is that of an Atwood machine, which we analyzed in terms of forces and acceleration. But when the unknown is final speed, it is natural to use energy conservation.

Analyze: As the system choose the two blocks *A* and *B*, the light string that does not stretch, the light pulley, and the Earth.

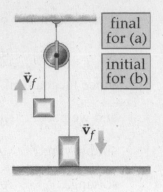

(a)   Choose the initial point before release and the final point, which we code with the subscript *fa*, just before the larger object hits the floor. No external forces do work on the system and no friction acts within the system. Then total mechanical energy of the system remains constant and the energy version of the isolated system model gives

$$\left(K_A + K_B + U_g\right)_i = \left(K_A + K_B + U_g\right)_{fa}$$

At the initial point K_{Ai} and K_{Bi} are zero and we define the gravitational potential energy of the system as zero. Thus the total initial energy is zero and we have

$$0 = \tfrac{1}{2}(m_1 + m_2)v_{fa}^2 + m_2gh + m_1g(-h)$$

Here we have used the fact that because the cord does not stretch, the two blocks have the same speed. The heavier mass moves down, losing gravitational energy, as the lighter mass moves up, gaining gravitational energy.

Simplifying, $(m_1 - m_2)gh = \tfrac{1}{2}(m_1 + m_2)v_{fa}^2$

$$v_{fa} = [2(m_1 - m_2)gh/(m_1 + m_2)]^{\frac{1}{2}}$$

$$= [2(2.00 \text{ kg})(9.80 \text{ m/s}^2)(4.00 \text{ m})/8.00 \text{ kg}]^{\frac{1}{2}}$$

and the solution is $v_{fa} = 4.43 \text{ m/s}$ ∎

(b)   Now the string goes slack. The 3.00-kg object becomes a projectile. We focus now on the system of the 3.00-kg object and the Earth. Take the initial point at the previous final point, and the new final point at its maximum height:

Energy is again conserved: $\left(K + U_g\right)_{fa} = \left(K + U_g\right)_{fb}$

Making the equation apply to this situation gives

$$\tfrac{1}{2}m_2v_{fa}^2 + m_2gh = m_2gy_b$$

$$y_b = h + v_{fa}^2/2g = 4.00 \text{ m} + (4.43 \text{ m/s})^2/2(9.80 \text{ m/s}^2)$$

so $y_b = 5.00 \text{ m}$ ∎

or 1.00 m above the height of the 5.00-kg mass when it was released.

Finalize: Energy is not matter, but you can visualize the energy as starting as gravitational energy in the 5-kg object and partly flowing along the string to become gravitational energy in the 3-kg object while some of it becomes kinetic energy in both. Then most of it suddenly turns into internal energy in the heavier block and the table while the lighter block by itself keeps constant energy as it rises to maximum height.

12. A sled of mass m is given a kick on a frozen pond. The kick imparts to the sled an initial speed of 2.00 m/s. The coefficient of kinetic friction between sled and ice is 0.100. Use energy considerations to find the distance the sled moves before it stops.

Solution

Conceptualize: Since the sled's initial speed of 2 m/s (~ 4 mi/h) is reasonable for a moderate kick, we might expect the sled to travel several meters before coming to rest.

Categorize: We could solve this problem using Newton's second law, but we will use the nonisolated system energy model, here written as $-f_k d = K_f - K_i$, where the kinetic energy change of the sled after the kick results only from the friction between the sled and ice. The weight and normal force both act at 90° to the motion, and therefore do no work on the sled.

Analyze: The friction force is $f_k = \mu_k n = \mu_k mg$

Since the final kinetic energy is zero, we have $-f_k d = -K_i$

or $\tfrac{1}{2}mv_i^2 = \mu_k mgd$

Thus, $d = \dfrac{mv_i^2}{2f_k} = \dfrac{mv_i^2}{2\mu_k mg} = \dfrac{v_i^2}{2\mu_k g} = \dfrac{(2.00 \text{ m/s})^2}{2(0.100)(9.80 \text{ m/s}^2)} = 2.04 \text{ m}$ ∎

Finalize: The distance agrees with the prediction. It is interesting that the distance does not depend on the mass and is proportional to the square of the initial velocity. This means that a small car and a massive truck should be able to stop within the same distance if they both skid to a stop from the same initial speed. Also, doubling the speed requires four times as much stopping distance, which is consistent with advice given by transportation safety officers who suggest that drivers maintain at least a 2 second gap between vehicles (rather than a fixed distance like 100 feet).

---

14. A 40.0-kg box initially at rest is pushed 5.00 m along a rough, horizontal floor with a constant applied horizontal force of 130 N. The coefficient of friction between box and floor is 0.300. Find (a) the work done by the applied force, (b) the increase in internal energy in the box–floor system as a result of friction, (c) the work done by the normal force, (d) the work done by the gravitational force, (e) the change in kinetic energy of the box, and (f) the final speed of the box.

Solution

Conceptualize: We expect perhaps hundreds of joules for answers (a), (b), and (e), zero for (c) and (d), and a few meters per second for (f).

Categorize: The final speed could come from a Newton's-second-law analysis, but we can get away without ever thinking about the value of the acceleration. We use the energy version of the nonisolated system model.

Analyze:

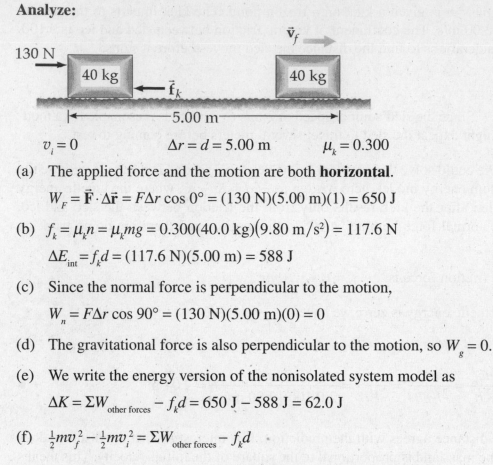

$v_i = 0$ $\qquad\qquad$ $\Delta r = d = 5.00$ m $\qquad$ $\mu_k = 0.300$

(a) The applied force and the motion are both **horizontal**.

$$W_F = \vec{\mathbf{F}} \cdot \Delta \vec{\mathbf{r}} = F\Delta r \cos 0° = (130 \text{ N})(5.00 \text{ m})(1) = 650 \text{ J}$$ ∎

(b) $f_k = \mu_k n = \mu_k mg = 0.300(40.0 \text{ kg})(9.80 \text{ m/s}^2) = 117.6 \text{ N}$

$$\Delta E_{int} = f_k d = (117.6 \text{ N})(5.00 \text{ m}) = 588 \text{ J}$$ ∎

(c) Since the normal force is perpendicular to the motion,

$$W_n = F\Delta r \cos 90° = (130 \text{ N})(5.00 \text{ m})(0) = 0$$ ∎

(d) The gravitational force is also perpendicular to the motion, so $W_g = 0$.

(e) We write the energy version of the nonisolated system model as

$$\Delta K = \Sigma W_{\text{other forces}} - f_k d = 650 \text{ J} - 588 \text{ J} = 62.0 \text{ J}$$ ∎

(f) $\frac{1}{2}mv_f^2 - \frac{1}{2}mv_i^2 = \Sigma W_{\text{other forces}} - f_k d$

$$v_f = \sqrt{\frac{2}{m}\left[\Delta K + \frac{1}{2}mv_i^2\right]} = \sqrt{\left(\frac{2}{40.0 \text{ kg}}\right)\left[62.0 \text{ J} + \frac{1}{2}(40.0 \text{ kg})(0)\right]} = 1.76 \text{ m/s}$$ ∎

Finalize: Most of the input work goes into increasing the internal energy of the floor and block by friction. We needed Newton's second law after all to identify the normal force as equal to mg, but we did avoid calculating acceleration.

———————————————

16. A crate of mass 10.0 kg is pulled up a rough incline with an initial speed of 1.50 m/s. The pulling force is 100 N parallel to the incline, which makes an angle of 20.0° with the horizontal. The coefficient of kinetic friction is 0.400, and the crate is pulled 5.00 m. (a) How much work is done by the gravitational force on the crate? (b) Determine the increase in internal energy of the crate–incline system owing to friction. (c) How much work is done by the 100-N force on the crate? (d) What is the change in kinetic energy of the crate? (e) What is the speed of the crate after being pulled 5.00 m?

Solution

Conceptualize: The gravitational force will do negative work of some tens of joules on the crate. We expect some hundreds of joules for answer (c), the input work. It should be larger than (b) and (d), which are two of the final forms of energy that the input work feeds into. The final speed should be a few meters per second.

Categorize: We could use Newton's second law to find the crate's acceleration, but using ideas of work and energy gets us more directly to the final speed. Finding the increase in internal energy would be a step toward finding the temperature increase of the rubbing surfaces.

Analyze:

(a) The force of gravitation is $(10.0 \text{ kg})(9.80 \text{ m/s}^2) = 98.0 \text{ N}$ straight down, at an angle of $(90.0° + 20.0°) = 110.0°$ with the motion. The work done by the gravitational force on the crate is

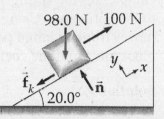

$$W_g = \vec{\mathbf{F}} \cdot \Delta \vec{\mathbf{r}} = (98.0 \text{ N})(5.00 \text{ m}) \cos 110.0° = -168 \text{ J} \quad \blacksquare$$

(b) We set the x and y axes parallel and perpendicular to the incline.

From $\qquad \sum F_y = ma_y$ we have $+n - (98.0 \text{ N}) \cos 20.0° = 0 \qquad$ so $\qquad n = 92.1 \text{ N}$

and $\qquad f_k = \mu_k n = 0.400 \ (92.1 \text{ N}) = 36.8 \text{ N}$

Therefore, $\qquad \Delta E_{int} = f_k d = (36.8 \text{ N})(5.00 \text{ m}) = 184 \text{ J} \qquad \blacksquare$

(c) $W = \vec{\mathbf{F}} \cdot \Delta \vec{\mathbf{r}} = 100 \text{ N}(5.00 \text{ m}) \cos 0° = +500 \text{ J} \qquad \blacksquare$

(d) We use the energy version of the nonisolated system model.

$$\Delta K = -f_k d + \sum W_{other forces}$$

$$\Delta K = -f_k d + W_g + W_{applied force} + W_n$$

The normal force does zero work, because it is at 90° to the motion.

$$\Delta K = -184 \text{ J} - 168 \text{ J} + 500 \text{ J} + 0 = 148 \text{ J} \qquad \blacksquare$$

(e) Again, $K_f - K_i = -f_k d + \sum W_{other forces}$

$$\tfrac{1}{2} m v_f^2 - \tfrac{1}{2} m v_i^2 = \sum W_{other forces} - f_k d$$

$$v_f = \sqrt{\frac{2}{m} \left[\Delta K + \tfrac{1}{2} m v_i^2 \right]} = \sqrt{\left(\frac{2}{10.0 \text{ kg}}\right)\left[148 \text{ J} + \tfrac{1}{2}(10.0 \text{ kg})(1.50 \text{ m/s})^2\right]}$$

Evaluating, $\qquad v_f = \sqrt{\dfrac{2\left(159 \text{ kg·m}^2/\text{s}^2\right)}{10.0 \text{ kg}}} = 5.65 \text{ m/s} \qquad \blacksquare$

Finalize: As an alternative to including the effect of gravitation as −168 J of work in $\Delta K = \sum W_{other forces} - f_k d$, we could have included its effect as +168 J of increase in gravitational energy in $\Delta K + \Delta U_g = W_{applied force} - f_k d$. We could say that the input work of 500 J goes into three places: +168 J becomes extra gravitational energy, +184 J becomes extra internal energy, and 148 J becomes extra kinetic energy on top of the 11 J of original kinetic energy.

Why do we do so many 'block and board' problems? Because the answers can be directly tested in the student laboratory and proven true. A very powerful course organization is theorize–predict–verify–consider another case.

======

21. A 200-g block is pressed against a spring of force constant 1.40 kN/m until the block compresses the spring 10.0 cm. The spring rests at the bottom of a ramp inclined at 60.0° to the horizontal. Using energy considerations, determine how far up the incline the block moves from its initial position before it stops (a) if the ramp exerts no friction force on the block and (b) if the coefficient of kinetic friction is 0.400.

Solution

Conceptualize: A good laboratory situation. We expect a distance of a few meters in (a) and perhaps half as much in (b).

Categorize: Even if the problem did not tell us explicitly to think about energy, we would have to do so to take account of the changing spring force by computing the spring energy before and after launch. We will use the nonisolated system energy model.

Analyze: We consider gravitational energy, elastic energy, kinetic energy, and in (b) internal energy.

(a) Apply the energy model between the release point and the point of maximum travel up the incline. In traveling distance d up along the incline (which includes distance $x = 0.1$ m) the block rises vertically by $d \sin 60°$. The initial and final states of the block are stationary, so $K_i = K_f = 0$. The elastic potential energy decreases in the firing process, and the gravitational energy increases in the uphill slide. Then

$$\Delta K + \Delta U_g + \Delta U_s = W_{\text{other forces}} - f_k d$$

becomes $\quad 0 + mgd \sin 60° - \tfrac{1}{2}kx_i^2 = 0$

Now d is the only unknown, so substituting gives

$$d = \frac{kx_i^2}{2mg \sin 60°} = \frac{(1\ 400\ \text{N/m})(0.100\ \text{m})^2}{2(0.200\ \text{kg})(9.80\ \text{m/s}^2)(0.866)} = 4.12\ \text{m} \qquad \blacksquare$$

(b) The weight of 0.200 kg is $mg = 1.96$ N. Taking our vertical axis to be perpendicular to the incline, we note

$$\Sigma F_y = ma_y: \qquad +n - (1.96\ \text{N})(\cos 60.0°) = 0 \qquad \text{and} \qquad n = 0.980\ \text{N}$$

Then the force of friction is $f_k = \mu_k n = (0.400)(0.980\ \text{N}) = 0.392$ N

In the energy equation we include one more term for the loss of mechanical energy due to friction:

$$\Delta K + \Delta U_g + \Delta U_s = W_{\text{other forces}} - f_k d$$

becomes $0 + mgd \sin 60° - \frac{1}{2}kx_i^2 = 0 - f_k d$

$$(mg \sin 60° + f_k)d = \frac{1}{2}kx_i^2$$

$$d = \frac{\frac{1}{2}(1\ 400\ \text{N/m})(0.100\ \text{m})^2}{1.96\ \text{N}(0.866) + 0.392\ \text{N}} = \frac{7.00\ \text{J}}{1.70\ \text{N} + 0.392\ \text{N}} = 3.35\ \text{m}$$ ∎

Finalize: Part (a) especially turns out to be simple to calculate, but only after you recognize what contains energy and what does not. Momentarily or permanently stationary objects have no kinetic energy and undistorted springs have no spring energy. In part (b), every student needs practice in reasoning through gravitational force, normal force, friction force, and friction-force-times-distance. The last of these is the term in the energy equation representing internal energy that appears as the molecules in the rubbing surfaces start vibrating faster.

---

22. The coefficient of friction between the block of mass $m_1 = 3.00$ kg and the surface in Figure P8.22 is $\mu_k = 0.400$. The system starts from rest. What is the speed of the ball of mass $m_2 = 5.00$ kg when it has fallen a distance $h = 1.50$ m?

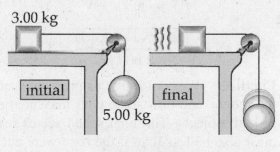

Figure P8.22

Solution

Conceptualize: Assuming that the block does not reach the pulley within the 1.50 m distance, a reasonable speed for the ball might be somewhere between 1 and 10 m/s, based on common experience.

Categorize: We could solve this problem by using $\Sigma F = ma$ to give a pair of simultaneous equations in the unknown acceleration and tension; after solving them we would still have to solve a motion problem to find the final speed. We find it easier to solve using the nonisolated system energy model.

Analyze: For the Earth plus objects 1 (block) and 2 (ball), we write the energy model equation as

$$\left(K_1 + K_2 + U_1 + U_2\right)_f - \left(K_1 + K_2 + U_1 + U_2\right)_i = \Sigma W_{\text{other forces}} - f_k d$$

Choose the initial point before release and the final point after each block has moved 1.50 m. Choose $U = 0$ with the 3.00-kg block on the tabletop and the 5.00-kg block in its final position.

So $K_{1i} = K_{2i} = U_{1i} = U_{1f} = U_{2f} = 0$

We have chosen to include the Earth in our system, so gravitation is an internal force. Because the only external forces are friction and normal forces exerted by the table and the pulley at right angles to the motion, $\Sigma W_{\text{other forces}} = 0$

We now have $\frac{1}{2}m_1 v_f^2 + \frac{1}{2}m_2 v_f^2 + 0 + 0 - 0 - 0 - 0 - m_2 g y_{2i} = 0 - f_k d$

Where the friction force is $f_k = \mu_k n = \mu_k m_A g$

The friction force causes a negative change in mechanical energy because the force opposes the motion. Since all of the variables are known except for v_f, we can substitute and solve for the final speed.

$$\frac{1}{2}m_1 v_f^2 + \frac{1}{2}m_2 v_f^2 - m_2 g y_{2i} = -f_k d$$

$$v_f = \sqrt{\frac{2}{m_1 + m_2}\left[m_2 gh - f_k d\right]}$$

$$= \sqrt{\left(\frac{2}{8 \text{ kg}}\right)\left[(5 \text{ kg})(9.8 \text{ m/s}^2)(1.5 \text{ m}) - (0.4)(3 \text{ kg})(9.8 \text{ m/s}^2)(1.5 \text{ m})\right]}$$

or $v_f = \sqrt{\dfrac{2(73.5 \text{ J} - 17.6 \text{ J})}{8.00 \text{ kg}}} = 3.74 \text{ m/s}$ ∎

Finalize: We could say this problem is about 73.5 J of energy, originally gravitational in the sphere, and how some goes into internal energy while the rest becomes kinetic energy in both objects. The final speed seems reasonable based on our expectation. This speed must also be less than if the rope were cut and the ball simply fell, in which case its final speed would be

$$v_f' = \sqrt{2gy} = \sqrt{2(9.80 \text{ m/s}^2)(1.50 \text{ m})} = 5.42 \text{ m/s}$$

---

23. A 5.00-kg block is set into motion up an inclined plane with an initial speed of $v_i = 8.00$ m/s (Fig. P8.23). The block comes to rest after traveling $d = 3.00$ m along the plane, which is inclined at an angle of $\theta = 30.0°$ to the horizontal. For this motion, determine (a) the change in the block's kinetic energy, (b) the change in the potential energy of the block–Earth system, and (c) the friction force exerted on the block (assumed to be constant). (d) What is the coefficient of kinetic friction?

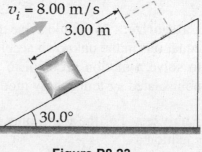

Figure P8.23

Solution

Conceptualize: The kinetic energy change is negative and on the order of –100 J. The gravitational energy change is positive and smaller in magnitude. We expect a friction force of several newtons and a friction coefficient of perhaps 0.8.

Categorize: There is enough information to find the block's acceleration, but we do not need to. We can think about energy on the way to finding the friction force and friction coefficient.

Analyze: We consider the block-plane-planet system between an initial point just after the block has been given its shove and a final point when the block comes to rest.

(a) The change in kinetic energy is

$$\Delta K = K_f - K_i = \tfrac{1}{2}mv_f^2 - \tfrac{1}{2}mv_i^2 = 0 - \tfrac{1}{2}(5.00 \text{ kg})(8.00 \text{ m/s})^2 = -160 \text{ J}$$ ∎

(b) The change in gravitational energy is

$$\Delta U = U_{gf} - U_{gi} = mgy_f - 0 = (5 \text{ kg})(9.8 \text{ m/s}^2)(3 \text{ m}) \sin 30.0° = 73.5 \text{ J}$$ ∎

(c) The nonisolated system energy model we write as

$$\Delta K + \Delta U = \Sigma W_{\text{other forces}} - f_k d = 0 - f_k d$$

The force of friction is the only unknown in it, so we may find it from

$$f_k = (-\Delta K - \Delta U)/d = (+160 \text{ J} - 73.5 \text{ J})/3.00 \text{ m}$$

and $\quad f_k = \dfrac{86.5 \text{ J}}{3.00 \text{ m}} = 28.8 \text{ N}$ ∎

(d) The forces perpendicular to the incline must add to zero.

$$\Sigma F_y = 0: \qquad +n - mg \cos 30.0° = 0$$

Evaluating, $\quad n = (5.00 \text{ kg})(9.80 \text{ m/s}^2) \cos 30.0° = 42.4 \text{ N}$

Now, $\quad f_k = \mu_k n \quad$ gives $\quad \mu_k = \dfrac{f_k}{n} = \dfrac{28.8 \text{ N}}{42.4 \text{ N}} = 0.679$ ∎

Finalize: Our estimates in the Conceptualize step were OK. In the equation representing the energy version of the nonisolated system model, different problems ask for very different unknowns. We can think of the energy equation as a precise set of prescriptions directing you to think about kinds of energy present in the pictures at the initial and final points you choose, and about whether energy is conserved.

The problem says "potential energy of the block-Earth system" and here we designate the same quantity as "gravitational energy in the block-plane-planet system." Calling it gravitational energy just emphasizes that it is not elastic [potential] energy or electrical [potential] energy.

---

29. An 820-N Marine in basic training climbs a 12.0-m vertical rope at a constant speed in 8.00 s. What is his power output?

Solution

Conceptualize: We expect some hundreds of watts.

Categorize: We must evaluate the mechanical work he does and use the definition of power.

Analyze: The Marine must exert an 820-N upward force, opposite the gravitational force, to lift his body at constant speed. Then his muscles do work

$$W = \vec{F} \cdot \Delta \vec{r} = (820\ \hat{\mathbf{j}}\ \text{N}) \cdot (12.0\ \hat{\mathbf{j}}\ \text{m}) = 9\,840\ \text{J}$$

The mechanical power he puts out is $P = \dfrac{W}{\Delta t} = \dfrac{9\,840\ \text{J}}{8.00\ \text{s}} = 1\,230\ \text{W}$ ■

Finalize: It is more than we expected, and the Marine also puts out energy by heat. His rate of converting chemical energy might be four thousand watts while he is climbing the rope, but much lower averaged over a day. We will use the symbol P for power and also for pressure, while p represents momentum and dipole moment. Do not forget that ρ [rho] stands for density. Do not mix them up yourself.

---

37.   A 3.50-kN piano is lifted by three workers at constant speed, to an apartment 25.0 m above the street using a pulley system fastened to the roof of the building. Each worker is able to deliver 165 W of power, and the pulley system is 75.0% efficient (so that 25.0% of the mechanical energy is transformed to other forms due to friction in the pulley). Neglecting the mass of the pulley, find the time required to lift the piano from the street to the apartment.

Solution

Conceptualize: Exerting roughly 4 000 N over 25 m, the workers must deliver on the order of a hundred thousand joules of gravitational energy to the piano. It should take them a few minutes.

Categorize: The piano has also some small kinetic energy on the way up, but not at the moment just after it stops at the destination. So we can compute the energy input to the piano just by thinking of gravitational energy. The definition of efficiency will then tell us the energy that must be put into the pulley system. Finally, the definition of power will tell us the time the workers require.

Analyze: The energy output of the pulley system is the final energy of the piano,

$$U_\text{g} = (mg)y = (3\,500\ \text{N})(25.0\ \text{m}) = 87\,500\ \text{J}$$

We can write the definition of efficiency as

$$\text{efficiency} = \frac{\text{useful energy output}}{\text{total energy input}}$$

Then the energy input to the pulley system is

$$\text{total energy input} = \frac{\text{useful energy output}}{\text{efficiency}} = \frac{87\,500\ \text{J}}{0.750} = 1.17 \times 10^5\ \text{J}$$

This input to the pulleys is the work done by the three laborers in

$$P = W/\Delta t, \text{ so } \quad \Delta t = \frac{W}{P} = \frac{1.17 \times 10^5 \text{ J}}{3 \times 165 \text{ W}} \left(\frac{1 \text{ W}}{1 \text{ J/s}}\right) = 236 \text{ s}$$ ■

Finalize: This is 3.93 minutes, as we estimated, and much less than the time required to rig the pulley system or to take it down again. If you find the "output" and "input" terminology confusing, you can draw an energy-chain diagram like this:

work by laborers	→ pulleys →	gravitational energy in piano

↘
| internal energy in pulleys and surroundings |

---

42.   What If? The block of mass $m = 200$ g described in Problem 41 (Fig. P8.41) is released from rest at point Ⓐ, and the surface of the bowl is rough. The block's speed at point Ⓑ is 1.50 m/s. (a) What is its kinetic energy at point Ⓑ? (b) How much mechanical energy is transformed into internal energy as the block moves from point Ⓐ to point Ⓑ? (c) Is it possible to determine the coefficient of friction from these results in any simple manner? (d) Explain your answer to part (c).

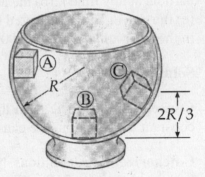

Figure P8.41

Solution

Conceptualize: The fraction of a joule of kinetic energy at Ⓓ will be less than the gravitational energy at Ⓐ because of friction.

Categorize: We calculate energy at Ⓐ, energy at Ⓑ, and take the difference. Problem 8.41 gives the radius of the bowl as 0.300 m.

Analyze: Let us take $U = 0$ for the particle-bowl-Earth system when the particle is at Ⓑ.

(a)   Since $v_B = 1.50$ m/s and $m = 200$ g,

$$K_B = \tfrac{1}{2}mv_B^2 = \tfrac{1}{2}(0.200 \text{ kg})(1.50 \text{ m/s})^2 = 0.225 \text{ J}$$ ■

(b)   At Ⓐ, $v_i = 0$, $K_A = 0$, and the whole energy at Ⓐ is $U_A = mgR$

$$E_i = K_A + U_A = 0 + mgR = (0.200 \text{ kg})(9.80 \text{ m/s}^2)(0.300 \text{ m}) = 0.588 \text{ J}$$

At Ⓑ, $E_f = K_B + U_B = 0.225 \text{ J} + 0$

The decrease in mechanical energy is equal to the increase in internal energy.
$$E_{mech,i} + \Delta E_{int} = E_{mech,f}$$
The energy transformed is

$$\Delta E_{int} = -\Delta E_{mech} = E_i - E_f = 0.588 \text{ J} - 0.225 \text{ J} = 0.363 \text{ J}$$ ■

(c)   There is no easy way to find the coefficient of friction.   ∎

(d)   Even though the energy transformed is known, both the normal force and the friction force change with position as the block slides on the inside of the bowl.   ∎

Finalize: Energy is not matter, but you can visualize the energy as 588 little pellets in the object at its starting point, with more than half turning into molecular-extra-vibration energy on the way down and 225 still in the object at the bottom as kinetic energy. We did not need to know the friction force or to put a number on the distance slid when we found $\Delta E_{int} = f_k d$.

---

47.   A 4.00-kg particle moves along the x axis. Its position varies with time according to $x = t + 2.0t^3$, where x is in meters and t is in seconds. Find (a) the kinetic energy of the particle at any time t, (b) the acceleration of the particle and the force acting on it at time t, (c) the power being delivered to the particle at time t, and (d) the work done on the particle in the interval $t = 0$ to $t = 2.00$ s.

Solution

Conceptualize: The term in time cubed indicates that the particle does not move with constant acceleration. We can use fundamental principles, namely...

Categorize: ...definitions, Newton's second law, and the work-kinetic energy theorem to find the answers, which will be functions of time.

Analyze:

(a)   Given $m = 4.00$ kg and $x = t + 2.0t^3$, we find the velocity by differentiating:

$$v = \frac{dx}{dt} = \frac{d}{dt}\left(t + 2t^3\right) = 1 + 6t^2$$

Then the kinetic energy from its definition is

$$K = \tfrac{1}{2}mv^2 = \tfrac{1}{2}(4.00)\left(1 + 6t^2\right)^2 = 2 + 24t^2 + 72t^4 \text{ where } K \text{ is in J and } t \text{ is in s.}   ∎$$

(b)   Acceleration is the measure of how fast velocity is changing:

$$a = \frac{dv}{dt} = \frac{d}{dt}\left(1 + 6t^2\right) = 12t \text{ where } a \text{ is in m/s}^2 \text{ and } t \text{ is in s.}$$

Newton's second law gives the total force exerted on the particle by the rest of the universe:

$$\Sigma F = ma = (4.00 \text{ kg})(12t) = 48t \quad \text{where } F \text{ is in N and } t \text{ is in s.}   ∎$$

(c)   Power is how fast work is done to increase the object's kinetic energy:

$$P = \frac{dW}{dt} = \frac{dK}{dt} = \frac{d}{dt}\left(2.00 + 24t^2 + 72t^4\right) = 48t + 288t^3 \text{ here } P \text{ is in W [watts]}$$

and t is in s.

Alternatively, we could use $P = Fv = 48t(1.00 + 6.0t^2)$

(d) The work-kinetic energy theorem $\Delta K = \Sigma W$ lets us find the work done on the object between $t_i = 0$ and $t_f = 2.00$ s. At $t_i = 0$ we have $K_i = 2.00$ J. At $t_f = 2.00$ s,

$$K_f = \left[2 + 24(2.00 \text{ s})^2 + 72(2.00 \text{ s})^4\right] = 1\,250 \text{ J}$$

Therefore the work input is $W = K_f - K_i = 1\,248 \text{ J} = 1.25 \times 10^3 \text{ J}$

Alternatively, we could start from $W = \int_{t_i}^{t_f} P\,dt = \int_0^2 \left(48t + 288t^3\right)dt$

Finalize: The force growing linearly in time makes this particle really take off. In 151 seconds the power delivered to it will be a million kilowatts. Notice how simple the equations expressing general principles are; *general* of course means that we can use them in any situation. For a contrasting example, $(\frac{1}{2})kx^2$ applies only to a spring described by Hooke's law, with x measured from one particular point.

---

63. A 10.0-kg block is released from rest at point Ⓐ in Figure P8.63. The track is frictionless except for the portion between points Ⓑ and Ⓒ, which has a length of 6.00 m. The block travels down the track, hits a spring of force constant 2 250 N/m, and compresses the spring 0.300 m from its equilibrium position before coming to rest momentarily. Determine the coefficient of kinetic friction between the block and the rough surface between points Ⓑ and Ⓒ.

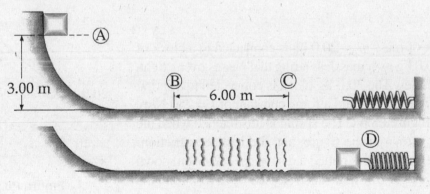

Figure P8.63

Solution

Conceptualize: We should expect the coefficient of friction to be somewhere between 0 and 1 since this is the range of typical μ_k values. It is possible that μ_k could be greater than 1, but it can never be less than 0.

Categorize: The easiest way to solve this problem about a chain-reaction process is by considering the energy changes experienced by the block between the point of release (initial) and the point of full compression of the spring (final). Recall that the change in potential energy (gravitational and elastic) plus the change in kinetic energy must equal the work done on the block by non-conservative forces. We choose the gravitational energy to be zero along the flat portion of the track.

Analyze: There is zero spring energy in situation Ⓐ and zero gravitational energy in situation Ⓓ. Putting the energy equation into symbols:

$$K_D - K_A - U_{gA} + U_{sD} = -f_k d_{BC}$$

Expanding into specific variables:

$$0 - 0 - mgy_A + \tfrac{1}{2}kx_s^2 = -f_k d_{BC}$$

The friction force is
$$f_k = \mu_k mg \quad \text{so} \quad mgy_A - \tfrac{1}{2}kx^2 = \mu_k mgd$$

Solving for the unknown variable μ_k gives
$$\mu_k = \frac{y_A}{d} - \frac{kx^2}{2mgd}$$

Substituting:
$$\mu_k = \frac{3.00\ \text{m}}{6.00\ \text{m}} - \frac{(2\,250\ \text{N/m})(0.300\ \text{m})^2}{2(10.0\ \text{kg})(9.80\ \text{m/s}^2)(6.00\ \text{m})} = 0.328 \quad ■$$

Finalize: Our calculated value seems reasonable based on the text's tabulation of coefficients of friction. The most important aspect to solving these energy problems is considering how the energy is transferred from the initial to final energy states and remembering to add or subtract the energy change resulting from any non-conservative forces. For friction the minus sign in $-f_k d$ is built in. If this problem were about a teenager on a toboggan run in *America's Funniest Home Videos*, some big dog might push forward on the block partway down, and the energy equation would contain an extra positive term $+W_{\text{other forces}}$ in $E_f - E_i = W_{\text{other forces}} - f_k d$.

---

64. A block of mass $m_1 = 20.0$ kg is connected to a block of mass $m_2 = 30.0$ kg by a massless string that passes over a light, frictionless pulley. The 30.0-kg block is connected to a spring that has negligible mass and a force constant of $k = 250$ N/m, as shown in Figure P8.64. The spring is unstretched when the system is as shown in the figure, and the incline is frictionless. The 20.0-kg block is pulled a distance $h = 20.0$ cm down the incline of angle $\theta = 40.0°$ and released from rest. Find the speed of each block when the spring is again unstretched.

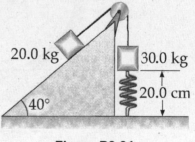

Figure P8.64

Solution

Conceptualize: The 30.0-kg block starts its motion at 40.0 cm above the floor, and we are to find its speed when it is 20.0 cm above the floor. The two blocks, joined by the inextensible string, will have the same speed of a few meters per second. The extra weight of the 30-kg block and the spring force are both reasons for both of the blocks to gain speed.

Categorize: We could not solve by Newton's second law, because the spring force is variable. Even if we could, we would have simultaneous equations to puzzle through and a subsequent motion problem to solve. The energy model takes account of two objects just by having two kinetic energy and two gravitational energy terms in the initial and final states.

Analyze: Let x represent the distance the spring is stretched from equilibrium ($x = 0.200$ m), which corresponds to the upward displacement of the 30.0-kg mass 1. Also let $U_g = 0$ be measured with respect to the lowest position of the 20.0-kg mass 2 when the system is released from rest. Finally, define v as the speed of both blocks at the moment the spring passes through its unstretched position. In

$$\Delta K_1 + \Delta K_2 + \Delta U_1 + \Delta U_2 + \Delta U_S = W_{\text{other forces}} - f_k d$$

since all forces are conservative, conservation of energy yields

$$\left(K_1 + K_2 + U_1 + U_2 + U_S \right)_i = \left(K_1 + K_2 + U_1 + U_2 + U_S \right)_f$$

Particularizing to the problem situation,

$$0 + 0 + m_2 g x + 0 + \tfrac{1}{2} k x^2 = \tfrac{1}{2}(m_1 + m_2)v^2 + 0 + m_1 g x \sin\theta + 0$$

Solving,
$$v = \left[\frac{2\left(m_2 g x + \tfrac{1}{2} k x^2 - m_1 g x \sin\theta \right)}{m_1 + m_2} \right]^{\tfrac{1}{2}}$$

Substituting,

$$v = \left[\frac{2\left\{ (30 \text{ kg})(9.8 \text{ m/s}^2)(0.2 \text{ m}) + \tfrac{1}{2}(250 \text{ N/m})(0.2 \text{ m})^2 - (20)(9.8)(0.2 \text{ J}) \sin 40° \right\}}{30 \text{ kg} + 20 \text{ kg}} \right]^{\tfrac{1}{2}}$$

$$v = \left[\frac{2\left\{ 58.8 \text{ J} + 5.00 \text{ J} - 25.2 \text{ J} \right\}}{50.0 \text{ kg}} \right]^{\tfrac{1}{2}}$$

and
$$v = 1.24 \text{ m/s} \qquad \blacksquare$$

Finalize: If you need to, you can turn your paper sideways to write down the long equation. Draw the original picture above the initial side of the equation and the final picture above the final side. The equation can be thought of as a story of the total energy 63.8 joules and what happens to it. It starts out as elastic energy in the spring plus gravitational energy in one object; then it becomes gravitational energy in the other object and kinetic energy in both.

---

65. A block of mass 0.500 kg is pushed against a horizontal spring of negligible mass until the spring is compressed a distance x (Fig. P8.65). The force constant of the spring is 450 N/m. When it is released, the block travels along a frictionless, horizontal surface to point Ⓐ, the bottom of a vertical circular track of radius $R = 1.00$ m, and continues to move up the track. The block's speed at the bottom of the track is $v_A = 12.0$ m/s, and the block experiences an average friction force of 7.00 N while sliding up the track. (a) What is x? (b) If the block were to reach the top of the track, what would be its speed at that point? (c) Does the block actually reach the top of the track, or does it fall off before reaching the top?

Solution

Conceptualize: This could be a model for a great roller-coaster ride! That loop is taller than you are. The compression will be several centimeters and the speed at the top much less than 12 m/s. We cannot guess in advance whether the block will fall from the track.

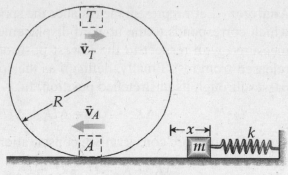

Figure P8.65

Categorize: We think of energy conservation in the firing process to find x, then a nonisolated energy system to find the speed at the top, and then forces (not centrifugal!) to decide whether the track has to be there to push down on the block at the top.

Analyze:

(a) The energy of the block-spring system is conserved in the firing of the block. Let point O represent the block and spring before the spring is released. We have $(K + U_s)_O = (K + U_s)_A$

$$0 + \tfrac{1}{2} kx_O^2 = \tfrac{1}{2} mv_A^2 + 0$$

Solving, $\quad x_O = [m/k]^{1/2} v_A = [0.500 \text{ kg}/(450 \text{ N/m})]^{1/2}(12.0 \text{ m/s})$

Thus, $\quad x_O = 0.400 \text{ m}$ ∎

(b) To find speed of block at the top, which we call point T, we consider the block-Earth system.

$$\left(K + U_g\right)_T - \left(K + U_g\right)_A = -f_k d$$

$$mgy_T + \tfrac{1}{2} mv_T^2 - 0 - \tfrac{1}{2} mv_A^2 = -f(\pi R)$$

$$v_T = \left[\frac{2\left(\tfrac{1}{2} mv_A^2 - f\pi R - mgy_T\right)}{m}\right]^{1/2}$$

$$= \left[\frac{2\left(\tfrac{1}{2}(0.5 \text{ kg})(12 \text{ m/s})^2 - (7 \text{ N})\pi(1 \text{ m}) - (0.5 \text{ kg})(9.8 \text{ m/s}^2)(2 \text{ m})\right)}{0.5 \text{ kg}}\right]^{1/2}$$

$$= \left[\frac{2(36.0 \text{ J} - 22.0 \text{ J} - 9.80 \text{ J})}{0.500 \text{ kg}}\right]^{1/2} = \left[\frac{2(4.21 \text{ J})}{0.500 \text{ kg}}\right]^{1/2}$$

and $\quad v_T = 4.10 \text{ m/s}$ ∎

(c) If its centripetal acceleration is greater than the free-fall acceleration, the block will require the track pushing down on it to stay on the circle, and it will not have fallen from the track. So we evaluate

$$a_c = \frac{v_T^2}{R} = \frac{(4.10 \text{ m/s})^2}{1.00 \text{ m}} = 16.8 \text{ m/s}^2$$

We see $a_c > g$. Some downward normal force is required along with the block's weight to provide the centripetal acceleration, and the block stays on the track. ■

Finalize: This is a good problem to review before an exam. It contained reasoning based on Newton's second law in part (c). Notice that the energy converted by friction (to internal energy) is the friction force times *distance*, but the energy converted by gravitation (to gravitational energy) is the weight times upward *displacement*. The *displacement* is the diameter of the circle. Identifying the *distance* over which friction acts as half the circumference is not tough once you look at the picture. The average force of friction would be hard to relate to a coefficient of friction, because the normal force depends strongly on the speed of the block.

---

69. A ball whirls around in a vertical circle at the end of a string. The other end of the string is fixed at the center of the circle. Assuming the total energy of the ball–Earth system remains constant, show that the tension in the string at the bottom is greater than the tension at the top by six times the ball's weight.

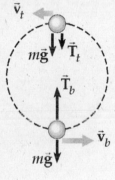

Solution

Conceptualize: The ball is going faster at the bottom and the string at that point must support its weight as well as provide the centripetal acceleration, so the tension at the bottom should be a lot larger than at the top.

Categorize: We relate speeds at top and bottom by energy conservation, and also think about the ball as a particle under net force.

Analyze: Applying the vertical component equation of Newton's second law at the bottom (*b*) and top (*t*) of the circular path gives

$$+T_b - mg = \frac{mv_b^2}{R} \tag{1}$$

$$-T_t - mg = -\frac{mv_t^2}{R} \tag{2}$$

Adding equations [2] and [1] gives

$$T_b = T_t + 2mg + \frac{m(v_b^2 - v_t^2)}{R} \tag{3}$$

Also, the string tension is perpendicular to the motion at every point and does no work on the ball, so energy of the ball-Earth system must be conserved; that is,

$$K_b + U_b = K_t + U_t$$

So,

$$\tfrac{1}{2}mv_b^2 + 0 = \tfrac{1}{2}mv_t^2 + 2mgR$$

or

$$\frac{v_b^2 - v_t^2}{R} = 4g \qquad [4]$$

Substituting [4] into [3] gives

$$T_b = T_t + 2mg + m4g$$

or

$$T_b = T_t + 6mg \qquad \blacksquare$$

Finalize: Observe how the derived equation completes the proof. This is a nice theorem to test with a laboratory force sensor connected to the string. If a roller coaster sent people around a circular loop in a vertical plane, they would be intensely uncomfortable and would risk serious injury at the bottom of the loop. So looping roller coasters are always built in some different way.

9

Linear Momentum and Collisions

EQUATIONS AND CONCEPTS

The **linear momentum** $\vec{\mathbf{p}}$ of a particle is defined as the product of its mass m and its velocity $\vec{\mathbf{v}}$. *This vector equation is equivalent to three component scalar equations, one along each of the coordinate axes.*

$$\vec{\mathbf{p}} \equiv m\vec{\mathbf{v}} \tag{9.2}$$

$$p_x = mv_x \quad p_y = mv_y \quad p_z = mv_z$$

The **net force acting on a particle** is equal to the time rate of change of the linear momentum of the particle. *This is a more general form of Newton's second law.*

$$\sum \vec{\mathbf{F}} = \frac{d\vec{\mathbf{p}}}{dt} \tag{9.3}$$

The **law of conservation of linear momentum states**: The total momentum of an isolated system (no external forces acting on the system) remains constant. There may be *internal forces* acting between particles within the system and the momentum of an *individual particle* may not be conserved. Although Equation 9.5 is written for a pair of particles, the conservation law holds for a system of any number of particles. *This fundamental law is especially useful in treating problems involving collisions between two bodies.*

$$\vec{\mathbf{p}}_{1i} + \vec{\mathbf{p}}_{2i} = \vec{\mathbf{p}}_{1f} + \vec{\mathbf{p}}_{2f} \ \left(\text{when} \sum \vec{\mathbf{F}} = 0\right) \tag{9.5}$$

Impulse is a vector quantity, defined by Equation 9.9. The magnitude of an impulse caused by a force is equal to the area under the force-time curve.

$$\vec{\mathbf{I}} \equiv \int_{t_i}^{t_f} \sum \vec{\mathbf{F}} \, dt \tag{9.9}$$

The **impulse-momentum theorem**, Equation 9.8, states that the impulse of a force $\vec{\mathbf{F}}$ acting on a particle equals the change in the momentum of the particle.

$$\Delta \vec{\mathbf{p}} = \vec{\mathbf{p}}_f - \vec{\mathbf{p}}_i = \int_{t_i}^{t_f} \sum \vec{\mathbf{F}} \, dt \tag{9.8}$$

$$\Delta \vec{\mathbf{p}} = \vec{\mathbf{I}} \tag{9.10}$$

145

The **time-averaged force** $(\sum \vec{\mathbf{F}})_{avg}$ is defined as that constant force which would give the same impulse to a particle as an actual time-varying force over the same time interval Δt. See the area outlined by the dashed line in the figure. *The impulse approximation assumes that one of the forces acting on a particle acts for a short time and is much larger than any other force present.* This approximation is usually made in collision problems, where the force is the contact force between the particles during the collision.

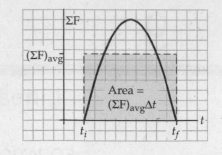

$$\vec{\mathbf{I}} = \sum \vec{\mathbf{F}}\,\Delta t \qquad (9.13)$$

An **elastic collision** is one in which both linear momentum and kinetic energy are conserved.

$$\left.\begin{array}{l} \vec{\mathbf{p}}_1 + \vec{\mathbf{p}}_2 = \text{const.} \\ K_1 + K_2 = \text{const.} \end{array}\right\} \text{ (Elastic collision)}$$

An **inelastic collision** is one in which only linear momentum is conserved. *A perfectly inelastic collision is an inelastic collision in which the two bodies stick together after the collision.*

$$\vec{\mathbf{p}}_1 + \vec{\mathbf{p}}_2 = \text{const.} \quad \text{(Inelastic collision)}$$

The **common velocity following a perfectly inelastic collision** between two bodies can be calculated in terms of the two mass values and the two initial velocities. *This result is a consequence of conservation of linear momentum.*

$$\vec{\mathbf{v}}_f = \frac{m_1 \vec{\mathbf{v}}_{1i} + m_2 \vec{\mathbf{v}}_{2i}}{m_1 + m_2} \qquad (9.15)$$

In an **elastic collision**, both momentum and kinetic energy are conserved. *Equations 9.16 and 9.17 apply in the case of a one-dimensional ("head-on") collision.*

$$m_1 v_{1i} + m_2 v_{2i} = m_1 v_{1f} + m_2 v_{2f} \qquad (9.16)$$

$$\tfrac{1}{2} m_1 v_{1i}^2 + \tfrac{1}{2} m_2 v_{2i}^2 = \tfrac{1}{2} m_1 v_{1f}^2 + \tfrac{1}{2} m_2 v_{2f}^2 \quad (9.17)$$

The **relative velocity** before a perfectly elastic collision between two bodies equals the negative of the relative velocity of the two bodies following the collision. Equation 9.20 can be combined with Equation 9.16 to solve problems involving one-dimensional perfectly elastic collisions.

$$v_{1i} - v_{2i} = -\left(v_{1f} - v_{2f}\right) \qquad (9.20)$$

The **final velocities following a one-dimensional elastic collision** between two particles can be calculated when the masses and initial velocities of both particles are known. *Remember the appropriate algebraic signs (designating direction) must be included for v_{1i} and v_{2i}.*

$$v_{1f} = \left(\frac{m_1 - m_2}{m_1 + m_2}\right)v_{1i} + \left(\frac{2m_2}{m_1 + m_2}\right)v_{2i}$$

$$(9.21)$$

$$v_{2f} = \left(\frac{2m_1}{m_1 + m_2}\right)v_{1i} + \left(\frac{m_2 - m_1}{m_1 + m_2}\right)v_{2i}$$

$$(9.22)$$

An **important special case** occurs when the second particle (m_2, the "target") is initially at rest.

$$v_{1f} = \left(\frac{m_1 - m_2}{m_1 + m_2}\right)v_{1i}$$ $$(9.23)$$

$$v_{2f} = \left(\frac{2m_1}{m_1 + m_2}\right)v_{1i}$$ $$(9.24)$$

A **two-dimensional elastic collision** in which an object m_1 moves along the x-axis and collides with m_2 initially at rest is illustrated in the figure below. *Momentum is conserved along each direction.* Angles in Equations 9.25 and 9.26 are defined in the diagram.

x-component:
$$m_1 v_{1i} + 0 = m_1 v_{1f}\cos\theta + m_2 v_{2f}\cos\phi$$

$$(9.25)$$

y-component:
$$0 + 0 = m_1 v_{1f}\sin\theta - m_2 v_{2f}\sin\phi \qquad (9.26)$$

The equation of conservation of kinetic energy for an elastic collision also applies.

$$\tfrac{1}{2}m_1\left(v_{1i}^2\right) = \tfrac{1}{2}m_1\left(v_{1f}^2\right) + \tfrac{1}{2}m_2\left(v_{2f}^2\right) \qquad (9.27)$$

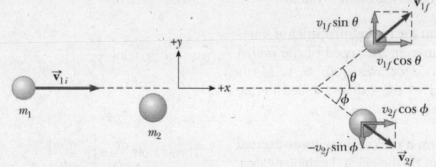

The **coordinates of the center of mass** of n particles with individual coordinates of (x_1, y_1, z_1), (x_2, y_2, z_2), (x_3, y_3, z_3) . . . and masses of m_1, m_2, m_3 . . . are given by Equations 9.29 and 9.30. The total mass is designated by M, and is found by summing over all n particles. *The center of mass of a homogeneous, symmetric body must lie on an axis of symmetry and on any plane of symmetry.*

$$x_{CM} \equiv \frac{1}{M}\sum_i m_i x_i \qquad (9.29)$$

$$y_{CM} \equiv \frac{1}{M}\sum_i m_i y_i \qquad (9.30)$$

$$z_{CM} \equiv \frac{1}{M}\sum_i m_i z_i$$

The **center of mass of an extended object** can be calculated by integrating over total length, area, or volume which includes the total mass M.

$$\vec{r}_{CM} \equiv \frac{1}{M}\int \vec{r}\,dm \qquad (9.34)$$

The **velocity of the center of mass of a system of particles** is given by Equation 9.35. In this equation, $\vec{v}_i$ is the velocity of the ith particle and M is the total mass of the system and remains constant.

$$\vec{v}_{CM} = \frac{d\vec{r}_{CM}}{dt} = \frac{1}{M}\sum_i m_i \vec{v}_i \qquad (9.35)$$

The **acceleration of the center of mass** of a system of particles depends on the value of the acceleration for each of the individual particles.

$$\vec{a}_{CM} = \frac{d\vec{v}_{CM}}{dt} = \frac{1}{M}\sum_i m_i \vec{a}_i \qquad (9.37)$$

The **net external force acting on a system of particles** equals the product of the total mass of the system and the acceleration of the center of mass. *The center of mass of a system of particles of combined mass M moves like an equivalent single particle of mass M would move if acted on by the same external force.*

$$\sum \vec{F}_{ext} = M\vec{a}_{CM} \qquad (9.39)$$

The **total momentum of a system of particles** is conserved when there is no net external force acting on the system.

$$M\vec{v}_{CM} = \vec{P}_{tot} = \text{constant} \quad (\text{when } \sum \vec{F}_{ext} = 0) \qquad (9.41)$$

The **equation for rocket propulsion** states that the change in the speed of the rocket (as the mass decreases from M_i to M_f) is proportional to the exhaust speed of the ejected gases.

$$v_f - v_i = v_e \ln\left(\frac{M_i}{M_f}\right) \qquad (9.43)$$

The **thrust on a rocket** is the force exerted on the rocket by the ejected exhaust gases. *The thrust increases as the exhaust speed increases and as the burn rate increases.*

$$\text{Thrust} = M\frac{dv}{dt} = \left| v_e \frac{dM}{dt} \right| \qquad (9.44)$$

SUGGESTIONS, SKILLS, AND STRATEGIES

The following procedure is recommended when dealing with problems involving collisions between two objects:

- Set up a coordinate system and show the velocities with respect to that system. That is, objects moving in the direction selected as the positive direction of the x axis are considered as having a positive velocity and negative if moving in the negative x direction. It is convenient to have the x axis coincide with one of the initial velocities.

- In your sketch of the coordinate system, draw all velocity vectors with labels and include all the given information.

- Write an equation which states that the total momentum of the system (including each mass) before the collision equals the total momentum of the system after the collision. Remember to include the appropriate signs for the velocity vectors. **In two-dimensional collision problems, write separate equations for the x and y components of momentum before and after the collision.**

- Remember: It is the momentum of the **system** (the two colliding objects) that is conserved, not the momentum of the individual objects.

- If the collision is **inelastic**, you should then proceed to solve the momentum equations for the unknown quantities.

- If the collision is **elastic**, kinetic energy is also conserved, so you can equate the total kinetic energy before the collision to the total kinetic energy after the collision. This gives an additional relationship between the various velocities. The conservation of kinetic energy for elastic collisions leads to the expression $v_{1i} - v_{2i} = -(v_{1f} - v_{2f})$. This equation can be used, together with the equation requiring conservation of momentum, to solve elastic collision problems.

REVIEW CHECKLIST

You should be able to:

- Apply the law of conservation of linear momentum to a two-body system. (Sections 9.1 and 9.2) *Remember: The momentum of any isolated system (one for which the net external force is zero) is conserved, regardless of the nature of the forces between the masses which comprise the system.*

- Determine the impulse delivered by a time-varying force from the area under the Force vs. Time curve. (Section 9.3) *Remember: The impulse of a force acting on a particle during some time interval equals the **change in momentum** of the particle.*

- Calculate the common velocity following an inelastic collision and the individual velocities following a one or two dimensional elastic collision between two objects. (Sections 9.4 and 9.5)

- Determine the center of mass of a system of discrete point-masses and, by integration, for a homogenous solid of sufficient symmetry. Determine also the motion (coordinate, velocity, and acceleration) of the center of mass of a system of objects. (Sections 9.6 and 9.7)

- Make calculations using the basic equations of rocket propulsion. (Section 9.9)

ANSWERS TO SELECTED OBJECTIVE QUESTIONS

9. If two particles have equal momenta, are their kinetic energies equal? (a) yes, always (b) no, never (c) no, except when their speeds are the same (d) yes, as long as they move along parallel lines

Answer (c) Their momenta will not be equal unless their speeds are equal. Equal momenta means that mv is the same for both. If they are moving and their speeds are different, then their masses must also be different. Let the speed of the lighter particle be larger by the factor α. Then the mass of the lighter particle must be smaller by the factor $1/\alpha$. The kinetic energy takes the increased speed into account twice, not once like the momentum. The less massive particle will have a kinetic energy larger by the factor $(1/\alpha)\alpha^2 = \alpha$.

14. A basketball is tossed up into the air, falls freely, and bounces from the wooden floor. From the moment after the player releases it until the ball reaches the top of its bounce, what is the smallest system for which momentum is conserved? (a) the ball (b) the ball plus player (c) the ball plus floor (d) the ball plus the Earth (e) momentum is not conserved for any system.

Answer (d) Momentum is not conserved for the ball, because the ball downward momentum increases as it accelerates downward. A larger upward momentum change occurs when the ball touches the floor and rebounds. The outside forces of gravitation and the normal force inject impulses to change the ball's momentum. The momentum changes of the ball mean that the momentum values of the ball-plus-player system and the ball-plus-floor system also change.

On the other hand, if we think of ball-and-Earth-together as our system, the gravitational and normal forces are internal and do not change the total momentum. It is conserved, if we neglect the curvature of the Earth's orbit. As the ball falls down, the Earth lurches up to meet it, on the order of 10^{25} times more slowly, but with an equal magnitude of momentum. Then, ball and planet bounce off each other and separate. In other words, while you dribble a ball, you also dribble the Earth.

ANSWERS TO SELECTED CONCEPTUAL QUESTIONS

6. A sharpshooter fires a rifle while standing with the butt of the gun against her shoulder. If the forward momentum of a bullet is the same as the backward momentum of the gun, why isn't it as dangerous to be hit by the gun as by the bullet?

Answer It is the product mv which is the same in magnitude for both the bullet and the gun. The bullet has a large velocity and a small mass, while the gun has a small velocity and a large mass. Its small velocity makes it safer. The bullet carries much more kinetic energy than the gun. Further, the wide area of the butt of the rifle reduces the pressure on the marksman's shoulder.

□   □   □   □

9.   (a) Does the center of mass of a rocket in free space accelerate? Explain. (b) Can the speed of a rocket exceed the exhaust speed of the fuel? Explain.

Answer   (a) The center of mass of a rocket, plus its exhaust, does not accelerate: system momentum must be conserved. However, if you consider the "rocket" to be the mechanical system plus the unexpended fuel, then it is clear that the center of mass of that "rocket" does accelerate.

Basically, a rocket can be modeled as two different rubber balls that are pressed together, and released. One springs in one direction with one velocity; the other springs in the other direction with a velocity equal to $v_2 = v_1(m_1/m_2)$.

(b) In a similar manner, the mass of the rocket body times its speed must equal the sum of the masses of all parts of the exhaust, times their speeds. After a sufficient quantity of fuel has been exhausted the ratio m_i/m_f will be greater than Euler's number e; then the speed of the rocket body **can** exceed the speed of the exhaust.

□   □   □   □

SOLUTIONS TO SELECTED END-OF-CHAPTER PROBLEMS

6.   A 45.0-kg girl is standing on a 150-kg plank. Both are originally at rest, on a frozen lake that constitutes a frictionless, flat surface. The girl begins to walk along the plank at a constant velocity of $1.50\hat{\mathbf{i}}$ m/s relative to the plank. (a) What is the velocity of the plank relative to the ice surface? (b) What is the girl's velocity relative to the ice surface?

Solution

Conceptualize: The plank slips backward, so the girl will move forward relative to the ice at a bit more than 1 m/s, and the plank backward at less than 0.5 m/s.

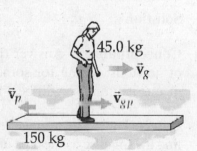

Categorize: Conservation of system momentum is the key.

Analyze: Define symbols: $\vec{\mathbf{v}}_g$ = velocity of the girl relative to the ice

$\vec{\mathbf{v}}_{gp}$ = velocity of the girl relative to the plank

$\vec{\mathbf{v}}_p$ = velocity of the plank relative to the ice

The girl and the plank exert forces on each other, but the ice isolates them from outside horizontal forces. Therefore, the net momentum is zero for the combined girl-plus-plank system, as it was before motion began:

$$0 = m_g \vec{v}_g + m_p \vec{v}_p$$

Further, the relation among relative velocities can be written

$$\vec{v}_g = \vec{v}_{gp} + \vec{v}_p \qquad \text{or} \qquad \vec{v}_g = 1.50\hat{i} \text{ m/s} + \vec{v}_p$$

We substitute into the momentum equation: $0 = (45.0 \text{ kg})(1.50\hat{i} \text{ m/s} + \vec{v}_p) + (150 \text{ kg}) \vec{v}_p$

and gather like terms: $(195 \text{ kg}) \vec{v}_p = -(45.0 \text{ kg})(1.50\hat{i} \text{ m/s})$

Because we happened to eliminate the speed of the girl by substitution, the velocity of the plank, answer (a) comes out first:

(a) $\vec{v}_p = -0.346\hat{i}$ m/s   ∎

Now substituting it back reveals

(b) $\vec{v}_g = 1.50\hat{i} - 0.346\hat{i}$ m/s $= 1.15\hat{i}$ m/s   ∎

Finalize: Our estimates were right. We can check them by comparing the momentum magnitudes of girl and plank, (45 kg)(1.15 m/s) and (150 kg)(0.346 m/s). They are equal.

---

11. An estimated force–time curve for a baseball struck by a bat is shown in P9.11. From this curve, determine (a) the magnitude of the impulse delivered to the ball and (b) the average force exerted on the ball.

Solution

Conceptualize: Answer (b) will be some thousands of newtons; acting for some thousandths of a second means its impulse will be a few newton-seconds.

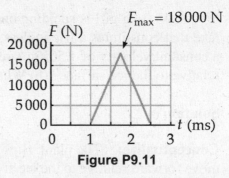

Figure P9.11

Categorize: The average force must be about half of the maximum force of 18 kN. We can find the impulse as the area under the graph line, and then think of an equivalent-area rectangle to get the average force.

Analyze:

(a)   The impulse delivered to the ball is equal to the area under the *F-t* graph. We have a triangle and so to get its area we multiply half its height times its width:

$$\text{Impulse} = \tfrac{1}{2}(0 + 18\,000\text{ N})(2.5 - 1.0) \times 10^{-3}\text{ s} = 9\,000\text{ N }(0.001\,5\text{ s})$$

$$= 13.5\text{ N}\cdot\text{s}$$

■

(b) $\quad F_{\text{avg}} = \dfrac{\int F\,dt}{\Delta t} = \dfrac{13.5\text{ N}\cdot\text{s}}{(2.5 - 1.0)\times10^{-3}\text{ s}} = 9.00\text{ kN}$

■

Finalize: Our estimates were good. Consider with care the plus sign in figuring the average height of the area and the minus sign in figuring its width.

---

19. A 10.0-g bullet is fired into a stationary block of wood having mass $m = 5.00$ kg. The bullet imbeds into the block. The speed of the bullet-plus-wood combination immediately after the collision is 0.600 m/s. What was the original speed of the bullet?

Solution

Conceptualize: A reasonable speed of a bullet should be somewhere between 100 and 1 000 m/s.

Categorize: We can find the initial speed of the bullet from conservation of momentum. We are told that the block of wood was originally stationary.

Analyze: Since there is no external force on the block and bullet system, the total momentum of the system is constant so that $\Delta\vec{p} = 0$.

That is,

$$m_1\vec{v}_{1i} + m_2\vec{v}_{2i} = m_1\vec{v}_{1f} + m_2\vec{v}_{2f}$$

We know all the quantities except for $\vec{v}_{1i}$, so we solve for it

$$\vec{v}_{1i} = \frac{m_1\vec{v}_{1f} + m_2\vec{v}_{2f} - m_2\vec{v}_{2i}}{m_1} = \frac{(m_1 + m_2)\vec{v}_f - 0}{m_1}$$

Substituting gives

$$\vec{v}_{1i} = \frac{(5.01\text{kg})(0.600\text{m/s})\,\hat{\textbf{i}}}{0.0100\text{kg}} = 301\hat{\textbf{i}}\text{ m/s}$$

■

Finalize: The speed seems reasonable, and is in fact just under the speed of sound in air (343 m/s at 20°C).

---

21. A neutron in a nuclear reactor makes an elastic head-on collision with the nucleus of a carbon atom initially at rest. (a) What fraction of the neutron's kinetic energy is transferred to the carbon nucleus? (b) The initial kinetic energy of the neutron is 1.60×10^{-13} J. Find its final kinetic energy and the kinetic energy of the carbon nucleus after the collision. (The mass of the carbon nucleus is nearly 12.0 times the mass of the neutron.)

Solution

Conceptualize: Visualize the light projectile as a rubber ball. It will bounce back at a bit less than its original speed, giving perhaps only ten percent of its energy to the target, and keeping the other ninety percent.

Categorize: Momentum and energy are both conserved for the neutron-nucleus system and this is a head-on collision, so we can use...

Analyze: ... the relative velocity equation $v_{1i} - v_{2i} = -(v_{1f} - v_{2f})$

(a) Let object 1 be the neutron, and object 2 be the carbon nucleus, with $m_2 = 12m_1$. Since $v_{2i} = 0$, we have $v_{2f} = v_{1i} + v_{1f}$

Now, by conservation of momentum for the system of colliding particles,

$$m_1 v_{1i} + m_2 v_{2i} = m_1 v_{1f} + m_2 v_{2f}$$

or $\qquad\qquad\qquad\qquad\qquad\qquad m_1 v_{1i} = m_1 v_{1f} + 12 m_1 v_{2f}$

The neutron mass divides out: $\qquad\qquad\qquad v_{1i} = v_{1f} + 12 v_{2f}$

Substituting our velocity equation, $\qquad\qquad v_{1i} = v_{1f} + 12(v_{1i} + v_{1f})$

We solve: $\qquad -11 v_{1i} = 13 v_{1f} \qquad\qquad v_{1f} = -\frac{11}{13} v_{1i}$

and $\qquad\qquad\qquad\qquad\qquad\qquad v_{2f} = v_{1i} - \frac{11}{13} v_{1i} = \frac{2}{13} v_{1i}$

The neutron's original kinetic energy is $\qquad \frac{1}{2} m_1 v_{1i}^2$

The carbon's final kinetic energy is $\qquad \frac{1}{2} m_2 v_{2f}^2 = \frac{1}{2}(12 m_1)\left(\frac{2}{13}\right)^2 v_{1i}^2 = \left(\frac{48}{169}\right)\left(\frac{1}{2}\right) m_1 v_{1i}^2$

So,

$$\frac{48}{169} = 0.284 \qquad \text{or} \qquad 28.4\% \text{ of the total energy is transferred.} \qquad\blacksquare$$

(b) For the carbon nucleus, $\qquad K_{2f} = (0.284)(1.60 \times 10^{-13}\,\text{J}) = 45.4\,\text{fJ} \qquad\blacksquare$

The collision is elastic, so the neutron retains the rest of the energy,

$$K_{1f} = (1.60 - 0.454) \times 10^{-13}\,\text{J} = 1.15 \times 10^{-13}\,\text{J} = 115\,\text{fJ} \qquad\blacksquare$$

Finalize: We could have written a conservation of kinetic energy equation in the same symbols and solved using it instead of the relative-velocity equation, along with conservation of momentum. Our ten-percent estimate was low. If the collision were not head-on, less than 28.4% of the neutron's energy would go into the carbon nucleus.

23. A 12.0-g wad of sticky clay is hurled horizontally at a 100-g wooden block initially at rest on a horizontal surface. The clay sticks to the block. After impact, the block slides 7.50 m before coming to rest. If the coefficient of friction between the block and the surface is 0.650, what was the speed of the clay immediately before impact?

Solution

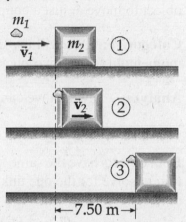

Conceptualize: 7.5 meter is a long way, so the clay-block combination must have started its motion at several meters per second, and the light wad of clay at on the order of a hundred m/s.

Categorize: We use the momentum version of the isolated system model to analyze the collision, and the energy version of the isolated system model to analyze the subsequent sliding process.

Analyze: The collision, for which figures (1) and (2) are before and after pictures, is **totally inelastic**, and momentum is conserved for the system of clay and block:

$$m_1 v_1 = (m_1 + m_2)v_2$$

In the sliding process occurring between figures (2) and (3), the original kinetic energy of the surface, block, and clay is equal to the increase in internal energy of the system due to friction:

$$\tfrac{1}{2}(m_1 + m_2)v_2^2 = fL$$

$$\tfrac{1}{2}(m_1 + m_2)v_2^2 = \mu(m_1 + m_2)gL$$

Solving for v_2 gives

$$v_2 = \sqrt{2\mu L g} = \sqrt{2(0.650)(7.50 \text{ m})(9.80 \text{ m/s}^2)} = 9.77 \text{ m/s}$$

Now from the momentum conservation equation,

$$v_1 = \left(\frac{m_1 + m_2}{m_1}\right)v_2 = \frac{112 \text{ g}}{12.0 \text{ g}}(9.77 \text{ m/s}) = 91.2 \text{ m/s} \quad \blacksquare$$

Finalize: This is a good review problem involving both conservation of momentum in the collision, and ideas about energy to describe the sliding process. We even needed to think about Newton's second law to identify the normal force as equal to $(m_1 + m_2)g$. Note well that the kinetic energy of the clay does not all become internal energy in the floor: $\tfrac{1}{2}m_1 v_1^2 = fL$ is not true.

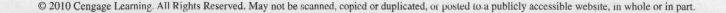

29. An object of mass 3.00 kg, moving with an initial velocity of $5.00\,\hat{\mathbf{i}}$ m/s, collides with and sticks to an object of mass 2.00 kg with an initial velocity of $-3.00\,\hat{\mathbf{j}}$ m/s. Find the final velocity of the composite object.

Solution

Conceptualize: Visualize an air-hockey game with magnetic pucks that latch together. The second object has less mass and is moving slower than the first, so we expect the 5-kg object to move at just a couple of m/s into the fourth quadrant, close to the x axis.

Categorize: We will use conservation of both the x component and the y component of momentum for the two-puck system, ...

Analyze: ...which we can write as a single vector equation.

$$m_1 \vec{\mathbf{v}}_{1i} + m_2 \vec{\mathbf{v}}_{2i} = m_1 \vec{\mathbf{v}}_{1f} + m_2 \vec{\mathbf{v}}_{2f}$$

Both objects have the same final velocity, which we call $\vec{\mathbf{v}}_f$. Doing the algebra and substitution to solve for the one unknown gives

$$\vec{\mathbf{v}}_f = \frac{m_1 \vec{\mathbf{v}}_{1i} + m_2 \vec{\mathbf{v}}_{2i}}{m_1 + m_2} = \frac{(3.00\,\text{kg})(5.00\hat{\mathbf{i}}\,\text{m/s}) + (2.00\,\text{kg})(-3.00\hat{\mathbf{j}}\,\text{m/s})}{3.00\,\text{kg} + 2.00\,\text{kg}}$$

and calculation gives $\vec{\mathbf{v}}_f = \dfrac{15.0\hat{\mathbf{i}} - 6.00\hat{\mathbf{j}}}{5.00}\,\text{m/s} = (3.00\hat{\mathbf{i}} - 1.20\hat{\mathbf{j}})\,\text{m/s}$ ∎

Finalize: Let us compute the kinetic energy of the system both before and after the collision and see whether kinetic energy is conserved.

$$K_{1i} + K_{2i} = \tfrac{1}{2}(3.00\text{ kg})(5.00\text{ m/s})^2 + \tfrac{1}{2}(2.00\text{ kg})(3.00\text{ m/s})^2 = 46.5\,\text{J}$$

$$K_{1f} + K_{2f} = \tfrac{1}{2}(5.00\text{ kg})[(3.00\text{ m/s})^2 + (1.20\text{ m/s})^2] = 26.1\,\text{J}$$

Thus the collision is inelastic; mechanical energy is not constant.

---

33. A billiard ball moving at 5.00 m/s strikes a stationary ball of the same mass. After the collision, the first ball moves at 4.33 m/s, at an angle of 30.0° with respect to the original line of motion. Assuming an elastic collision (and ignoring friction and rotational motion), find the struck ball's velocity after the collision.

Solution

Conceptualize: We guess a bit less than 3 m/s, at about −60°.

Categorize: We apply conservation of both x and y components of momentum to the two-ball system.

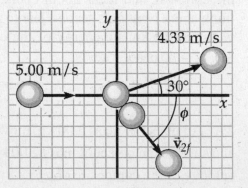

Analyze: We are given plenty of information:
$v_{1ix} = 5.00$ m/s, $v_{1iy} = 0$, $v_{2ix} = 0$, $v_{2iy} = 0$,

$v_{1fx} = (4.33 \text{ m/s}) \cos 30.0°$, $v_{1fy} = (4.33 \text{ m/s}) \sin 30.0°$. Call each mass m, and let $\vec{\mathbf{v}}_{2f}$ represent the velocity of the second ball after the collision, as in the figure.

In the x direction, conservation of momentum for the two-ball system reads

$$mv_{1ix} + mv_{2ix} = mv_{1fx} + mv_{2fx}$$

so $\quad v_{2fx} = v_{1ix} + v_{2ix} - v_{1fx} = 5.00 \text{ m/s} + 0 - (4.33 \text{ m/s}) \cos 30°$

giving $\quad v_{2fx} = 1.25 \text{ m/s}$

In the y direction we have similarly $\quad mv_{1iy} + mv_{2iy} = mv_{1fy} + mv_{2fy}$

so $\quad v_{2fy} = v_{1iy} + v_{2iy} - v_{1fy} = 0 + 0 - (4.33 \text{ m/s}) \sin 30°$

and $\quad v_{2fy} = -2.17 \text{ m/s}$

In summary, $\quad \vec{\mathbf{v}}_{2f} = (1.25\hat{\mathbf{i}} - 2.17\,\hat{\mathbf{j}}) \text{ m/s}$

$$= (1.25^2 + 2.17^2)^{1/2} \text{ m/s} \text{ at } \tan^{-1}(-2.17/1.25) = 2.50 \text{ m/s at } -60.0° \qquad \blacksquare$$

Finalize: We did not have to use the fact that the collision is elastic. We could prove that it is elastic by evaluating the kinetic energy before and after. For this special case of equal masses with one initially at rest, conservation of energy reduces to the true equation about speeds $5^2 = 4.33^2 + 2.5^2$. And this equation in turn is a Pythagorean theorem, a way of stating the condition that the velocities after the collision are perpendicular to each other.

---

35. An unstable atomic nucleus of mass 17.0×10^{-27} kg initially at rest disintegrates into three particles. One of the particles, of mass 5.00×10^{-27} kg, moves in the y direction with a speed of 6.00×10^6 m/s. Another particle, of mass 8.40×10^{-27} kg, moves in the x direction with a speed of 4.00×10^6 m/s. Find (a) the velocity of the third particle and (b) the total kinetic energy increase in the process.

Solution

Conceptualize: The third particle has small mass, so we expect a velocity of more than 6 million m/s into the third quadrant.

Categorize: Total vector momentum for the three-particle system is conserved.

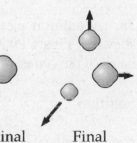

Original   Final

Analyze:

(a) With three particles, the total final momentum of the system is $\quad m_1\vec{\mathbf{v}}_{1f} + m_2\vec{\mathbf{v}}_{2f} + m_3\vec{\mathbf{v}}_{3f}$

and it must be zero to equal the original momentum.

The mass of the third particle is $m_3 = (17.0 - 5.00 - 8.40) \times 10^{-27} \, \text{kg}$

or $m_3 = 3.60 \times 10^{-27} \, \text{kg}$

Solving $m_1 \vec{\mathbf{v}}_{1f} + m_2 \vec{\mathbf{v}}_{2f} + m_3 \vec{\mathbf{v}}_{3f} = 0$

for $\vec{\mathbf{v}}_{3f}$ gives $\vec{\mathbf{v}}_{3f} = -\dfrac{m_1 \vec{\mathbf{v}}_{1f} + m_2 \vec{\mathbf{v}}_{2f}}{m_3}$

$$\vec{\mathbf{v}}_{3f} = -\frac{(3.00\hat{\mathbf{j}} + 3.36\hat{\mathbf{i}}) \times 10^{-20} \, \text{kg} \cdot \text{m/s}}{3.60 \times 10^{-27} \, \text{kg}} = (-9.33\hat{\mathbf{i}} - 8.33\hat{\mathbf{j}}) \, \text{Mm/s} \qquad \blacksquare$$

(b)   The original kinetic energy of the system is zero.

The final kinetic energy is $K = K_{1f} + K_{2f} + K_{3f}$

The terms are $K_{1f} = \frac{1}{2}(5.00 \times 10^{-27} \, \text{kg})(6.00 \times 10^6 \, \text{m/s})^2 = 9.00 \times 10^{-14} \, \text{J}$

$$K_{2f} = \tfrac{1}{2}(8.40 \times 10^{-27} \, \text{kg})(4.00 \times 10^6 \, \text{m/s})^2 = 6.72 \times 10^{-14} \, \text{J}$$

$$K_{3f} = \tfrac{1}{2}(3.60 \times 10^{-27} \, \text{kg}) \times \left[(-9.33 \times 10^6 \, \text{m/s})^2 + (-8.33 \times 10^6 \, \text{m/s})^2\right] = 28.2 \times 10^{-14} \, \text{J}$$

Then the system kinetic energy is

$$K = 9.00 \times 10^{-14} \, \text{J} + 6.72 \times 10^{-14} \, \text{J} + 28.2 \times 10^{-14} \, \text{J} = 4.39 \times 10^{-13} \, \text{J} = 439 \, \text{fJ} \qquad \blacksquare$$

Finalize: The total kinetic energy in the final picture is positive, but the resultant momentum is zero—scalar addition is different from vector addition. If you rely on writing down formulas with just two objects mentioned, you deserve to get a three-object problem like this on your next exam.

---

38.   A uniform piece of sheet metal is shaped as shown in Figure P9.38. Compute the x and y coordinates of the center of mass of the piece.

Solution

Conceptualize: By inspection, it appears that the center of mass is located at about $(12\hat{\mathbf{i}} + 13\hat{\mathbf{j}})$ cm.

Categorize: We could analyze the object as nine squares, each represented by an equal-mass particle at its center. But we will have less writing to do if we

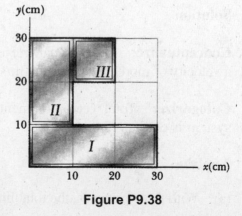

Figure P9.38

think of the sheet as composed of three sections, and consider the mass of each section to be at the geometric center of that section. Define the mass per unit area to be σ, and number the rectangles as shown. We can then calculate the mass and identify the center of mass of each section.

Analyze: $m_I = (30.0 \text{ cm})(10.0 \text{ cm})\sigma$ with $CM_I = (15.0 \text{ cm}, 5.00 \text{ cm})$.

$m_{II} = (10.0 \text{ cm})(20.0 \text{ cm})\sigma$ $CM_{II} = (5.00 \text{ cm}, 20.0 \text{ cm})$

$m_{III} = (10.0 \text{ cm})(10.0 \text{ cm})\sigma$ $CM_{III} = (15.0 \text{ cm}, 25.0 \text{ cm})$

The overall center of mass is at a point defined by the vector equation

$$\vec{r}_{CM} \equiv \left(\sum m_i \vec{r}_i \right) / \sum m_i$$

Substituting the appropriate values, $\vec{r}_{CM}$ is calculated to be:

$$\vec{r}_{CM} = \frac{\sigma[(300)(15.0\hat{i} + 5.00\hat{j}) + (200)(5.00\hat{i} + 20.0\hat{j}) + (100)(15.0\hat{i} + 25.0\hat{j})] \text{ cm}^3}{\sigma(300 \text{ cm}^2 + 200 \text{ cm}^2 + 100 \text{ cm}^2)}$$

Calculating, $\vec{r}_{CM} = \dfrac{45.0\hat{i} + 15.0\hat{j} + 10.0\hat{i} + 40.0\hat{j} + 15.0\hat{i} + 25.0\hat{j}}{6.00} \text{ cm}$

and evaluating, $\vec{r}_{CM} = (11.7\hat{i} + 13.3\hat{j}) \text{ cm}$ ∎

Finalize: The coordinates are close to our eyeball estimate. In solving this problem, we could have chosen to divide the original shape some other way, but the answer would be the same. This problem also shows that the center of mass can lie outside the boundary of the object.

---

41. A 2.00-kg particle has a velocity $(2.00\hat{i} - 3.00\hat{j})$ m/s, and a 3.00-kg particle has a velocity $(1.00\hat{i} + 6.00\hat{j})$ m/s. Find (a) the velocity of the center of mass and (b) the total momentum of the system.

Solution

Conceptualize: Visualize the center of mass as a certain point on a freely stretching straight line between the objects as they steadily move apart. The center of mass will have a constant velocity of a few meters per second into the first quadrant, and the total momentum is headed this way too.

Categorize: We will use $\vec{v}_{CM} = \dfrac{m_1\vec{v}_1 + m_2\vec{v}_2}{m_1 + m_2}$ and $\vec{p}_{CM} = (m_1 + m_2)\vec{v}_{CM}$

Analyze:

(a) by substitution,

$$\vec{v}_{CM} = \frac{(2.00 \text{ kg})[(2.00\hat{i} - 3.00\hat{j}) \text{ m/s}] + (3.00 \text{ kg})[(1.00\hat{i} + 6.00\hat{j}) \text{ m/s}]}{(2.00 \text{ kg} + 3.00 \text{ kg})}$$

$$\vec{v}_{CM} = (1.40\hat{i} + 2.40\hat{j}) \text{ m/s}$$ ∎

(b) $$\vec{p}_{CM} = (2.00 \text{ kg} + 3.00 \text{ kg})[(1.40\hat{i} + 2.40\hat{j}) \text{ m/s}]$$

$$= (7.00\hat{i} + 12.0\hat{j}) \text{ kg} \cdot \text{m/s}$$ ∎

Finalize: We can symbolize the resultant system momentum as $\vec{p}_{CM}$ because it is the momentum of a particle with the total system mass, moving at the center of mass. The chapter text proved this theorem and we see it again when we write

$$(m_1 + m_2)\vec{v}_{CM} = m_1\vec{v}_1 + m_2\vec{v}_2 = \Sigma\vec{p} = \vec{p}_{CM}$$

---

43. Romeo (77.0 kg) entertains Juliet (55.0 kg) by playing his guitar from the rear of their boat at rest in still water, 2.70 m away from Juliet, who is in the front of the boat. After the serenade, Juliet carefully moves to the rear of the boat (away from shore) to plant a kiss on Romeo's cheek. How far does the 80.0-kg boat move toward the shore it is facing?

Solution

Conceptualize: As the girl makes her way toward the rear of the boat, the boat and the boy glide forward, towards shore, by a distance smaller in proportion as their mass is larger than hers.

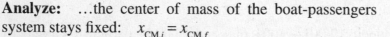

Categorize: No outside forces act on the boat-plus-lovers system, so its momentum is conserved at zero and…

Analyze: …the center of mass of the boat-passengers system stays fixed: $x_{CM,i} = x_{CM,f}$

Define K to be the point where they kiss, and Δx_J and Δx_b as shown in the figure. Since Romeo moves with the boat (and thus $\Delta x_{Romeo} = \Delta x_b$), let m_b be the combined mass of Romeo and the boat.

The front of the boat and the shore are to the right in this picture, and we take the positive x direction to the right.

Then, $$m_J\Delta x_J + m_b\Delta x_b = 0$$

Choosing the x axis to point toward the shore,

$$(55.0 \text{ kg})\Delta x_J + (77.0 \text{ kg} + 80.0 \text{ kg})\Delta x_b = 0$$

and $$\Delta x_J = -2.85\Delta x_b$$

As Juliet moves away from shore, the boat and Romeo glide toward the shore until the original 2.70 m gap between them is closed. We describe the relative motion with the equation

$$|\Delta x_J| + \Delta x_b = 2.70 \text{ m}$$

Here the first term needs absolute value signs because Juliet's change in position is toward the left. An equivalent equation is then

$$-\Delta x_J + \Delta x_b = 2.70 \text{ m}$$

Substituting, we find $+2.85 \, \Delta x_b + \Delta x_b = 2.70$ m

so $\Delta x_b = 2.70$ m/3.85 $= +0.700$ m or 0.700 m towards the shore. ∎

Finalize: This is physically a problem about conservation of momentum. A person inside the boat can do nothing to move the center of mass of the boat-people system. The equations we wrote could be about position rather than momentum because position is the integral of velocity.

---

45. For a technology project, a student has built a vehicle, of total mass 6.00 kg, that moves itself. As shown in Figure P9.45, it runs on four light wheels. A reel is attached to one of the axles, and a cord originally wound on the reel goes up over a pulley attached to the vehicle, to support an elevated load. After the vehicle is released from rest, the load descends very slowly, unwinding the cord to turn the axle and make the vehicle move forward (to the left in Fig. P9.45). Friction is negligible in the pulley and axle bearings. The wheels do not slip on the floor. The reel has been constructed with a conical shape so that the load descends at a constant low speed while the vehicle moves horizontally across the floor with constant acceleration, reaching a final velocity of $3.00\hat{\mathbf{i}}$ m/s. (a) Does the floor impart impulse to the vehicle? If so, how much? (b) Does the floor do work on the vehicle? If so, how much? (c) Does it make sense to say that the final momentum of the vehicle came from the floor? If not, where did it come from? (d) Does it make sense to say that the final kinetic energy of the vehicle came from the floor? If not, where did it come from? (e) Can we say that one particular force causes the forward acceleration of the vehicle? What does cause it?

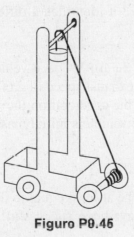

Figuro P9.45

Solution

Conceptualize.   How does your car set itself moving? The vehicle here can be considered a model for a gasoline automobile, or even for you starting to walk across the floor. Newton proposes that nothing can exert a force on itself—only another object can exert a force to accelerate the car. This is a good exercise to describe precisely how Newton is right, but still you must put gasoline into the car and not into some other object.

Categorize.   The calculations here will be short and simple. The answers to the questions can be considered simple as well, if we agree that different theories of motion can all be correct as they direct our attention to different things. We will use reasoning based on the nonisolated system (energy) model, the nonisolated system (momentum) model, and the particle under a net force model.

Analyze:

(a) Yes, the floor imparts impulse to the vehicle. The only horizontal force on the vehicle is the frictional force exerted by the floor, so it gives the vehicle all of its final momentum, $(6 \text{ kg})(3\,\hat{\mathbf{i}} \text{ m/s}) = 18.0\,\hat{\mathbf{i}} \text{ kg} \cdot \text{m/s}$ ∎

(b) No, the floor does no work on the vehicle. Each drive wheel rolls without slipping, with its lowest point stationary relative to the floor. The friction force exerted by the floor on the bottom of each wheel acts over no distance at the contact point, so it does zero work. ∎

(c) Yes, we could say that the final momentum of the cart came from the floor or from the planet through the floor, because the floor imparts impulse. ∎

(d) No, the floor does no work. The final kinetic energy came from the original gravitational energy of the Earth-load system, in amount $(1/2)(6 \text{ kg})(3 \text{ m/s})^2 = 27.0 \text{ J}$. ∎

(e) Yes, the forward acceleration is caused by one particular force. The acceleration is caused by the static friction force exerted by the floor on each drive wheel. This is the force that prevents the contact points on the wheels from slipping backward. ∎

Finalize.   The impulse-momentum theory of motion is quite different from the work-energy theory of motion. It identifies a different physical quantity as the cause for changes in motion, and in the situation here it identifies a different object as acting to make the motion of the vehicle change.

50. Review.   The first stage of a Saturn V space vehicle consumed fuel and oxidizer at the rate of 1.50×10^4 kg/s, with an exhaust speed of 2.60×10^3 m/s. (a) Calculate the thrust produced by this engine. (b) Find the acceleration the vehicle had just as it lifted off the launch pad on the Earth, taking the vehicle's initial mass as 3.00×10^6 kg.

Solution

Conceptualize: The thrust must be at least equal to the weight of the rocket (≈ 30 MN); otherwise the launch would not have been successful! However, since Saturn V rockets accelerate rather slowly compared to the acceleration of falling objects, the thrust should be less than about twice the rocket's weight, to meet the condition $0 < a < g$.

Categorize: Use Newton's second law to find the force and acceleration from the changing momentum.

Analyze:

(a) The thrust, F, is equal to the time rate of change of momentum as fuel is exhausted from the rocket.

$$F = \frac{dp}{dt} = \frac{d}{dt}(mv_e)$$

Since v_e is a constant exhaust velocity,

$$F = v_e(dm/dt) \quad \text{where} \quad dm/dt = 1.50 \times 10^4 \text{ kg/s}$$

$$\text{and} \quad v_e = 2.60 \times 10^3 \text{ m/s}$$

Then $\quad F = (2.60 \times 10^3 \text{ m/s})(1.50 \times 10^4 \text{ kg/s}) = 39.0 \text{ MN}$ ∎

(b) Applying $\sum F = ma$ gives

$$(3.90 \times 10^7 \text{ N}) - (3.00 \times 10^6 \text{ kg})(9.80 \text{ m/s}^2) = (3.00 \times 10^6 \text{ kg})a$$

$$a = \frac{(3.90 \times 10^7 \text{ N}) - (2.94 \times 10^7 \text{ N})}{3.00 \times 10^6 \text{ kg}} = 3.20 \text{ m/s}^2 \text{ up}$$ ∎

Finalize: As expected, the thrust is slightly greater than the weight of the rocket, and the acceleration is about $0.3g$, so the answers appear to be reasonable. This kind of rocket science is not so complicated after all!

---

55. A 3.00-kg steel ball strikes a wall with a speed of 10.0 m/s at an angle of $\theta = 60.0°$ with the surface. It bounces off with the same speed and angle (Fig. P9.55). If the ball is in contact with the wall for 0.200 s, what is the average force exerted by the wall on the ball?

Solution

Conceptualize: If we think about the angle as a variable and consider the limiting cases, then the force should be zero when the angle is 0° (no contact between the ball and the wall). When the angle is 90°, the force will be its maximum and can be found from the momentum-impulse equation as 3 kg (20 m/s)/0.2 s = 300 N. Then for the 60° angle we expect a force of less than 300 N. It must be directed to the left.

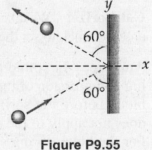

Figure P9.55

Categorize: We use the momentum-impulse equation to find the force, and carefully consider the direction of the velocity vectors by defining up and to the right as positive.

Analyze: In $\vec{\mathbf{F}}\Delta t = \Delta(m\vec{\mathbf{v}})$

one component gives $\Delta p_y = m(v_{fy} - v_{iy}) = m(v \cos 60.0° - v \cos 60.0°) = 0$

So the wall does not exert a force on the ball in the y direction. The other component gives

$$\Delta p_x = m(v_{fx} - v_{ix}) = m(-v \sin 60.0° - v \sin 60.0°) = -2mv \sin 60.0°$$

$$\Delta p_x = -2(3.00 \text{ kg})(10.0 \text{ m/s})(0.866) = -52.0 \text{ kg} \cdot \text{m/s}$$

So

$$\vec{\mathbf{F}} = \frac{\Delta\vec{\mathbf{p}}}{\Delta t} = \frac{\Delta p_x \hat{\mathbf{i}}}{\Delta t} = \frac{-52.0\hat{\mathbf{i}} \text{ kg} \cdot \text{m/s}}{0.200 \text{ s}} = -260\hat{\mathbf{i}} \text{ N} \qquad \blacksquare$$

Finalize: The force is to the left and has a magnitude less than 300 N as expected.

---

73. A 5.00-g bullet moving with an initial speed of $v_i = 400$ m/s is fired into and passes through a 1.00-kg block as shown in Figure P9.73. The block, initially at rest on a frictionless, horizontal surface, is connected to a spring with force constant 900 N/m. The block moves $d = 5.00$ cm to the right after impact before being brought to rest by the spring. Find (a) the speed at which the bullet emerges from the block and (b) the amount of initial kinetic energy of the bullet that is converted into internal energy in the bullet-block system during the collision.

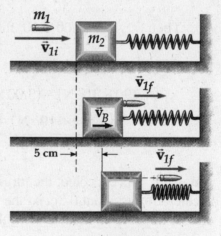

Figure P9.73 (modified)

Solution

Conceptualize: We expect a speed much less than 400 m/s because the bullet gave some momentum to the block. The massive block has only a low speed when it starts to compress the spring, so it has relatively little kinetic energy. Most of the bullet's original kinetic energy must disappear into internal energy.

Categorize: We take the process apart into two stages: the very brief collision of the bullet and the block to set the block in motion, and the subsequent compression of the spring.

Analyze: First find the initial velocity of the block, using conservation of energy during compression of the spring. Assume that the bullet has passed completely through the block before the spring has started to compress. Note that conservation of momentum does not apply to the spring-compression process, but conservation of energy does apply. For the block-spring system during the compression process (between the second of our pictures and the third) we have

$$(K + U)_{\text{before}} = (K + U)_{\text{after}}$$

$$\tfrac{1}{2}m_2 v_B^2 = \tfrac{1}{2}kx^2$$

becomes $\quad v_B = \sqrt{\dfrac{k}{m_2}}\,x = \sqrt{\dfrac{900\,\text{N}/\text{m}}{1.00\,\text{kg}}}\,0.0500\,\text{m} = 1.50\,\text{m}/\text{s}$

(a) When the bullet collides with the block, it is not energy but momentum that is conserved. For the bullet-block system we have

$$m_1 v_{1i} + m_2 v_{2i} = m_1 v_{1f} + m_2 v_B$$

so $\quad v_{1f} = (m_1 v_{1i} - m_2 v_B)/m_1$

$$v_{1f} = \frac{\left(5.00 \times 10^{-3}\,\text{kg}\right)(400\,\text{m}/\text{s}) - (1.00\,\text{kg})(1.50\,\text{m}/\text{s})}{5.00 \times 10^{-3}\,\text{kg}} = 100\,\text{m}/\text{s} \qquad \blacksquare$$

(b) We think about energy in an isolated system to find the mechanical energy converted into internal energy in the collision. Before the collision, the block is motionless, and the bullet's energy is

$$K_i = \tfrac{1}{2}mv_{1i}^2 = \tfrac{1}{2}(0.005\,00\,\text{kg})(400\,\text{m}/\text{s})^2 = 400\,\text{J}$$

After the collision, the mechanical energy is

$$K_f = \tfrac{1}{2}m_1 v_{1f}^2 + \tfrac{1}{2}m_2 v_B^2$$

$$K_f = \tfrac{1}{2}(0.005\,00\,\text{kg})\,(100\,\text{m}/\text{s})^2 + \tfrac{1}{2}(1.00\,\text{kg})(1.50\,\text{m}/\text{s})^2 = 26.1\,\text{J}$$

Therefore the energy converted from mechanical to internal in the collision is given by $\Delta K + \Delta E_{int} = 0$

$$\Delta E_{int} = -\Delta K = -K_f + K_i = -26.1\,\text{J} + 400\,\text{J} = 374\,\text{J} \qquad \blacksquare$$

Finalize: Energy and momentum are quite different things and follow quite different rules. The spring counts as exerting an external force on the block over time during the compression process, so it completely upsets momentum conservation—it makes the block stop moving. But the spring force is conservative, so the stopping process just converts a constant 1.12 J from kinetic into elastic energy. In the bullet-block collision, friction between them negates conservation of mechanical energy. But the friction forces are internal and must leave total momentum constant for the pair of objects.

10

Rotation of a Rigid Object About a Fixed Axis

EQUATIONS AND CONCEPTS

The **arc length**, s, is the distance traveled by a particle which moves along a circular path of radius r. *As the particle moves along the arc, the radius line sweeps out an angle θ (measured counterclockwise relative to a reference line).*

$$s = r\theta \qquad \qquad 10.1a$$

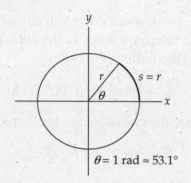

$\theta = 1 \text{ rad} \approx 53.1°$

The **angular displacement** θ is the ratio of two lengths (arc length to radius) and hence is a dimensionless quantity. However, it is common practice to refer to the angle theta as being in units of radians. In calculations, the relationship between radians and degrees is shown to the right.

$$\theta(\text{rad}) = \frac{\pi}{180°}\,\theta(\text{deg})$$

The **average angular speed** ω_{avg} of a particle or body rotating about a fixed axis equals the ratio of the angular displacement $\Delta\theta$ to the time interval Δt, where θ is measured in radians. *In the case of a rigid body, $\Delta\theta$ is associated with the entire body as well as with each element of the body.*

$$\omega_{avg} \equiv \frac{\theta_f - \theta_i}{t_f - t_i} = \frac{\Delta\theta}{\Delta t} \qquad (10.2)$$

The **instantaneous angular speed** ω is defined as the limit of the average angular speed as Δt approaches zero. *ω is considered positive when θ is increasing counterclockwise.*

$$\omega \equiv \lim_{\Delta t \to 0} \frac{\Delta\theta}{\Delta t} = \frac{d\theta}{dt} \qquad (10.3)$$

The **average angular acceleration** α_{avg} of a rotating body is defined as the ratio of the change in angular velocity to the time interval Δt during which the change occurs.

$$\alpha_{avg} \equiv \frac{\omega_f - \omega_i}{t_f - t_i} = \frac{\Delta\omega}{\Delta t} \qquad (10.4)$$

166

The **instantaneous angular acceleration** equals the limit of the average angular acceleration as Δt approaches zero. *When a rigid object is rotating about a fixed axis, every element of the object rotates through the same angle in a given time interval and has the same angular velocity and the same angular acceleration.*

$$\alpha \equiv \lim_{\Delta t \to 0} \frac{\Delta \omega}{\Delta t} = \frac{d\omega}{dt} \tag{10.5}$$

The **equations of rotational kinematics** apply when a particle or body rotates about a fixed axis with *constant angular acceleration*.

$$\omega_f = \omega_i + \alpha t \tag{10.6}$$

$$\theta_f = \theta_i + \omega_i t + \tfrac{1}{2}\alpha t^2 \tag{10.7}$$

$$\omega_f^2 = \omega_i^2 + 2\alpha(\theta_f - \theta_i) \tag{10.8}$$

$$\theta_f = \theta_i + \tfrac{1}{2}(\omega_i + \omega_f)t \tag{10.9}$$

The values of **tangential speed and tangential acceleration** are proportional to the corresponding angular values. All points on an object rotating about a fixed axis have the same values of angular speed and angular acceleration. However, the corresponding values of tangential speed and acceleration depend on the radial distance from the axis of rotation. The figure shows a disk rotating counterclockwise with negative angular acceleration about an axis through the center.

$$v = r\omega \tag{10.10}$$

$$a_t = r\alpha \tag{10.11}$$

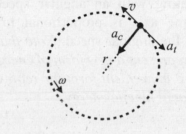

The **centripetal acceleration** is directed toward the center of rotation.

$$a_c = \frac{v^2}{r} = r\omega^2 \tag{10.12}$$

The **total linear acceleration** has both radial and tangential components and a magnitude given by Equation 10.13. *The magnitude of a_r equals a_c in Equation 10.12.*

$$a = \sqrt{a_t^2 + a_r^2} = r\sqrt{\alpha^2 + \omega^4} \tag{10.13}$$

The **moment of inertia of a system of particles** is defined by Equation 10.15, where m_i is the mass of the ith particle and r_i is its distance from a specified axis. Note that I has SI units of kg·m². *Moment of inertia is a measure of the resistance of an object to changes in its rotational motion.*

$$I \equiv \sum_i m_i r_i^2 \tag{10.15}$$

The **moment of inertia of an extended rigid object** can be determined by integrating over each mass element of the object. The integration can usually be simplified by introducing volumetric mass density, ρ ($dm = \rho dV$), surface mass density σ ($dm = \sigma dA$), and linear mass density λ ($dm = \lambda dx$). *The integration extends over the corresponding volume, area, or length of the object.*

$$I = \int r^2 dm \qquad (10.17)$$

The **parallel-axis theorem** can be used to calculate the moment of inertia of an object of mass M about an axis which is parallel to an axis through the center of mass and located a distance D away. See *Calculating Moments of Inertia in Suggestions, Skills, and Strategies* section.

$$I = I_{CM} + MD^2 \qquad (10.18)$$

The **rotational kinetic energy** of a rigid body rotating with an angular speed ω about some axis is proportional to the square of the angular speed. *Note that K_R does not represent a new form of energy. It is simply a convenient form for representing rotational kinetic energy.*

$$K_R = \tfrac{1}{2}I\omega^2 \qquad (10.16)$$

The **torque** due to an applied force has a magnitude given by the product of the force and its moment arm d, *where d equals the perpendicular distance from the rotation axis to the line of action of F (See figure at right). The line of action of a force is an imaginary line extending out of each end of the vector representing the force, which gives rise to the torque. Torque is a measure of the ability of a force to rotate a body about a specified axis and should not be confused with force.*

$$\tau \equiv rF \sin\phi = Fd \qquad (10.19)$$

where $d = r\sin\phi$

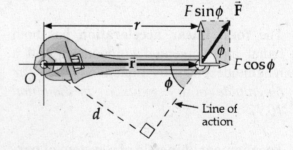

See *Calculating Torque in Suggestions, Skills, and Strategies* section.

The **rotational analog of Newton's second law** of motion is valid for a rigid body of arbitrary shape rotating about a fixed axis:

$$\sum \tau_{ext} = I\alpha \qquad (10.20)$$

the net torque is proportional to the angular acceleration. *Equation 10.20 applies when forces acting on a rigid body have radial as well as tangential components.*

The **instantaneous power** delivered by a force in rotating a rigid body about a fixed axis is proportional to the angular velocity. *Equation 10.23 is analogous to P = Fv in the case of linear motion.*

$$P = \frac{dW}{dt} = \tau\omega \tag{10.23}$$

The **work-kinetic energy theorem for rotational motion** states that the net work done by external forces in rotating a symmetric rigid object about a fixed axis equals the change in the object's rotational kinetic energy.

$$\sum W = \tfrac{1}{2}I\omega_f^2 - \tfrac{1}{2}I\omega_i^2 \tag{10.24}$$

For **pure rolling motion**, the linear speed of the center of mass and the magnitude of the linear acceleration of the center of mass are given by Equations 10.25 and 10.26. *Pure rolling motion occurs when a rigid object rolls without slipping.*

$$v_{CM} = R\omega \tag{10.25}$$

$$a_{CM} = R\alpha \tag{10.26}$$

The **total kinetic energy of a rolling object** is the sum of the rotational kinetic energy about the center of mass and the translational kinetic energy of the center of mass.

$$K = \tfrac{1}{2}I_{CM}\omega^2 + \tfrac{1}{2}Mv_{CM}^2 \tag{10.28}$$

$$K = \tfrac{1}{2}\left(\frac{I_{CM}}{R^2} + M\right)v_{CM}^2 \tag{10.29}$$

SUGGESTIONS, SKILLS, AND STRATEGIES

CALCULATING MOMENTS OF INERTIA

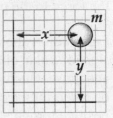

The method for calculating the moment of inertia of a system of particles about a specified axis is straightforward; apply Equation 10.15, $I = \sum m_i r_i^2$, where m_i is the mass of the ith particle and r_i is the distance from the axis of rotation to the particle. For mass m in the figure, the moment of inertia about the x axis is $I_x = my^2$ and about the y axis the moment of inertia is $I_y = mx^2$. A particle on the x axis would have a zero moment of inertia about that axis.

In using Equation 10.17, $I = \int r^2 dm$, to calculate the moment of inertia of a rigid body about an axis of symmetry, it is usually easier to change the integrand to sum over all volume elements. To do this let $dm = \rho dV$, where ρ is the volumetric mass density and dV is an element of volume. In the case of a plane sheet, $dm = \sigma dA$ where σ is the surface mass density and dA is an element of area; and for a long thin rod, $dm = \lambda dx$ where λ is the linear mass density and dx is an element of length. In the case of a uniform thin hoop or ring, it is easy to integrate directly over the mass elements since all mass elements are the same distance from the axis of symmetry. **Note:** If the density of the object is not constant over the volume, area, or length, then ρ, σ, or λ must be replaced with the expression showing the manner in which density varies.

If the moment of inertia of a rigid body about an axis through the center of mass I_{CM} is known, you can easily evaluate the moment of inertia about any axis parallel to the axis through the center of mass using the **parallel-axis theorem**:

$$I = I_{CM} + MD^2$$

where M is the mass of the body and D is the distance between the two axes.

For example, the moment of inertia of a solid cylinder about an axis through its center (the z axis in the figure) is given by

$$I_z = \tfrac{1}{2}MR^2$$

Hence, the moment of inertia about the z' axis located a distance $D = R$ from the z axis is:

$$I_{z'} = I_z + MR^2 = \tfrac{1}{2}MR^2 + MR^2 = \tfrac{3}{2}MR^2$$

CALCULATING TORQUE

Do not confuse force and torque. Both are vector quantities. Torque is the effect of a force in causing or changing the rotation of a rigid body. The magnitude of a torque is the product of the force and the moment arm of the force (the perpendicular distance from the line of action of the force to the pivot point).

For rotation about a fixed axis, the vector nature of the torque can be represented by using signed scalar quantities. Torques which produce counterclockwise rotations are positive and those which produce clockwise rotations are negative. The correct algebraic signs must be taken into account when using Equation 10.20, $\Sigma\tau = I\alpha$. In the figure, F_5 gives rise to a positive torque about an axis through O, force F_4 produces zero torque (the moment arm is zero) and the other forces produce negative torques. Consider the forces and moment arms shown in the figure; and note that if the magnitudes of the corresponding forces are equal, $|\tau_2| > |\tau_1| > |\tau_3|$.

REVIEW CHECKLIST

You should be able to:

- Apply the equations of rotational kinematics to make calculations of rotational displacement, angular velocity, and angular acceleration. (Section 10.2)

- Make calculations relating corresponding angular and linear quantities. (Section 10.3)

- Calculate the moment of inertia of a system of discrete point masses. (Section 10.4)

- Calculate the moment of inertia of solids which have a high degree of symmetry by integrating over the volume, area, or length. Use the parallel axis theorem. (Sections 10.4 and 10.5)

- Calculate the torque due to a force acting on a rigid body and determine the angular acceleration due to the net force. (Sections 10.6 and 10.7)

- Make calculations of rotational kinetic energy and apply the work-kinetic energy theorem to rotating bodies. Calculate the power delivered by a tangential force. (Sections 10.8 and 10.4)

- Calculate the kinetic energy of a rigid body in pure rolling motion. (Section 10.9)

Some important points to remember:

- Quantitatively, the angular displacement, speed, and acceleration for a point on a rigid body in rotational motion are related to the distance traveled, tangential speed, and tangential acceleration. The linear quantity is calculated by multiplying the angular quantity by the radius arm for the point in question.

- If a body rotates about a fixed axis, every element of the body has the same angular speed and angular acceleration. The kinematic equations which describe angular motion are analogous to the corresponding set of equations pertaining to linear motion. The expressions for Newton's second law, kinetic energy, and power also have analogous forms for linear and rotational motion.

- The value of the moment of inertia I of a system of particles or a rigid body depends on the manner in which the mass is distributed relative to the axis of rotation. For example, the moment of inertia of a disk is less than that of a ring of equal mass and radius when calculated about an axis perpendicular to their planes and through the centers of mass. The parallel-axis theorem is useful for calculating I about an axis parallel to one that goes through the center of mass.

- Torque should not be confused with force. The magnitude of a torque associated with a force depends on the magnitude of the force, the point of application of the force, and the angle between the force vector and the line joining the pivot point and the point where the force is applied.

- The description of the motion of an object in pure rolling motion can be simplified by considering the motion (displacement, velocity, and acceleration) of the center of mass.

ANSWER TO AN OBJECTIVE QUESTION

11. A basketball rolls across a classroom floor without slipping, with its center of mass moving at a certain speed. A block of ice of the same mass is set sliding across the floor with the same speed along a parallel line. Which object has more (i) kinetic energy and (ii) momentum? (a) The basketball does. (b) The ice does. (c) The two quantities are equal. (iii) The two objects encounter a ramp sloping upward. Which object will travel farther up the ramp? (a) The basketball will. (b) The ice will. (c) They will travel equally far up the ramp.

Answer   (i) (a) The basketball has more kinetic energy. It possesses rotational kinetic energy which adds to its translational kinetic energy.
(ii) (c) The rolling ball and the sliding ice have equal quantities of momentum.
(iii) (a) The basketball starts with more kinetic energy, due to its rotational kinetic energy. The initial kinetic energy of both the ice block and the basketball has transformed entirely into gravitational potential energy of the object-Earth systems when the objects momentarily come to rest at their highest points on the ramp. The basketball will be at a higher location, corresponding to its larger gravitational potential energy.

ANSWERS TO SELECTED CONCEPTUAL QUESTIONS

1.   (a) What is the angular speed of the second hand of a clock? (b) What is the direction of $\vec{\omega}$ as you view a clock hanging on a vertical wall? (c) What is the magnitude of the angular acceleration vector $\vec{\alpha}$ of the second hand?

Answer   The second hand of a clock turns at one revolution per minute, so

$$\omega = \frac{2\pi \text{ rad}}{60 \text{ sec}} = 0.105 \text{ rad/s}$$

The motion is clockwise, so the direction of the vector angular velocity is away from you. (b) The second hand turns steadily, so ω is constant, and α is zero.

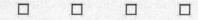

11.   If you see an object rotating, is there necessarily a net torque acting on it?

Answer   No. A rigid object rotates with constant angular velocity when zero total torque acts on it. For example, consider the Earth; it rotates at a constant rate of once per day, but there is no net torque acting on it.

SOLUTIONS TO SELECTED END-OF-CHAPTER PROBLEMS

5. A wheel starts from rest and rotates with constant angular acceleration to reach an angular speed of 12.0 rad/s in 3.00 s. Find (a) the magnitude of the angular acceleration of the wheel and (b) the angle in radians through which it rotates in this time interval.

Solution

Conceptualize: We expect on the order of 10 rad/s² for the angular acceleration. The angular displacement must be greater than zero and less than (12 rad/s)(3 s) = 36 rad, because the rotor is always turning slower than 12 rad/s.

Categorize: This is a straightforward motion-with-constant-α problem. We identify the symbols for the given information

$$\omega_i = 0 \quad \omega_f = 12 \text{ rad/s} \quad \theta_i = 0 \quad t = 3 \text{ s} \quad \theta_f = ?? \quad \alpha = ?$$

and choose one of the four standard equations to solve each part.

Analyze:

(a) $\omega_f = \omega_i + \alpha t$

So $\alpha = \dfrac{\omega_f - \omega_i}{t} = \dfrac{(12.0 - 0) \text{ rad/s}}{3.00 \text{ s}} = 4.00 \text{ rad/s}^2$ ∎

(b) $\theta_f = \omega_i t + \frac{1}{2}\alpha t^2 = \frac{1}{2}(4.00 \text{ rad/s}^2)(3.00 \text{ s})^2 = 18.0 \text{ rad}$ ∎

Finalize: A rotating object does not have to be round. Think of a square rotating platform with a motor that makes it speed up steadily. This problem is not about motion through space with constant acceleration a. Different points on the rotating object move through different distances with different accelerations. Points on the axis do not move through space at all.

---

7. An electric motor rotating a workshop grinding wheel at 1.00×10^2 rev/min is switched off. Assume the wheel has a constant negative angular acceleration of magnitude 2.00 rad/s². (a) How long does it take the grinding wheel to stop? (b) Through how many radians has the wheel turned during the time interval found in part (a)?

Solution

Conceptualize: We estimate several seconds for the slowing process, and many radians for the angular displacement.

Categorize: We use the rigid body under constant angular acceleration model. We will use one of the four equations about motion with constant angular acceleration for each part.

Analyze: We are given $\alpha = -2.00 \text{ rad/s}^2$ $\omega_f = 0$

and

$$\omega_i = 100 \frac{\text{rev}}{\text{min}} \left(2\pi \frac{\text{rad}}{\text{rev}} \right) \left(\frac{1 \text{ min}}{60.0 \text{ s}} \right) = 10.5 \text{ rad/s}$$

(a) From $\omega_f = \omega_i + \alpha t$

we have

$$t = \frac{\omega_f - \omega_i}{\alpha} = \frac{0 - (10.5 \text{ rad/s})}{-2.00 \text{ rad/s}^2} = 5.24 \text{ s} \qquad \blacksquare$$

(b) From $\omega_f^2 - \omega_i^2 = 2\alpha(\theta_f - \theta_i)$

we have

$$\theta_f - \theta_i = \frac{\omega_f^2 - \omega_i^2}{2\alpha} = \frac{0 - (10.5 \text{ rad/s})^2}{2(-2.00 \text{ rad/s}^2)} = 27.4 \text{ rad} \qquad \blacksquare$$

Finalize: Note also in part (b) that since a constant angular acceleration persists for time t, we could have found the answer from

$$\theta_f - \theta_i = \tfrac{1}{2}\left(\omega_i + \omega_f\right)t = \frac{(10.5 + 0 \text{ rad/s})}{2}(5.24 \text{ s}) = 27.4 \text{ rad}$$

That is, the set of four equations is more than complete.

---

17. A wheel 2.00 m in diameter lies in a vertical plane and rotates about its central axis with a constant angular acceleration of 4.00 rad/s². The wheel starts at rest at $t = 0$, and the radius vector of a certain point P on the rim makes an angle of 57.3° with the horizontal at this time. At $t = 2.00$ s, find (a) the angular speed of the wheel and, for point P, (b) the tangential speed, (c) the total acceleration, and (d) the angular position.

Solution

Conceptualize: Visualize a large carnival roulette wheel. It will be spinning at several radians per second. A point on the rim will have a rather large centripetal acceleration as well as a smaller tangential acceleration.

Categorize: Parts (a) and (d) are 'pure Greek-letter' exercises about describing rotation with constant α. They are about the kinematics of rotation. Parts (b) and (c) make us relate rotation at one instant to the linear motion of a point on the object, through radian measure.

Analyze: We are given

$$\alpha = 4 \text{ rad/s}^2, \ \omega_i = 0, \ \theta_i = 57.3° = 1 \text{ rad, and } t = 2 \text{ s}$$

(a)   We choose the equation $\omega_f = \omega_i + \alpha t = 0 + \alpha t$

and compute $\omega_f = (4.00 \text{ rad}/\text{s}^2)(2.00 \text{ s}) = 8.00 \text{ rad/s}$ ∎

(b)   We are also given $r = 1.00$ m. The speed of a point P on the rim is

$$v = r\omega = (1.00 \text{ m})(8.00 \text{ rad/s}) = 8.00 \text{ m/s}$$ ∎

(c)   The components of the point's acceleration are

$$a_c = r\omega^2 = (1.00 \text{ m})(8.00 \text{ rad/s})^2 = 64.0 \text{ m/s}^2$$ ∎

$$a_t = r\alpha = (1.00 \text{ m})(4.00 \text{ rad/s}^2) = 4.00 \text{ m/s}^2$$ ∎

The magnitude of the total acceleration is:

$$a = \sqrt{a_c^2 + a_t^2} = \sqrt{(64.0 \text{ m/s}^2)^2 + (4.00 \text{ m/s}^2)^2} = 64.1 \text{ m/s}^2$$ ∎

The direction of the total acceleration vector makes an angle ϕ forward with respect to the radius to point P:

$$\phi = \tan^{-1}\left(\frac{a_t}{a_c}\right) = \tan^{-1}\left(\frac{4.00 \text{ m/s}^2}{64.0 \text{ m/s}^2}\right) = 3.58°$$ ∎

(d) $\theta_f = \theta_i + \omega_i t + \frac{1}{2}\alpha t^2 = (1.00 \text{ rad}) + \frac{1}{2}(4.00 \text{ rad/s}^2)(2.00 \text{ s})^2 = 9.00 \text{ rad}$ ∎

$\theta_f - \theta_i$ is the total angle through which point P has passed, and is greater than one revolution. The position we see for point P is found by subtracting one revolution from θ_f. Therefore P is at

$$9.00 \text{ rad} - 2\pi \text{ rad} = 2.72 \text{ rad}$$ ∎

Finalize: The acceleration is several times larger than the free-fall acceleration. Notice that you need not set your calculator for radian measure to do problems like this. The radian is just a unit.

---

19.   A disk 8.00 cm in radius rotates at a constant rate of 1 200 rev/min about its central axis. Determine (a) its angular speed in radians per second, (b) the tangential speed at a point 3.00 cm from its center, (c) the radial acceleration of a point on the rim, and (d) the total distance a point on the rim moves in 2.00 s.

Solution

Conceptualize: If this is a sanding disk driven by a workshop electric drill, you need to wear goggles. The acceleration of a point on the circumference is many times g.

Categorize: Every part is about using radian measure to relate rotation of the whole object to the linear motion of a point on the object.

Analyze:

(a) $\omega = 2\pi f = (2\pi \text{ rad/rev})\left(\dfrac{1\,200\text{ rev/min}}{60\text{ s/min}}\right) = 125.7\text{ rad/s} = 126\text{ rad/s}$ ∎

(b) $v = \omega r = (125.7\text{ rad/s})(0.030\,0\text{ m}) = 3.77\text{ m/s}$ ∎

(c) $a_c = v^2/r = \omega^2 r = (125.7\text{ rad/s})^2(0.080\,0\text{ m}) = 1.26 \times 10^3\text{ m/s}^2$ ∎

(d) $s = r\theta = r\omega t = (8.00 \times 10^{-2}\text{ m})(125.7\text{ rad/s})(2.00\text{ s}) = 20.1\text{ m}$ ∎

Finalize: The change-in-direction acceleration of the rim point is hundreds of times larger than g, while the other answers have more ordinary sizes. The units work out as $\text{rad} \cdot \text{m} = \text{m}$ because 1 rad = 1.

---

25. The four particles in Figure P10.25 are connected by rigid rods of negligible mass. The origin is at the center of the rectangle. The system rotates in the xy plane about the z axis with an angular speed of 6.00 rad/s. Calculate (a) the moment of inertia of the system about the z axis and (b) the rotational kinetic energy of the system.

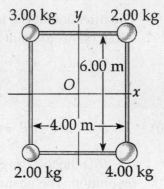

Figure P10.25

Solution

Conceptualize: Visualize a room-size framework wheeling like a runaway advertising sign. Its rotational inertia will be several kilogram-meter-squared's and its kinetic energy many joules.

Categorize: We use the definition of moment of inertia to add up the contributions of the four particles, and then the definition of rotational kinetic energy.

Analyze:

(a)   All four particles are at a distance r from the z axis,

with $r^2 = (3.00\text{ m})^2 + (2.00\text{ m})^2 = 13.00\text{ m}^2$

Thus the moment of inertia is

$I_z = \Sigma m_i r_i^2 = (3.00\text{ kg})(13.00\text{ m}^2) + (2.00\text{ kg})(13.00\text{ m}^2) + (4.00\text{ kg})(13.00\text{ m})$
$\qquad + (2.00\text{ kg})(13.00\text{ m})$

$I_z = 143\text{ kg} \cdot \text{m}^2$ ∎

(b) $K_R = \frac{1}{2}I_z\omega^2 = \frac{1}{2}(143 \text{ kg} \cdot \text{m}^2)(6.00 \text{ rad/s})^2 = 2.57 \text{ kJ}$ ■

Finalize: If we added up $(1/2)mv^2$ for all four particles, we would get the same answer.

---

35. Find the net torque on the wheel in Figure P10.35 about the axle through O, taking $a = 10.0$ cm and $b = 25.0$ cm.

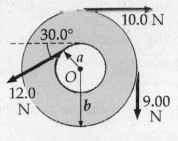

Figure P10.35

Solution

Conceptualize: By examining the magnitudes of the forces and their respective lever arms, it appears that the wheel will rotate clockwise, and the net torque appears to be about $5 \text{ N} \cdot \text{m}$ in magnitude.

Categorize: To find the net torque, we add the individual torques, remembering to apply the convention that a torque producing clockwise rotation is negative and a counterclockwise rotation is positive.

Analyze:

$$\sum\tau = \sum Fd = +(12.0 \text{ N})(0.100 \text{ m}) - (10.0 \text{ N})(0.250 \text{ m}) - (9.00 \text{ N})(0.250 \text{ m})$$

$$\sum\tau = -3.55 \text{ N} \cdot \text{m}$$ ■

This is $3.55 \text{ N} \cdot \text{m}$ into the plane of the page, a clockwise torque.

Finalize: The resultant torque has a reasonable magnitude and produces clockwise rotation as expected. Note that the $30°$ angle was not required for the solution since each force acted perpendicular to its lever arm. The 10-N force is to the right, but its torque is negative: that is, it is clockwise, just like the torque of the downward 9-N force.

---

39. A model airplane with mass 0.750 kg is tethered to the ground by a wire so that it flies in a horizontal circle 30.0 m in radius. The airplane engine provides a net thrust of 0.800 N perpendicular to the tethering wire. (a) Find the torque the net thrust produces about the center of the circle. (b) Find the angular acceleration of the airplane. (c) Find the translational acceleration of the airplane tangent to its flight path.

Solution

Conceptualize: The torque will be several newton-meters. The linear acceleration will be about one meter per second squared. By contrast, the angular acceleration is a small fraction of one radian per second squared.

Categorize: We use the definition of torque in part (a), Newton's second law for rotation in (b), and the definition of radian measure in (c). Each is a short equation involving new ideas.

Analyze:

(a) $\tau = Fd$: $\tau = (0.800 \text{ N})(30.0 \text{ m}) = 24.0 \text{ N} \cdot \text{m}$

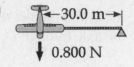
←30.0 m→

↓ 0.800 N

(b) The airplane presents rotational inertia

$$I = mr^2: \qquad I = (0.750 \text{ kg})(30.0 \text{ m})^2 = 675 \text{ kg} \cdot \text{m}^2$$

The angular acceleration is found from

$$\sum \tau = I\alpha: \qquad \alpha = \frac{\sum \tau}{I} = \frac{24.0 \text{ N} \cdot \text{m}}{675 \text{ kg} \cdot \text{m}^2} = 0.035\ 6 \text{ rad/s}^2$$

(c) $a = r\alpha$: $a = (30.0 \text{ m})(0.035\ 6/\text{s}^2) = 1.07 \text{ m/s}^2$

Finalize: In part (c), we could also find the tangential acceleration from

$$\sum F_t = ma_t: \qquad a_t = \frac{\sum F_t}{m} = \frac{0.800 \text{ N}}{0.750 \text{ kg}} = 1.07 \text{ m/s}^2$$

Note that the airplane in flight will also have a centripetal acceleration, large if it is moving fast, associated with the string tension.

---

51. Review. An object with a mass of $m = 5.10$ kg is attached to the free end of a light string wrapped around a reel of radius $R = 0.250$ m and mass $M = 3.00$ kg. The reel is a solid disk, free to rotate in a vertical plane about the horizontal axis passing through its center. The suspended object is released from rest 6.00 m above the floor. Determine (a) the tension in the string, (b) the acceleration of the object, and (c) the speed with which the object hits the floor. (d) Verify your answer to (c) by using the isolated system (energy) model.

Solution

Conceptualize: Since the rotational inertia of the reel will slow the fall of the weight, we should expect the downward acceleration to be less than g. If the reel did not rotate, the tension in the string would be equal to the weight of the object; and if the reel disappeared, the tension would be zero. Therefore, $T < 50$ N for the given problem. With similar reasoning, the final speed must be less than if the weight were to fall freely:

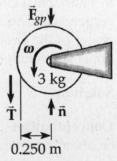

$$v_f < \sqrt{2g\Delta y} \approx 11 \text{ m/s}$$

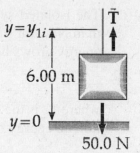

Categorize: We can find the acceleration and tension using the rotational form of Newton's second law. The final speed can be found from the kinematics equation stated above and from conservation of energy. Force diagrams will greatly assist in analyzing the forces.

Analyze: The gravitational force exerted on the reel is

$$mg = (5.10 \text{ kg})(9.80 \text{ m/s}^2) = 50.0 \text{ N down.}$$

(a)   Use $\Sigma\tau = I\alpha$ to find T and a.

First find I for the reel, which we know is a uniform disk.

$$I = \tfrac{1}{2}MR^2 = \tfrac{1}{2}3.00 \text{ kg } (0.250 \text{ m})^2 = 0.093\ 8 \text{ kg} \cdot \text{m}^2$$

The forces on it are shown, including a normal force exerted by its axle. From the diagram, we can see that the tension is the only force that produces a torque causing the reel to rotate.

$$\Sigma\tau = I\alpha \qquad \text{becomes} \qquad n(0) + F_{gp}(0) + T(0.250 \text{ m})$$
$$= (0.093\ 8 \text{ kg} \cdot \text{m}^2)(a/0.250 \text{ m}) \qquad \text{[1]}$$

where we have applied $a_t = r\alpha$ to the point of contact between string and reel.

For the object that moves down,

$$\Sigma F_y = ma_y \qquad \text{becomes} \qquad 50.0 \text{ N} - T = (5.10 \text{ kg})a \qquad \text{[2]}$$

Note that we have defined downwards to be positive, so that positive linear acceleration of the object corresponds to positive angular acceleration of the reel. We now have our two equations in the unknowns T and a for the two connected objects. Substituting T from equation [2] into equation [1], we have

$$[50.0 \text{ N} - (5.10 \text{ kg})a](0.250 \text{ m}) = (0.093\ 8 \text{ kg} \cdot \text{m}^2)(a/0.250 \text{ m})$$

$$12.5 \text{ N} \cdot \text{m} - (1.28 \text{ kg} \cdot \text{m})a = (0.375 \text{ kg} \cdot \text{m})a$$

(b) $$12.5 \text{ N} \cdot \text{m} = a(1.65 \text{ kg} \cdot \text{m}) \quad \text{or} \quad a = 7.57 \text{ m/s}^2 \qquad \blacksquare$$

Because we eliminated T in solving the simultaneous equations, the answer for a, required for part (b), emerged first. No matter—we can now substitute back to get the answer to part (a).

(a) $$T = 50.0 \text{ N} - 5.10 \text{ kg } (7.57 \text{ m/s}^2) = 11.4 \text{ N} \qquad \blacksquare$$

(c)   For the motion of the hanging weight,

$$v_f^2 = v_i^2 + 2a(y_f - y_i) = 0^2 + 2(7.57 \text{ m/s}^2)(6.00 \text{ m})$$

$$v_f = 9.53 \text{ m/s (down)} \qquad \blacksquare$$

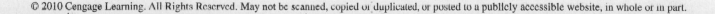

(d) The isolated-system energy model can take account of multiple objects more easily than Newton's second law. Like your bratty cousins, the equation for conservation of energy grows between visits. Now it reads for the counterweight-reel-Earth system:

$$\left(K_1 + K_2 + U_g\right)_i = \left(K_1 + K_2 + U_g\right)_f$$

where K_1 is the translational kinetic energy of the falling object and K_2 is the rotational kinetic energy of the reel.

$$0 + 0 + m_1 g y_{1i} = \tfrac{1}{2} m_1 v_{1f}^2 + \tfrac{1}{2} I_2 \omega_{2f}^2 + 0$$

Now note that $\omega = v/r$ as the string unwinds from the reel.

$$mgy_i = \tfrac{1}{2}mv^2 + \tfrac{1}{2}I\omega^2$$

$$2mgy_i = mv^2 + I\left(\frac{v^2}{R^2}\right) = v^2\left(m + \frac{I}{R^2}\right)$$

$$v = \sqrt{\frac{2mgy_i}{m + \left(I/R^2\right)}} = \sqrt{\frac{2(5.10\,\text{kg})(9.80\,\text{m/s}^2)(6.00\,\text{m})}{5.10\,\text{kg} + (0.093\,8/0.25^2)\,\text{kg}}} = 9.53\,\text{m/s} \quad\blacksquare$$

Finalize: As we should expect, both methods give the same final speed for the falling object, but the energy method is simpler. The energy method does not require solving simultaneous equations. The acceleration is less than g, and the tension is less than the object's weight, as we predicted. Now that we understand the effect of the reel's moment of inertia, this problem solution could be applied to solve other real-world pulley systems with masses that should not be ignored.

---

52. This problem describes one experimental method for determining the moment of inertia of an irregularly shaped object such as the payload for a satellite. Figure P10.52 shows a counterweight of mass m suspended by a cord wound around a spool of radius r, forming part of a turntable supporting the object. The turntable can rotate without friction. When the counterweight is released from rest, it descends through a distance h, acquiring a speed v. Show that the moment of inertia I of the rotating apparatus (including the turntable) is $mr^2(2gh/v^2 - 1)$.

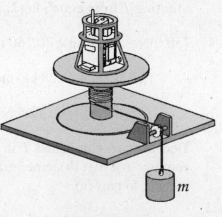

Figure P10.52

Solution

Conceptualize: Think of the counterweight unreeling like a bucket on a rope, falling down a well. Observing its motion gives us information about how hard-to-set-rotating the reel is.

Categorize: If the friction is negligible, then the energy of the counterweight-payload-turntable-Earth system is conserved as the counterweight unwinds.

Analyze: Each point on the cord moves at a linear speed of $v = \omega r$, where r is the radius of the spool. The energy conservation equation for the counterweight-turntable-Earth system is:

$$\left(K_1 + K_2 + U_g\right)_i + W_{\text{other}} = \left(K_1 + K_2 + U_g\right)_f$$

Specializing, we have
$$0 + 0 + mgh + 0 = \tfrac{1}{2}mv^2 + \tfrac{1}{2}I\omega^2 + 0$$

$$mgh = \tfrac{1}{2}mv^2 + \tfrac{1}{2}\frac{Iv^2}{r^2}$$

$$2\,mgh - mv^2 = I\frac{v^2}{r^2}$$

and finally,
$$I = mr^2\!\left(\frac{2gh}{v^2} - 1\right)$$ ∎

Finalize: We could have used Newton's laws for linear motion and for rotation, and solved simultaneous equations for the two objects, but the energy method is easier. According to the equation we have proved, a larger counterweight mass and a lower impact speed on the floor are both evidence for a larger moment of inertia for the rotor.

---

53. A uniform solid disk of radius R and mass M is free to rotate on a frictionless pivot through a point on its rim (Fig. P10.53). If the disk is released from rest in the position shown by the shaded circle, (a) what is the speed of its center of mass when the disk reaches the position indicated by the dashed circle? (b) What is the speed of the lowest point on the disk in the dashed position? (c) **What If?** Repeat part (a) using a uniform hoop.

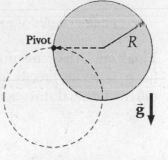

Figure P10.53

Solution

Conceptualize: It is tempting to identify the angular displacement as $\pi/2$ radians and the angular acceleration as due to gravitation, and then to use the equation $\omega^2 - \omega_i^2 = 2\alpha(\pi/2)$ to find ω_f. This idea is quite wrong, because α is not constant.

Categorize: Instead, we use conservation of energy.

Analyze: To identify the change in gravitational energy, think of the height through which the center of mass falls. From the parallel axis theorem, the moment of inertia of the disk about the pivot point on the circumference is:

$$I = I_{\text{CM}} + MD^2 = \tfrac{1}{2}MR^2 + MR^2 = \tfrac{3}{2}MR^2$$

The pivot point is fixed, so the kinetic energy is entirely rotational around the pivot:

The equation for the isolated system (energy) model

$$(K + U)_i = (K + U)_f$$

for the disk-Earth system becomes $0 + MgR = \frac{1}{2}\left(\frac{3}{2}MR^2\right)\omega^2 + 0$

Solving for ω, $\omega = \sqrt{\dfrac{4g}{3R}}$

(a) At the center of mass, $v = R\omega = 2\sqrt{\dfrac{Rg}{3}}$ ∎

(b) At the lowest point on the rim, $v = 2R\omega = 4\sqrt{\dfrac{Rg}{3}}$ ∎

(c) For a hoop, $I_{CM} = MR^2$ and $I_{rim} = 2MR^2$

By conservation of energy for the hoop-Earth system, then

$$MgR = \frac{1}{2}\left(2MR^2\right)\omega^2 + 0$$

so $\omega = \sqrt{\dfrac{g}{R}}$

and the center of mass moves at $v_{CM} = R\omega = \sqrt{gR}$, slower than the disk.   ∎

Finalize: The mass divides out, but the difference in rotational inertia between disk and hoop shows up directly in the different speeds. The moments of inertia are different because of different mass distributions. This could be a good laboratory investigation.

---

55. A cylinder of mass 10.0 kg rolls without slipping on a horizontal surface. At a certain instant its center of mass has a speed of 10.0 m/s. Determine (a) the translational kinetic energy of its center of mass, (b) the rotational kinetic energy about its center of mass, and (c) its total energy.

Solution

Conceptualize: Think of a keg of beer, frozen solid, rolling fast on *America's Funniest Home Videos*. It will have many joules of kinetic energy, which we can calculate separately for translation and for rotation.

Categorize: We substitute into the two kinetic energy 'formulas.'

Analyze:

(a) The kinetic energy of translation is

$$K_{trans} = \tfrac{1}{2}mv_{CM}^2 = \tfrac{1}{2}(10.0 \text{ kg})(10.0 \text{ m/s})^2 = 500 \text{ J} \quad \blacksquare$$

(b) Call the radius of the cylinder R. An observer at the center sees the rough surface and the circumference of the cylinder moving at 10.0 m/s, so the angular speed of the cylinder is

$$\omega = \frac{v_{CM}}{R} = \frac{10.0 \text{ m/s}}{R}$$

The moment of inertia about an axis through the center of mass is $\quad I_{CM} = \tfrac{1}{2}mR^2$

so $\quad K_{rot} = \tfrac{1}{2}I_{CM}\omega^2 = \left(\tfrac{1}{2}\right)\left[\tfrac{1}{2}(10.0 \text{ kg})R^2\right]\left(\frac{10.0 \text{ m/s}}{R}\right)^2 = 250 \text{ J} \quad \blacksquare$

(c) We can now add up the total energy:

$$K_{tot} = 500 \text{ J} + 250 \text{ J} = 750 \text{ J} \quad \blacksquare$$

Finalize: We could use the parallel-axis theorem to find the moment of inertia about the cylinder's line of contact with the ground. Then the whole 750 J could be calculated as kinetic energy of rotation.

---

57. (a) Determine the acceleration of the center of mass of a uniform solid disk rolling down an incline making angle θ with the horizontal. (b) Compare the acceleration found in (a) with that of a uniform hoop. (c) What is the minimum coefficient of friction required to maintain pure rolling motion for the disk?

Solution

Conceptualize: The acceleration of the disk will depend on the angle of the incline. In fact, it should be proportional to $g \sin\theta$ since the disk should not accelerate when the incline angle is zero. The acceleration of the disk should also be greater than a hoop since the mass of the disk is closer to its center, giving it less rotational inertia so that it can roll faster than the hoop. The required coefficient of friction is difficult to predict, but is probably between 0 and 1 since this is a typical range for μ.

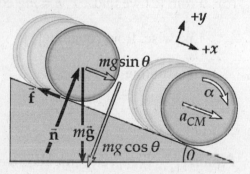

Categorize: We can find the acceleration by applying Newton's second law and considering both the linear and rotational motion. A force diagram, shown for the left-hand disk, and a motion diagram, on the right-hand disk, will greatly assist us in defining our variables and seeing how the forces are related.

Analyze:

$$\sum F_x = ma_x \quad \text{becomes} \quad mg\sin\theta - f = ma_{CM} \qquad \textbf{[1]}$$

$$\sum F_y = ma_y \quad \text{yields} \quad n - mg\cos\theta = 0 \qquad \textbf{[2]}$$

$$\sum \tau = I_{CM}\alpha \quad \text{gives} \quad fr = I_{CM}a_{CM}/r \qquad \textbf{[3]}$$

(a) For a disk, $\qquad (I_{CM})_{disk} = \frac{1}{2}mr^2$

From [3] we find $\qquad f = \left(\frac{1}{2}mr^2\right)\left(a_{CM}\right)/r^2 = \frac{1}{2}ma_{CM}$

Substituting this into [1], $\qquad mg\sin\theta - \frac{1}{2}ma_{CM} = ma_{CM}$

so that $\qquad \left(a_{CM}\right)_{disk} = \frac{2}{3}g\sin\theta$ ∎

(b) For a hoop, $\qquad (I_{CM})_{hoop} = mr^2$

From [3], $\qquad f = mr^2 a_{CM}/r^2 = ma_{CM}$

Substituting this into [1], $\qquad mg\sin\theta - ma_{CM} = ma_{CM}$

so $\qquad \left(a_{CM}\right)_{hoop} = \frac{1}{2}g\sin\theta$

Therefore, $\qquad \dfrac{\left(a_{CM}\right)_{disk}}{\left(a_{CM}\right)_{hoop}} = \dfrac{\frac{2}{3}g\sin\theta}{\frac{1}{2}g\sin\theta} = \dfrac{4}{3}$

The acceleration of the disk is larger by the factor 4/3 ∎

(c) From [2] we find $\quad n = mg\cos\theta: \qquad f \le \mu n = \mu mg\cos\theta$

Likewise, from equation [1], $\qquad f = (mg\sin\theta) - ma_{CM}$

Substituting for f gives $\qquad \mu_s mg\cos\theta \ge mg\sin\theta - \frac{2}{3}mg\sin\theta$

so $\qquad \mu_s \ge \frac{1}{3}\left(\dfrac{\sin\theta}{\cos\theta}\right) = \frac{1}{3}\tan\theta$ ∎

Finalize: As expected, the acceleration of the disk is proportional to $g\sin\theta$ and is slightly greater than the acceleration of the hoop. The minimum coefficient of friction result is similar to the result found for a block on an incline plane, where $\mu = \tan\theta$ is required to avoid slipping. However, μ is not always between 0 and 1 as predicted. For angles greater than 72° the coefficient of friction must be larger than 1. For angles greater than 80°, μ must be extremely large to make the disk roll without slipping.

63. A 4.00-m length of light nylon cord is wound around a uniform cylindrical spool of radius 0.500 m and mass 1.00 kg. The spool is mounted on a frictionless axle and is initially at rest. The cord is pulled from the spool with a constant acceleration of magnitude 2.50 m/s². (a) How much work has been done on the spool when it reaches an angular speed of 8.00 rad/s? (b) How long does it take the spool to reach this angular speed? (c) How much cord is left on the spool when it reaches this angular speed?

Solution

Conceptualize: The numbers could describe a tabletop laboratory apparatus. We expect a few joules of work and a few seconds of motion.

Categorize: We will use the work-kinetic energy theorem in (a) and simple ideas of radian measure in (b). To answer the question in (c), we can calculate the linear displacement of a point on the surface of the spool and compare it to 4 m.

Analyze:

(a) $W = \Delta K_R = \frac{1}{2}I\omega_f^2 - \frac{1}{2}I\omega_i^2 = \frac{1}{2}I(\omega_f^2 - \omega_i^2)$ where $I = \frac{1}{2}mR^2$

$W = \left(\frac{1}{2}\right)\left(\frac{1}{2}\right)(1.00 \text{ kg})(0.500 \text{ m})^2[(8.00 \text{ rad/s})^2 - 0] = 4.00 \text{ J}$ ∎

(b) $\omega_f = \omega_i + \alpha t$ where $\alpha = \dfrac{a}{r} = \dfrac{2.50 \text{ m/s}^2}{0.500 \text{ m}} = 5.00 \text{ rad/s}^2$

$t = \dfrac{\omega_f - \omega_i}{\alpha} = \dfrac{8.00 \text{ rad/s} - 0}{5.00 \text{ rad/s}^2} = 1.60 \text{ s}$ ∎

(c) The spool turns through angular displacement

$$\theta_f = \theta_i + \omega_i t + \frac{1}{2}\alpha t^2$$

$$\theta_f = 0 + 0 + \frac{1}{2}(5.00 \text{ rad/s}^2)(1.60 \text{ s})^2 = 6.40 \text{ rad}$$

The length pulled from the spool is

$$s = r\theta = (0.500 \text{ m})(6.40 \text{ rad}) = 3.20 \text{ m}$$

When the spool reaches an angular velocity of 8.00 rad/s, 1.60 s will have elapsed and 3.20 m of cord will have been removed from the spool. Remaining on the spool will be 0.800 m. ∎

Finalize: What would be wrong with saying the string tension is $F = ma = (1 \text{ kg})(2.5 \text{ m/s}^2) = 2.5$ N and the work is $F\Delta x = 2.5$ N$(3.2$ m$)$? This is wrong because the acceleration a of the object as a whole is zero. Its axis does not move through space. The reel is in rotational rather than in translational motion.

67. A long, uniform rod of length L and mass M is pivoted about a frictionless, horizontal pin through one end. The rod is nudged from rest in a vertical position as shown in Figure P10.67. At the instant the rod is horizontal, find (a) its angular speed, (b) the magnitude of its angular acceleration, (c) the x and y components of the acceleration of its center of mass, and (d) the components of the reaction force at the pivot.

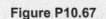
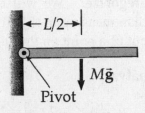

Figure P10.67

Solution

Conceptualize: The downward acceleration of the center of mass will be proportional to the free-fall acceleration. But it will be smaller in magnitude because of the upward force exerted by the hinge. And the hinge will have to pull to the left on the bar to produce centripetal acceleration of the bar.

Categorize: In (a) we use conservation of energy. In (b), (c), and (d) we use Newton's second laws for translation and rotation.

Analyze:

(a) Since only conservative forces are acting on the bar, we have conservation of energy of the bar-Earth system:

$$K_i + U_i = K_f + U_f$$

For evaluation of its gravitational energy, a rigid body can be modeled as a particle at its center of mass. Take the zero configuration for potential energy with the bar horizontal.

Under these conditions $\quad U_f = 0 \quad$ and $\quad U_i = MgL/2$

Using the equation above, $\quad 0 + \frac{1}{2}MgL = \frac{1}{2}I\omega_f^2 \quad$ and $\quad \omega_f = \sqrt{MgL/I}$

For a bar rotating about an axis through one end, $\quad I = ML^2/3$

Therefore, $\quad \omega_f = \sqrt{(MgL)/\frac{1}{3}ML^2} = \sqrt{3g/L}$ ∎

Note that we have chosen clockwise rotation as positive.

(b) $\sum \tau = I\alpha$: $\quad Mg(L/2) = \left(\frac{1}{3}ML^2\right)\alpha \quad$ and $\quad \alpha = 3g/2L$ ∎

(c) $a_x = -a_c = -r\omega_f^2 = -\left(\dfrac{L}{2}\right)\left(\dfrac{3g}{L}\right) = -\dfrac{3g}{2}$ ∎

Since this is **centripetal** acceleration, it is directed along the **negative** horizontal.

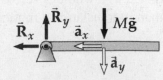

$$a_y = -a_t = -r\alpha = -\frac{L}{2}\alpha = -\frac{3g}{4}$$ ■

(d) Using $\qquad \Sigma\vec{\Gamma} - m\vec{a}$

we have $\qquad R_x = Ma_x = -3Mg/2$ in the **negative** x direction ■

and $\qquad R_y - Mg = Ma_y \qquad$ so $\qquad R_y = M(g + a_y) = M(g - 3g/4) = Mg/4$ ■

Finalize: The components of the hinge force are quite real. The horizontal component is surprisingly large. But they do not affect the angular acceleration of the rod. We needed the result of part (a) about speed to find the centripetal acceleration of the center of mass.

---

71. Review. As shown in Figure P10.71, two blocks are connected by a string of negligible mass passing over a pulley of radius $r = 0.250$ m and moment of inertia I. The block on the frictionless incline is moving with a constant acceleration of magnitude $a = 2.00$ m/s². From this information, we wish to find the moment of inertia of the pulley. (a) What analysis model is appropriate for the blocks? (b) What analysis model is appropriate for the pulley? (c) From the analysis model in part (a), find the tension T_1. (d) Similarly, find the tension T_2. (e) From the analysis model in part (b), find a symbolic expression for the moment of inertia of the pulley in terms of the tensions T_1 and T_2, the pulley radius r, and the acceleration a. (f) Find the numerical value of the moment of inertia of the pulley.

Solution

Conceptualize: In earlier problems, we assumed that the tension in a string was the same on either side of a pulley. Here we see that the moment of inertia changes that assumption, but we should still expect the tensions to be similar in magnitude (about the weight of each mass ~150 N), and $T_2 > T_1$ for the pulley to rotate clockwise as shown.

If we knew the mass of the pulley, we could calculate its moment of inertia, but since we only know the acceleration, it is difficult to estimate I. We at least know that I must have units of kg·m², and a 50-cm disk probably has a mass less than 10 kg, so I is probably less than 0.3 kg·m².

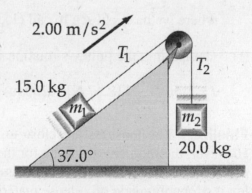

Figure P10.71

Categorize: For each block, we know its mass and acceleration, so we can use Newton's second law to find the net force, and from it the tension. The difference in the two tensions causes the pulley to rotate, so this net torque and the resulting angular acceleration can be used to find the pulley's moment of inertia.

Analyze:

(a) We apply the particle-under-a-net force model to each block.   ■

(b) We apply the rigid-body-under-a-net torque model to the pulley.   ■

(c) We use $\Sigma F = ma$ for each block to find each string tension. The forces acting on the 15-kg block are its weight, the normal support from the incline, and T_1. Taking the positive x axis as directed up the incline,

$\Sigma F_x = ma_x$ yields: $-(m_1 g)_x + T_1 = m_1(+a)$

Solving and substituting known values, we have

$T_1 = m_1(+a) + (m_1 g)_x$

$\quad = (15.0 \text{ kg})(2.00 \text{ m/s}^2) + (15.0 \text{ kg})(9.80 \text{ m/s}^2)\sin 37° = 118 \text{ N}$ ■

(d) Similarly, for the counterweight, we have

$\sum F_y = ma_y$ or $T_2 - m_2 g = m_2(-a)$

$T_2 = m_2 g + m_2(-a)$

$\quad\quad = (20.0 \text{ kg})(9.80 \text{ m/s}^2) + (20.0 \text{ kg})(-2.00 \text{ m/s}^2) = 156 \text{ N}$ ■

(e) Now for the pulley, $\Sigma \tau = r(T_2 - T_1) = I\alpha = I\,a/r$

so $I = (T_2 - T_1)r^2/a$ ■

where we have chosen to call clockwise positive.

(f) Computing, the pulley's rotational inertia is

$I = [156 \text{ N} - 118.47 \text{ N}](0.250 \text{ m})^2/(2.00 \text{ m/s}^2) = 1.17 \text{ kg} \cdot \text{m}^2$ ■

Finalize: The tensions are close to the weight of each mass and $T_2 > T_1$ as expected. However, the moment of inertia for the pulley is about 4 times greater than expected. Our result means that the pulley has a mass of 37.4 kg (about 80 lb), which means that the pulley is probably made of a dense material, like steel. This is certainly not a problem where the mass of the pulley can be ignored since the pulley has more mass than the combination of the two blocks!

73. Review.   A string is wound around a uniform disk of radius R and mass M. The disk is released from rest with the string vertical and its top end tied to a fixed bar (Fig. P10.73). Show that (a) the tension in the string is one third of the weight of the disk, (b) the magnitude of the acceleration of the center of mass is $2g/3$, and (c) the speed of the center of mass is $(4gh/3)^{1/2}$ after the disk has descended through distance h. (d) Verify your answer to part (c) using the energy approach.

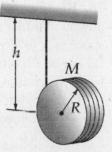

Figure P10.73

Solution

Conceptualize: It does not have a small-diameter axle, but you can think of the disk as a model yo-yo. A gravitational force pulls down on the disk and string tension up, so it is reasonable that its downward acceleration is less than g and the string tension is less than Mg.

Categorize. Newton's second laws for translation and rotation will give us simultaneous equations in acceleration, angular acceleration, and tension. The answers will be in terms of the known quantities M, g, R, and h.

Analyze: Choosing positive linear quantities to be downwards and positive angular quantities to be clockwise,

$$\sum F_y = ma_y \qquad \text{yields} \qquad \sum F = Mg - T = Ma \quad \text{or} \quad a = \frac{Mg - T}{M}$$

$$\sum \tau = I\alpha \qquad \text{becomes} \qquad \sum \tau = TR = I\alpha = \tfrac{1}{2}MR^2\left(\frac{a}{R}\right) \quad \text{so} \quad a = \frac{2T}{M}$$

(a) Setting these two expressions equal, $\quad \dfrac{Mg - T}{M} = \dfrac{2T}{M} \quad$ and $\quad T = Mg/3 \quad$ ∎

(b) Substituting back, $\quad a = \dfrac{2T}{M} = \dfrac{2Mg}{3M} \quad$ or $\quad a = \tfrac{2}{3}g \quad$ ∎

(c) Since $v_i = 0$ and $a = \tfrac{2}{3}g$, $\qquad v_f^2 = v_i^2 + 2ah$

gives us $\qquad v_f^2 = 0 + 2\left(\tfrac{2}{3}g\right)h$

or $\qquad v_f = \sqrt{4gh/3} \quad$ ∎

(d) Now we verify this answer. Requiring conservation of mechanical energy for the disk-Earth system, we have $U_i + K_{\text{rot i}} + K_{\text{trans i}} = U_f + K_{\text{rot f}} + K_{\text{trans f}}$

$$mgh + 0 + 0 = 0 + \tfrac{1}{2}I\omega^2 + \tfrac{1}{2}mv^2$$

$$mgh = \tfrac{1}{2}\left(\tfrac{1}{2}MR^2\right)\omega^2 + \tfrac{1}{2}Mv^2$$

When there is no slipping, $\qquad \omega = \dfrac{v}{R} \quad$ and $\quad v = \sqrt{\dfrac{4gh}{3}} \quad$ ∎

Finalize: Conservation of energy gives the same final speed as the Newton's-second-laws-plural approach. In both approaches we have to think about the rotational inertia and use an equation like $\alpha = a/R$ or $\omega = v/R$. With energy, we do not have to deal with torque or simultaneous equations.

76. Review. A spool of wire of mass M and radius R is unwound under a constant force $\vec{F}$ (Fig. P10.76). Assuming the spool is a uniform solid cylinder that doesn't slip, show that (a) the acceleration of the center of mass is $4\vec{F}/3M$ and (b) the force of friction is to the *right* and equal in magnitude to $F/3$. (c) If the cylinder starts from rest and rolls without slipping, what is the speed of its center of mass after it has rolled through a distance d?

Solution

Figure P10.76

Conceptualize: Borrow a spool of thread and try it yourself. With a sharp tug, the spool is set into rotation that has its contact point slipping backward over the table. To prevent this slipping, friction must act forward on the spool and help to produce its extra-large acceleration greater than F/M.

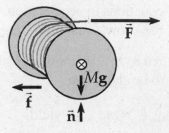

Categorize: We use Newton's second laws for translation and rotation.

Analyze: To keep the spool from slipping, there must be static friction acting at its contact point with the floor. To illustrate the power of our problem-solving approach, let us assume friction is directed to the left. Then the full set of Newton's second-law equations are:

$$\sum F_x = ma_x: \qquad F - f = Ma$$

$$\sum F_y = ma_y: \qquad -Mg + n = 0$$

$$\sum \tau_{CM} = I_{CM}\alpha_{CM}: \qquad -FR + Mg(0) + n(0) - fR = \tfrac{1}{2}MR^2\left(-\frac{a}{R}\right)$$

We have written $\alpha_{CM} = -a/R$ to describe the clockwise rotation. With F, M, and R regarded as known, the first and third of these equations allow us to solve for f and a:

$$f = F - Ma \qquad \text{and} \qquad -F - f = -\tfrac{1}{2}Ma$$

(a) Substituting for f, $\qquad F + F - Ma = \tfrac{1}{2}Ma$

Thus, $\qquad\qquad 2F = \dfrac{3}{2}Ma \qquad$ and $\qquad a = \dfrac{4F}{3M}$ ∎

(b) Then solving for f, $\qquad f = F - M\left(\dfrac{4F}{3M}\right) = F - \dfrac{4F}{3} = -\dfrac{F}{3}$ ∎

The negative sign means that the force of friction is opposite the direction we assumed. That is, the friction force exerted by the table on the spool is to the right, so the spool does not spin like the wheel of a car stuck in the snow.

(c) Since a is constant, we can use

$$v_f^2 = v_i^2 + 2a(x_f - x_i) = 0 + 2\left(\frac{4F}{3M}\right)d$$

to find $v_f = \sqrt{\dfrac{8Fd}{3M}}$ ∎

Finalize: Note that only one force diagram lets us write down the equations about force and also about torques. When the solution comes out with a minus sign to indicate that a force is in the opposite direction to that originally assumed, you should keep the original assumption and the minus sign in any subsequent steps.

---

83. As a result of friction, the angular speed of a wheel changes with time according to

$$\frac{d\theta}{dt} = \omega_0 e^{-\sigma t}$$

where ω_0 and σ are constants. The angular speed changes from 3.50 rad/s at $t = 0$ to 2.00 rad/s at $t = 9.30$ s. (a) Use this information to determine σ and ω_0. Then determine (b) the magnitude of the angular acceleration at $t = 3.00$ s, (c) the number of revolutions the wheel makes in the first 2.50 s, and (d) the number of revolutions it makes before coming to rest.

Solution

Conceptualize: The exponential decay of the angular speed might be produced by fluid friction torque on the wheel, proportional to its angular speed. The decay process goes on with positive values for ω continuing forever in time, but we will see whether the net angular displacement is infinite or finite.

Categorize: We will use the definitions of angular acceleration as the time derivative of angular speed and of angular speed as the time derivative of angular position.

Analyze: (a) When $t = 0$, $\omega = 3.50$ rad/s and $e^0 = 1$

so $\omega = \omega_0 e^{-\sigma t}$ becomes $\omega_{t=0} = \omega_0 e^{-\sigma 0} = 3.50$ rad/s

which gives $\omega_0 = 3.50$ rad/s ∎

We now calculate σ: To solve $\omega = \omega_0 e^{-\sigma t}$ for σ, we recall that the natural logarithm function is the inverse of the exponential function.

$\omega/\omega_0 = e^{-\sigma t}$ becomes $\ln(\omega/\omega_0) = -\sigma t$ or $\ln(\omega_0/\omega) = +\sigma t$

so $\sigma = (1/t)\ln(\omega_0/\omega) = (1/9.30\text{ s})\ln(3.50/2.00) = 0.560/(9.30\text{ s})$

and $\sigma = 0.060\,2$ s^{-1} ∎

(b) At all times $\quad \alpha = \dfrac{d\omega}{dt} = \dfrac{d}{dt}\left[\omega_0 e^{-\sigma t}\right] = -\sigma\omega_0 e^{-\sigma t}$

$\qquad$ At $t = 3.00$ s, $\qquad \alpha = -(0.060\ 2\ \text{s}^{-1})\ (3.50\ \text{rad/s})e^{-0.181} = -0.176\ \text{rad/s}^2 \qquad \blacksquare$

(c) From the given equation, we have $\ d\theta = \omega_0 e^{-\sigma t}\ dt$

$\qquad$ and $\quad \theta = \displaystyle\int_{0\ \text{s}}^{2.50\ \text{s}} \omega_0 e^{-\sigma t}\ dt = \dfrac{\omega_0}{-\sigma}e^{-\sigma t}\Big|_{0\ \text{s}}^{2.50\ \text{s}} = \dfrac{\omega_0}{-\sigma}\left(e^{-2.50\sigma} -1\right)$

$\qquad$ Substituting and solving, $\theta = -58.2(0.860 - 1)\ \text{rad} = 8.12\ \text{rad}$

$\qquad$ or $\quad \theta = (8.12\ \text{rad})\left(\dfrac{1\ \text{rev}}{2\pi\ \text{rad}}\right) = 1.29\ \text{rev} \qquad \blacksquare$

(d) The motion continues to a finite limit, as ω approaches zero and t goes to infinity. From part (c), the total angular displacement is

$$\theta = \int_0^\infty \omega_0 e^{-\sigma t}\ dt = \dfrac{\omega_0}{-\sigma}e^{-\sigma t}\Big|_0^\infty = \dfrac{\omega_0}{-\sigma}\ (0 - 1) = \dfrac{\omega_0}{\sigma}$$

$\qquad$ Substituting, $\quad \theta = 58.2\ \text{rad} \qquad$ or $\qquad \theta = \left(\dfrac{1\ \text{rev}}{2\pi\ \text{rad}}\right)(58.2\ \text{rad}) = 9.26\ \text{rev} \qquad \blacksquare$

Finalize: This problem used differential and integral calculus, but it was not a 'derivation' problem to prove a symbolic result. It has numerical answers for each part, like any 'plug and chug' problem. $\alpha = d\omega/dt$ and $\omega = d\theta/dt$ are general equations (the most general!) that are also prescriptions for calculation.

11

Angular Momentum

EQUATIONS AND CONCEPTS

The **vector torque** due to an applied force can be expressed as a vector product. The position vector $\vec{r}$ locates the point of application of the force. The force $\vec{F}$ will tend to produce a rotation about an axis perpendicular to the plane formed by $\vec{r}$ and $\vec{F}$. *Torque depends on the choice of the origin and has SI units of N·m.*

$$\vec{\tau} \equiv \vec{r} \times \vec{F} \tag{11.1}$$

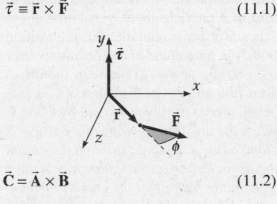

The **vector product or cross product** of any two vectors $\vec{A}$ and $\vec{B}$ is a vector $\vec{C}$ whose magnitude is given by $C = AB\sin\theta$. The direction of $\vec{C}$ is perpendicular to the plane formed by $\vec{A}$ and $\vec{B}$; and θ is the angle between $\vec{A}$ and $\vec{B}$. The cross product of $\vec{A}$ and $\vec{B}$ is written $\vec{A} \times \vec{B}$ and is read "A cross B." *The sense of $\vec{C}$ can be determined from the right-hand rule.* See the figure in *Suggestions, Skills, and Strategies.*

$$\vec{C} = \vec{A} \times \vec{B} \tag{11.2}$$

$$C = AB\sin\theta \tag{11.3}$$

The **vector or cross product of two vectors** can be expressed in terms of the components of the vectors and unit vectors.

$$\vec{A} \times \vec{B} = \left(A_y B_z - A_z B_y\right)\hat{i} + \left(A_z B_x - A_x B_z\right)\hat{j}$$
$$+ \left(A_x B_y - A_y B_x\right)\hat{k} \tag{11.8}$$

The **angular momentum of a particle** relative to the origin is defined as the cross product of the instantaneous position vector and the instantaneous momentum. The SI unit of angular momentum is kg·m²/s. *Note that both the magnitude and direction of $\vec{L}$ depend on the choice of origin.*

$$\vec{L} \equiv \vec{r} \times \vec{p} \tag{11.10}$$

A **particle moving in a circular path** of radius, r, has angular momentum of magnitude $L = mvr$.

A **change in the angular momentum** of a particle is caused by a torque. Compare Equation 11.11 and Equation 9.3; and note that torque causes a change in angular momentum and force causes a change in linear momentum. *Equation 11.11 is valid for any origin fixed in an inertial frame when the same origin is used to define $\vec{L}$ and $\vec{\tau}$.*

$$\sum \vec{\tau} = \frac{d\vec{L}}{dt} \tag{11.11}$$

The **magnitude of the angular momentum of a particle** about a specified origin depends on the magnitude of its momentum (mv), the magnitude of the position vector (r), and the angle (ϕ) between the directions of $\vec{v}$ and $\vec{r}$. The direction of $\vec{L}$ is perpendicular to the plane formed by $\vec{r}$ and $\vec{v}$. *When the linear momentum of a particle is along a line that passes through the origin, the particle has zero angular momentum. Note in the figure that the correct angle to use in Equation 11.12 is larger than 90°.*

$$L = mvr \sin\phi \tag{11.12}$$

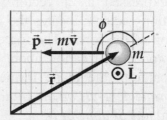

In the figure, $\vec{L}$ is directed out of the page.

The **net external torque acting on a system of particles** equals the time rate of change of the total angular momentum of the system. The angular momentum of a system of particles is obtained by taking the vector sum of the individual angular momenta about some point in an inertial frame. *Equation 11.13 is valid regardless of the motion of the center of mass.*

$$\sum \vec{\tau}_{ext} = \frac{d\vec{L}_{tot}}{dt} \tag{11.13}$$

The **magnitude of the angular momentum of a rigid body** rotating in the xy plane about a *fixed axis* (the z axis) is given by the product $I\omega$, where I is the moment of inertia about the axis of rotation and ω is the angular speed. *The angular velocity and angular momentum are along the z axis (perpendicular to the xy plane).*

$$L_z = I\omega \tag{11.14}$$

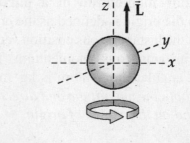

The **rotational form of Newton's second law** states that the net torque acting on a rigid object rotating about a fixed axis equals the product of the moment of inertia about the axis of rotation and the angular

$$\sum \tau_{ext} = I\alpha \tag{11.16}$$

acceleration relative to that axis. *This equation is also valid for a rigid object rotating about a moving axis if the axis of rotation is an axis of symmetry and also passes through the center of mass.*

The **law of conservation of angular momentum** states that if the resultant external torque acting on a system is zero, the total angular momentum is constant in both magnitude and direction.

$$\text{If}\quad \sum \vec{\tau}_{ext} = \frac{d\vec{L}_{tot}}{dt} = 0 \qquad (11.17)$$

$$\text{then,}\quad \vec{L}_{tot} = \text{constant} \quad \text{or} \quad \vec{L}_i = \vec{L}_f \quad (11.18)$$

If a **redistribution of mass in an isolated system** results in a change in the moment of inertia of the system, conservation of angular momentum can be used to find the final angular speed in terms of the initial angular speed.

$$I_i\omega_i = I_f\omega_f = \text{constant} \qquad (11.19)$$

Conserved quantities in an isolated system are: total mechanical energy, linear momentum, and angular momentum.

$E_i = E_f$ (zero energy transfers)

$\vec{P}_i = \vec{P}_f$ (zero net external force)

$\vec{L}_i = \vec{L}_f$ (zero net external torque)

The **precessional frequency of a gyroscope** is the rate at which the axle of the gyroscope rotates about the vertical axis. The gyroscope has an angular velocity ω; and the center of mass of the gyroscope is a distance r from the pivot point. The axle of the gyroscope sweeps out an angle $d\phi$ in a time interval dt. **The direction of the angular momentum vector changes while the magnitude remains constant.** *Equation 11.20 is valid only when $\omega_p \ll \omega$. The precessional frequency decreases as the spinning rate increases.*

$$\omega_p = \frac{d\phi}{dt} = \frac{Mgr}{I\omega} \qquad (11.20)$$

Top view

SUGGESTIONS, SKILLS, AND STRATEGIES

The operation of the vector or cross product is used for the first time in this chapter. (Recall that the angular momentum $\vec{L}$ of a particle is defined as $\vec{L} \equiv \vec{r} \times \vec{p}$, while torque is defined by the expression $\vec{\tau} \equiv \vec{r} \times \vec{F}$.) Let us briefly review the cross-product operation and some of its properties.

If $\vec{A}$ and $\vec{B}$ are any two vectors, their cross product, written as $\vec{A} \times \vec{B}$, is also a vector.

That is, if $\vec{C} = \vec{A} \times \vec{B}$, the magnitude of $\vec{C}$ is given by $C = |\vec{C}| = AB \sin\theta$, and θ is the angle between $\vec{A}$ and $\vec{B}$ as in the figure below.

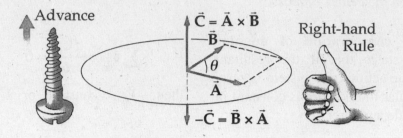

The direction of $\vec{C}$ is perpendicular to the plane formed by $\vec{A}$ and $\vec{B}$, and its sense is determined by the right-hand rule. You should practice this rule for various choices of vector pairs. Note that $\vec{B} \times \vec{A}$ is directed opposite to $\vec{A} \times \vec{B}$. That is, $\vec{A} \times \vec{B} = -(\vec{B} \times \vec{A})$. This follows from the right-hand rule. You should not confuse the cross product of two vectors, which is a vector quantity, with the dot product of two vectors, which is a scalar quantity. (Recall that the dot product is defined as $\vec{A} \cdot \vec{B} = AB \cos\theta$.) Note that the cross product of any vector with itself is zero. That is, $\vec{A} \times \vec{A} = 0$ since $\theta = 0$, and $\sin(0) = 0$.

Very often, vectors will be expressed in unit vector form, and it is convenient to make use of the multiplication table for unit vectors. Note that $\hat{i}$, $\hat{j}$, and $\hat{k}$ represent a set of orthogonal vectors as shown in the figure at right.

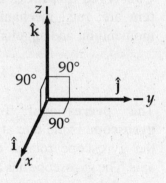

$$\hat{i} \times \hat{i} = \hat{j} \times \hat{j} = \hat{k} \times \hat{k} = 0$$

$$\hat{i} \times \hat{j} = -\hat{j} \times \hat{i} = \hat{k}$$

$$\hat{j} \times \hat{k} = -\hat{k} \times \hat{j} = \hat{i}$$

$$\hat{k} \times \hat{i} = -\hat{i} \times \hat{k} = \hat{j}$$

The cross product of two vectors can be expressed in determinant form. When the determinant is expanded, the cross product can be calculated in terms of the vector components:

$$\vec{A} \times \vec{B} = (A_y B_z - A_z B_y)\hat{i} + (A_z B_x - A_x B_z)\hat{j} + (A_x B_y - A_y B_x)\hat{k}$$

For example, consider two vectors in the *xy* plane:

$$\vec{A} = 3.00\hat{i} + 5.00\hat{j} \quad \text{and} \quad \vec{B} = 4.00\hat{j}$$

The cross product can be found using the expression above:

Let $\vec{C} = \vec{A} \times \vec{B}$

$$\vec{C} = [(5.00)(0) - (0)(4.00)]\hat{i} + [(0)(0) - (3.00)(0)]\hat{j} + [(3.00)(4.00) - (5.00)(0)]\hat{k}$$

$$\vec{C} = 12\hat{k}$$

As expected from the right hand rule, the direction of $\vec{C}$ is along $\hat{k}$, the *z* axis.

REVIEW CHECKLIST

You should be able to:

- Determine the magnitude and direction of the cross product of two vectors and find the angle between two vectors. (Section 11.1)

- Given the position vector of a particle, calculate the net torque on the particle due to several specified forces. (Section 11.1)

- Determine the angular momentum of a particle of mass, m: (i) rotating about a specified axis and (ii) moving with a given velocity relative to a specified origin. Note that both angular momentum and torque are quantities which depend on the choice of the origin since they involve the vector position of the particle. (Section 11.2)

- Apply Equation 11.13 to calculate the change in angular momentum of a system due to the resultant torque acting on the system. (Section 11.2)

- Calculate the angular acceleration of a rigid body about an axis due to an external torque. (Section 11.3)

- Apply the conservation of angular momentum principle to a body rotating about a fixed axis when the moment of inertia changes due to a change in the mass distribution. (Section 11.4)

- Calculate the precessional frequency of a gyroscope. (Section 11.5)

ANSWER TO AN OBJECTIVE QUESTION

3. Let us name three perpendicular directions as right, up, and toward you, as you might name them when you are facing a television screen that lies in a vertical plane. Unit vectors for these directions are $\hat{r}$, $\hat{u}$, and $\hat{t}$, respectively. Consider the quantity $(-3\hat{u} \times 2\hat{t})$. **(i)** Is the magnitude of this vector (a) 6, (b) 3, (c) 2, or (d) 0? **(ii)** Is the direction of this vector (a) down, (b) toward you, (c) up, (d) away from you, or (e) left?

Answer **(i)** (a). The magnitude of the cross product here is $(3)(2)\sin(90°)$. The angle between "down" and "toward you" is a right angle.

(ii) (e) We identify $-\hat{u}$ as down. The diagram shows that "down" cross "toward you" is "left."

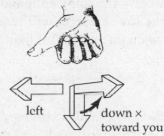

left   down ×
         toward you

ANSWERS TO SELECTED CONCEPTUAL QUESTIONS

3.   Why does a long pole help a tightrope walker stay balanced?

Answer   The long pole increases the tightrope walker's moment of inertia, and therefore decreases his angular acceleration, under any given torque. That gives him a greater time interval to respond, and in essence aids his reflexes.

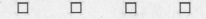

4.   Two children are playing with a roll of paper towels. One child holds the roll between the index fingers of her hands so that it is free to rotate, and the second child pulls at constant speed on the free end of the paper towels. As the child pulls the paper towels, the radius of the roll of remaining towels decreases. (a) How does the torque on the roll change with time? (b) How does the angular speed of the roll change in time? (c) If the child suddenly jerks the end paper towel with a large force, is the towel more likely to break from the others when it is being pulled from a nearly full roll or from a nearly empty roll?

Answer   (a) We assume that the first child holds the tubular cardboard spool firmly enough that the roll does not turn frictionlessly, but feels a significant frictional torque. The angular speed of the roll changes quite slowly in time, so its angular acceleration is nearly zero and the net torque on the roll is nearly zero. The torque exerted by the second child through the ribbon of towels very nearly balances the frictional torque. We suppose that it is nearly constant in time. The tension force exerted by the second child must then be slowly increasing as its moment arm decreases.

(b) As the source roll radius R shrinks, the roll's angular speed $\omega = v/R$ must increase to keep constant the speed v of the unrolling towels. But the biggest change is to the roll's moment of inertia. If it did not have a hollow core, the moment of inertia of the cylindrical roll would be $I = (\frac{1}{2}) MR^2$. The roll's mass is proportional to its base area πR^2. Then the biggest term in the moment of inertia is proportional to R^4. The moment of inertia decreases very rapidly as the roll shrinks.

(c) We have said that the tension in the ribbon of towels, as the children unroll it steadily, will be greatest when the roll is smallest. However, in the case of a sudden jerk on the towels, the rotational dynamics of the roll becomes important. If the roll is full, then its moment of inertia, proportional to R^4, will be so large that very high tension will be required to give the roll its angular acceleration. If the roll is nearly empty, the same acceleration of the towel ribbon will require smaller tension. We find that it is easier to snap a towel free when the roll is full than when it is nearly empty.

9.   If global warming continues over the next one hundred years, it is likely that some polar ice will melt and the water will be distributed closer to the equator. (a) How would that change the moment of inertia of the Earth? (b) Would the duration of the day (one revolution) increase or decrease?

Answer (a) The Earth already bulges slightly at the equator, and is slightly flat at the poles. If more mass moved towards the equator, it would essentially move the mass to a greater distance from the axis of rotation, and increase the moment of inertia. (b) Since conservation of angular momentum for the Earth as a system requires that $\omega_z I_z = $ constant, an increase in the moment of inertia would decrease the angular velocity, and slow down the spinning of the Earth. Thus, the length of each day would increase.

□ □ □ □

SOLUTIONS TO SELECTED END-OF-CHAPTER PROBLEMS

3. Two vectors are given by $\vec{\mathbf{A}} = \hat{\mathbf{i}} + 2\hat{\mathbf{j}}$ and $\vec{\mathbf{B}} = -2\hat{\mathbf{i}} + 3\hat{\mathbf{j}}$. Find (a) $\vec{\mathbf{A}} \times \vec{\mathbf{B}}$ and (b) the angle between $\vec{\mathbf{A}}$ and $\vec{\mathbf{B}}$.

Solution

Conceptualize: $\vec{\mathbf{A}}$ is in the first quadrant and $\vec{\mathbf{B}}$ points into the second quadrant. Their cross product then points toward you along the z axis. We can estimate its magnitude as somewhat less than $(2.5)(4) \sin 90° = 10$ units.

Categorize: We take the cross product of each term of $\vec{\mathbf{A}}$ with each term of $\vec{\mathbf{B}}$, using the cross-product multiplication table for unit vectors. Then we use the identification of the magnitude of the cross product as $AB \sin\theta$ to find θ.

Analyze: We assume the data are known to three significant digits.

(a) $\vec{\mathbf{A}} \times \vec{\mathbf{B}} = (1\hat{\mathbf{i}} + 2\hat{\mathbf{j}}) \times (-2\hat{\mathbf{i}} + 3\hat{\mathbf{j}})$

$\vec{\mathbf{A}} \times \vec{\mathbf{B}} = -2\hat{\mathbf{i}} \times \hat{\mathbf{i}} + 3\hat{\mathbf{i}} \times \hat{\mathbf{j}} - 4\hat{\mathbf{j}} \times \hat{\mathbf{i}} + 6\hat{\mathbf{j}} \times \hat{\mathbf{j}}$

$\qquad = 0 + 3\hat{\mathbf{k}} - 4(-\hat{\mathbf{k}}) + 0 = 7.00\hat{\mathbf{k}}$

(b) Since $|\vec{\mathbf{A}} \times \vec{\mathbf{B}}| = AB \sin\theta$

$\quad$ we have $\quad \theta = \sin^{-1}\dfrac{|\vec{\mathbf{A}} \times \vec{\mathbf{B}}|}{AB} = \sin^{-1}\left(\dfrac{7}{\sqrt{1^2 + 2^2}\sqrt{2^2 + 3^2}}\right) = 60.3°$

Finalize: To solidify your understanding, review taking the dot product of these same two vectors, and again find the angle between them.

$\vec{\mathbf{A}} \cdot \vec{\mathbf{B}} = (+1\hat{\mathbf{i}} + 2\hat{\mathbf{j}}) \cdot (-2\hat{\mathbf{i}} + 3\hat{\mathbf{j}}) = -2.00 + 6.00 = 4.00$

$\vec{\mathbf{A}} \cdot \vec{\mathbf{B}} = AB \cos\theta$

$\theta = \cos^{-1}\dfrac{\vec{\mathbf{A}} \cdot \vec{\mathbf{B}}}{AB} = \cos^{-1}\left(\dfrac{4.00}{\sqrt{1^2 + 2^2}\sqrt{2^2 + 3^2}}\right) = 60.3°$

The answers agree. The method based upon the dot product is more general. It gives an unambiguous answer when the angle between the vectors is greater than 90°.

7. If $|\vec{A} \times \vec{B}| = \vec{A} \cdot \vec{B}$, what is the angle between $\vec{A}$ and $\vec{B}$?

Solution

Conceptualize: We review here...

Categorize: ...the different definitions of the dot and cross products.

Analyze: We are given the condition $|\vec{A} \times \vec{B}| = \vec{A} \cdot \vec{B}$.

This says that $AB \sin\theta = AB \cos\theta$

so $\tan\theta = 1$

$\theta = 45.0°$ satisfies this condition.   ∎

Finalize: Would any other angle do? If either $|\vec{A}|$ or $|\vec{B}|$ were zero, the condition would be satisfied for any angle between the vectors, but the angle might not be definable for a zero vector. An angle of 135° or 225° will not work, because it would make the dot product negative but not the magnitude of the cross product. An angle of 315° or of 405° does work but is geometrically the same as our answer of 45°.

---

11.   A light, rigid rod of length $\ell = 1.00$ m joins two particles, with masses $m_1 = 4.00$ kg and $m_2 = 3.00$ kg, at its ends. The combination rotates in the xy plane about a pivot through the center of the rod (Fig. P11.11). Determine the angular momentum of the system about the origin when the speed of each particle is 5.00 m/s.

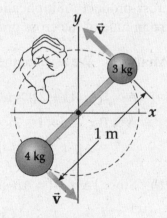

Solution

Conceptualize: This object is more like a weapon than a majorette's baton. We expect the angular momentum to be several kilogram-meter-squareds-per-second.

Figure P11.11

Categorize: We use the identification of angular momentum as moment of inertia times angular velocity.

Analyze: Taking the geometric center of the compound object to be the pivot, the angular speed and the moment of inertia are

$$\omega = v/r = (5.00 \text{ m/s})/0.500 \text{ m} = 10.0 \text{ rad/s} \text{ and}$$

$$I = \sum mr^2 = (4.00 \text{ kg})(0.500 \text{ m})^2 + (3.00 \text{ kg})(0.500 \text{ m})^2 = 1.75 \text{ kg} \cdot \text{m}^2$$

By the right-hand rule (shown in our version of the figure), we find that the angular velocity is directed out of the plane. So the object's angular momentum, with magnitude

$$L = I\omega = (1.75 \text{ kg} \cdot \text{m}^2)(10.0 \text{ rad/s})$$

is the vector $\vec{L} = 17.5$ kg $\cdot$ m^2/s out of the plane.   ∎

Finalize: Alternatively, we could solve this using $\vec{L} = \sum m \, \vec{r} \times \vec{v}$

$$L = (4.00 \text{ kg})(0.500 \text{ m})(5.00 \text{ m/s}) + (3.00 \text{ kg})(0.500 \text{ m})(5.00 \text{ m/s})$$

$$L = 10.0 \text{ kg} \cdot \text{m}^2/\text{s} + 7.50 \text{ kg} \cdot \text{m}^2/\text{s}$$

and $\quad \vec{L} = 17.5 \text{ kg} \cdot \text{m}^2/\text{s}$ out of the plane. ∎

---

15. The position vector of a particle of mass 2.00 kg as a function of time is given by $\vec{r} = (6.00\hat{i} + 5.00t \, \hat{j})$, where $\vec{r}$ is in meters and t is in seconds. Determine the angular momentum of the particle about the origin as a function of time.

Solution

Conceptualize: Think of plotting the particle's trajectory. It moves parallel to the y axis along the line $x = 6$ m, with speed 5 m/s. Its angular momentum will be $2(6)(5) = 60$ units pointing out of the plane parallel to the z axis.

Categorize: We will find the particle's velocity by differentiating its position, and then use $\vec{L} = m \, \vec{r} \times \vec{v}$

Analyze: The velocity of the particle is $\quad \vec{v} = \dfrac{d\vec{r}}{dt} = \dfrac{d}{dt}(6.00\hat{i} \text{ m} + 5.00 \, t\hat{j} \text{ m}) = 5.00\hat{j} \text{ m/s}$

The angular momentum is $\quad \vec{L} = \vec{r} \times \vec{p} = m \, \vec{r} \times \vec{v}$

$$\vec{L} = (2.00 \text{ kg})(6.00\hat{i} \text{ m} + 5.00 \, t\hat{j} \text{ m}) \times 5.00\hat{j} \text{ m/s}$$

$$\vec{L} = (60.0 \text{ kg} \cdot \text{m}^2/\text{s})\hat{i} \times \hat{j} + (50.0t \text{ kg} \cdot \text{m}^2/\text{s})\hat{j} \times \hat{j}$$

$$\vec{L} = (60.0\hat{k})\text{kg} \cdot \text{m}^2/\text{s}, \text{ constant in time}$$ ∎

Finalize: Make sure that you know that angular momentum is not a kind of momentum. Also, angular momentum does not need to have any obvious relation with rotation.

---

19. Review. A projectile of mass m is launched with an initial velocity $\vec{v}_i$ making an angle θ with the horizontal as shown in Figure P11.19. The projectile moves in the gravitational field of the Earth. Find the angular momentum of the projectile about the origin (a) when the projectile is at the origin, (b) when it is at the highest point of its trajectory, and (c) just before it hits the ground. (d) What torque causes its angular momentum to change?

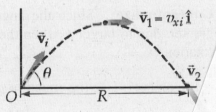

Figure P11.19

Solution

Conceptualize: The $\vec{L} = m \, \vec{r} \times \vec{v}$ will point perpendicularly into the plane of the It will grow in magnitude as time goes on and the particle gets farther from the

© 201

Categorize: We use $\vec{L} = m\,\vec{r} \times \vec{v}$ to find the angular momentum as a function of time, based on our knowledge of projectile motion.

Analyze:

(a) At the origin, $\vec{r} = 0$, and $\vec{L} = m\,\vec{r} \times \vec{v} = 0$ ∎

(b) At the highest point of the trajectory,

$$\vec{L} = \left(\frac{v_i^2 \sin 2\theta}{2g}\hat{i} + \frac{(v_i \sin\theta)^2}{2g}\hat{j} \right) \times mv_{xi}\hat{i} = \frac{-m(v_i \sin\theta)^2 v_i \cos\theta}{2g}\hat{k}$$ ∎

(c) $\vec{L} = R\hat{i} \times m\vec{v}_2$, where $R = \dfrac{v_i^2 \sin 2\theta}{g} = \dfrac{2v_i^2 \sin\theta\cos\theta}{g}$

$$\vec{L} = m\left[R\hat{i} \times (v_i \cos\theta\hat{i} - v_i \sin\theta\hat{j}) \right] = -mRv_i \sin\theta\,\hat{k} = \frac{-2mv_i^3 \sin^2\theta\cos\theta}{g}\hat{k}$$ ∎

(d) The downward gravitational force exerts a torque in the $-z$ direction. ∎

Finalize: The particle's angular momentum is perpendicular to its motion, into the plane of the paper. The angular momentum grows in time, and grows faster and faster as the lever arm of the gravitational force becomes larger and larger. The angular momentum starts from zero and is four times larger at the end of the aboveground trajectory than at the apex. That is, the angular momentum behaves very differently from the energy (constant in time) and from the momentum (changing steadily toward the downward direction).

---

27. A particle of mass 0.400 kg is attached to the 100-cm mark of a meterstick of mass 0.100 kg. The meterstick rotates on the surface of a frictionless horizontal table with an angular speed of 4.00 rad/s. Calculate the angular momentum of the system when the stick is pivoted about an axis (a) perpendicular to the table through the 50.0-cm mark and (b) perpendicular to the table through the 0-cm mark.

Solution

Conceptualize: Since the angular speed is constant, the angular momentum will be greater, due to larger I, when the center of mass of the system is farther from the axis of rotation.

Categorize: Use the equation $L = I\omega$ to find L for each I.

Analyze: Defining the distance from the pivot to the particle as d, we first find the rotational inertia of the system for each case, from the information $M = 0.100$ kg, $m = 0.400$ kg, and $D = 1.00$ m.

(a) For the meter stick rotated about its center, $I_m = \frac{1}{12}MD^2$

For the additional particle, $I_w = md^2 = m\left(\frac{1}{2}D\right)^2$

Together, $I = I_m + I_w = \left(\frac{1}{12}MD^2\right) + \left(\frac{1}{4}mD^2\right)$

$$I = \frac{(0.100\,\text{kg})(1.00\,\text{m})^2}{12} + \frac{(0.400\,\text{kg})(1.00\,\text{m})^2}{4} = 0.108\,\text{kg}\cdot\text{m}^2$$

And the angular momentum is

$$L = I\omega = (0.108\,\text{kg}\cdot\text{m}^2)(4.00\,\text{rad/s}) = 0.433\,\text{kg}\cdot\text{m}^2/\text{s} \qquad\blacksquare$$

(b) For a stick rotated about a point at one end,

$$I_m - \tfrac{1}{3}mD^2$$

$$I_m = \tfrac{1}{3}(0.100\,\text{kg})(1.00\,\text{m})^2 = 0.033\,3\,\text{kg}\cdot\text{m}^2$$

For a point mass, $\quad I_w = mD^2 = (0.400\,\text{kg})(1.00\,\text{m})^2 = 0.400\,\text{kg}\cdot\text{m}^2$

so together they have rotational inertia $\quad I = I_m + I_w = 0.433\,\text{kg}\cdot\text{m}^2$

and angular momentum

$$L = I\omega = (0.433\,\text{kg}\cdot\text{m}^2)(4.00\,\text{rad/s}) = 1.73\,\text{kg}\cdot\text{m}^2/\text{s} \qquad\blacksquare$$

Finalize: As we expected, the angular momentum is larger when the center of mass is further from the rotational axis. In fact, the angular momentum for part (b) is 4 times greater than for part (a). The units of $\text{kg}\cdot\text{m}^2/\text{s}$ also make sense since $[L] = [r] \times [p]$.

---

35. A 60.0-kg woman stands at the western rim of a horizontal turntable having a moment of inertia of $500\,\text{kg}\cdot\text{m}^2$ and a radius of 2.00 m. The turntable is initially at rest and is free to rotate about a frictionless, vertical axle through its center. The woman then starts walking around the rim clockwise (as viewed from above the system) at a constant speed of 1.50 m/s relative to the Earth. Consider the woman-turntable system as motion begins. (a) Is the mechanical energy of the system constant? (b) Is the momentum of the system constant? (c) Is the angular momentum of the system constant? (d) In what direction and with what angular speed does the turntable rotate? (e) How much chemical energy does the woman's body convert into mechanical energy of the woman-turntable system as the woman sets herself and the turntable into motion?

Solution

Conceptualize: The turntable starts to turn counterclockwise. We review thinking about the three quantities energy, momentum, and angular momentum that characterize motion.

Categorize: We use the angular momentum version of the isolated system model. We contrast it with what the momentum version of the nonisolated system model and the energy version of the isolated system model have to say.

Analyze:

We answer the questions in a different order. (c) The angular momentum of the woman-turntable system is conserved because the frictionless axle isolates this pair of objects from external torques. Only torques from outside the system can change the total angular momentum. $\qquad\blacksquare$

(a)   The mechanical energy of this system is not conserved because the internal forces, of the woman pushing backward on the turntable and of the turntable pushing forward on the woman, both do positive work, converting chemical into kinetic energy.   ∎

(b)   Momentum of the woman-turntable system is not conserved. The turntable's center of mass is always fixed. The turntable always has zero momentum. The woman starts walking north, gaining northward momentum. Where does it come from? She pushes south on the turntable. Its axle holds it still against linear motion by pushing north on it, and this outside force delivers northward linear momentum into the system.   ∎

(d)   From conservation of angular momentum for the woman-turntable system, we have
$L_f = L_i = 0$

$$L_f = I_w \omega_w + I_t \omega_t = 0 \qquad \text{and} \qquad \omega_t = -\frac{I_w}{I_t}\omega_w$$

Solving, $\omega_t = -\left(\dfrac{m_w r^2}{I_t}\right)\left(\dfrac{v_w}{r}\right) = -\dfrac{(60.0\ \text{kg})(2.00\ \text{m})(1.50\ \text{m/s})}{500\ \text{kg}\cdot\text{m}^2}$

$$\omega_t = -0.360\ \text{rad/s} = 0.360\ \text{rad/s counterclockwise}$$ ∎

(e)   The chemical energy converted becomes kinetic energy, and on the way it is work done $= \Delta K$:

$$W = K_f - 0 = \tfrac{1}{2}m_{\text{woman}}v^2_{\text{woman}} + \tfrac{1}{2}I_{\text{table}}\omega^2_{\text{table}}$$

$$W = \tfrac{1}{2}(60.0\ \text{kg})(1.50\ \text{m/s})^2 + \tfrac{1}{2}(500\ \text{kg}\cdot\text{m}^2)(0.360\ \text{rad/s})^2 = 99.9\ \text{J}$$ ∎

Finalize: We can think of living in a universe of matter, that interpenetrates with three universes of equally real immaterial things: energy, momentum, and angular momentum. Here system angular momentum is conserved while momentum changes, because the force of the turntable bearing causes no torque. The turntable bearing does no work, so no energy is exchanged with the environment. But energy exists in nonmechanical forms and processes within the system changed energy out of chemical and into kinetic. Note also that momentum and angular momentum could never be added together, but rotational kinetic energy and translational kinetic energy of the turntable and woman add up directly to tell us the mechanical energy.

=====

37.   A wooden block of mass M resting on a frictionless, horizontal surface is attached to a rigid rod of length ℓ and of negligible mass (Fig. P11.37). The rod is pivoted at the other end. A bullet of mass m traveling parallel to the horizontal surface and perpendicular to the rod with speed v hits the block and becomes embedded in it. (a) What is the angular momentum of the bullet-block system about a vertical axis through the pivot? (b) What fraction of the original kinetic energy of the bullet is converted into internal energy in the system during the collision?

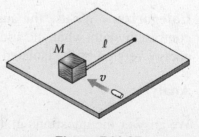

Figure P11.37

Solution

Conceptualize: Since there are no external torques acting on the bullet-block system, the angular momentum of the system will be constant and will simply be that of the bullet before it hits the block.

We should expect there to be a significant but not total loss of kinetic energy in this perfectly inelastic "angular collision," since the block and bullet do not bounce apart but move together after the collision with a small velocity that can be found from conservation of system angular momentum.

Categorize: We have practically solved this problem already! We just need to work out the details for the loss of kinetic energy.

Analyze: Taking the origin at the pivot point,

(a) Note that $\vec{r}$ is perpendicular to $\vec{v}$,

so $\sin\theta = 1$ and $L = rmv\,\sin\theta = \ell mv$ vertically down ∎

(b) Taking v_f to be the speed of the bullet and the block together, we first apply conservation of angular momentum.

$$L_i = L_f \quad \text{becomes} \quad \ell mv = \ell(m + M)v_f \quad \text{or} \quad v_f = \left(\frac{m}{m + M}\right)v$$

The total kinetic energies before and after the collision are respectively

$$K_i = \tfrac{1}{2}mv^2 \quad \text{and} \quad K_f = \tfrac{1}{2}(m + M)v_f^2 = \tfrac{1}{2}(m + M)\left(\frac{m}{m + M}\right)^2 v^2 = \tfrac{1}{2}\left(\frac{m^2}{m + M}\right)v^2$$

So the fraction of the kinetic energy that is converted into internal energy will be

$$\text{Fraction} = \frac{-\Delta K}{K_i} = \frac{K_i - K_f}{K_i} = \frac{\tfrac{1}{2}mv^2 - \tfrac{1}{2}\left(\frac{m^2}{m + M}\right)v^2}{\tfrac{1}{2}mv^2} = \frac{M}{m + M} \qquad ∎$$

Finalize: We could have also used conservation of system linear momentum to solve part (b), instead of conservation of angular momentum. The answer would be the same in this particular situation, but if the pivoting target had any shape other than a particle on a light rod, momentum conservation would not apply.

From the final equation, we can see that a larger fraction of kinetic energy is transformed for a smaller bullet. This is consistent with reasoning that if the bullet were so small that the block barely moved after the collision, then nearly all the initial kinetic energy would disappear into internal energy.

52. A puck of mass $m = 50.0$ g is attached to a taut cord passing through a small hole in a frictionless, horizontal surface (Fig. P11.52). The puck is initially orbiting with speed $v_i = 1.50$ m/s in a circle of radius $r_i = 0.300$ m. The cord is then slowly pulled from below, decreasing the radius of the circle to $r = 0.100$ m. (a) What is the puck's speed at the smaller radius? (b) Find the tension in the cord at the smaller radius. (c) How much work is done by the hand in pulling the cord so that the radius of the puck's motion changes from 0.300 m to 0.100 m?

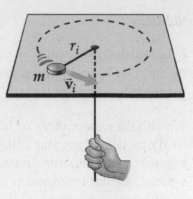

Figure P11.52

Solution

Conceptualize: Think of this as a model of water circling the bathtub drain, speeding up in linear speed and speeding up a lot in rate of rotation as it moves in. The string tension will increase steeply. The hand must do work as any point on the string, exerting downward or inward tension on the string above it, moves down or inward.

Categorize: The puck's linear momentum is always changing. Its mechanical energy changes as work is done on it. But its angular momentum stays constant because…

Analyze:

(a)   …although an external force (tension of rope) acts on the puck, no external torques act.

Therefore, $L = $ constant

and at any time $mvr = mv_i r_i$

giving us $v = \dfrac{v_i r_i}{r} = (1.50 \text{ m/s})(0.300 \text{ m})/(0.100 \text{ m}) = 4.50$ m/s   ∎

(b)   From Newton's second law, the tension is always $T = \dfrac{mv^2}{r}$

Substituting for v from (a), we find that

$$T = (0.050\ 0 \text{ kg})(4.50 \text{ m/s})^2/\ 0.100 \text{ m} = 10.1 \text{ N}$$ ∎

(c)   The work-kinetic energy theorem identifies the work as

$$W = \Delta K = \tfrac{1}{2}m\left(v^2 - v_i^2\right)$$

Substituting again for v from (a), we find

$$W = 0.5(0.050\ 0 \text{ kg})(4.50^2 - 1.50^2)(\text{m/s})^2 = 0.450 \text{ J}$$ ∎

Finalize: The puck's speed grows as the radius decreases in an inverse proportionality. The kinetic energy grows according to an inverse square law. But the tension is inversely proportional to the radius cubed.

55. Two astronauts (Fig. P11.55), each having a mass of 75.0 kg, are connected by a 10.0-m rope of negligible mass. They are isolated in space, orbiting their center of mass at speeds of 5.00 m/s. Treating the astronauts as particles, calculate (a) the magnitude of the angular momentum of the two-astronaut system and (b) the rotational energy of the system. By pulling on the rope, one astronaut shortens the distance between them to 5.00 m. (c) What is the new angular momentum of the system? (d) What are the astronauts' new speeds? (e) What is the new rotational energy of the system? (f) How much chemical potential energy in the body of the astronaut was converted to mechanical energy in the system when he shortened the rope?

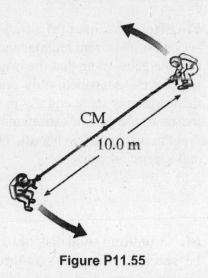

CM

10.0 m

Figure P11.55

Solution

Conceptualize: It takes real effort for one astronaut to pull hand over hand on the tether—it takes mechanical work. Both astronauts will speed up equally as…

Categorize: …angular momentum for the two-astronaut-rope system is conserved in the absence of external torques. We use this principle to find the new angular speed with the shorter tether. Standard equations will tell us the original amount of angular momentum and the original and final amounts of kinetic energy. Then the kinetic energy difference is the work.

Analyze:

(a) The angular momentum magnitude is $|\vec{L}| = m|\vec{r} \times \vec{v}|$. In this case, r and v are perpendicular, so the magnitude of L about the center of mass is

$$L = \sum mrv = 2(75.0 \text{ kg})(5.00 \text{ m})(5.00 \text{ m/s}) = 3.75 \times 10^3 \text{ kg} \cdot \text{m}^2/\text{s} \quad \blacksquare$$

(b) The original kinetic energy is

$$K = \tfrac{1}{2}mv^2 + \tfrac{1}{2}mv^2 = \tfrac{1}{2}(75.0 \text{ kg})(5.00 \text{ m/s})^2(2) = 1.88 \times 10^3 \text{ J} \quad \blacksquare$$

(c) With a lever arm of zero, the rope tension generates no torque about the center of mass. Thus, the angular momentum for the two-astronaut-rope system is unchanged:

$$L = 3.75 \times 10^3 \text{ kg} \cdot \text{m}^2/\text{s} \quad \blacksquare$$

(d) Again, $\quad L = 2mrv \quad$ so $\quad v = \dfrac{L}{2mr} = \dfrac{3.75 \times 10^3 \text{ kg} \cdot \text{m}^2/\text{s}}{2(75.0\text{kg})(2.50 \text{ m})}$

and $\quad v = 10.0$ m/s $\quad \blacksquare$

(e) The final kinetic energy is

$$K = 2\left(\tfrac{1}{2}mv^2\right) = 2\left(\tfrac{1}{2}\right)(75 \text{ kg})(10 \text{ m/s})^2 = 7.50 \times 10^3 \text{ J} \quad \blacksquare$$

(f) The energy converted by the astronaut is the work he does:

$$W_{nc} = K_f - K_i = 7.50 \times 10^3 \text{ J} - 1.88 \times 10^3 \text{ J}$$

$$W_{nc} = 5.62 \times 10^3 \text{ J} \quad \blacksquare$$

Finalize: Was part (c) a trick question? It is answered without calculation, and the conservation of system angular momentum is the idea from the start of the problem. Just for completeness, note that the original momentum is zero and the final momentum is zero. In the process described, only one astronaut converts chemical energy into mechanical, but the astronauts share equally in the kinetic energy. We could have identified the moment of inertia of the pair of astronauts as $2m(\ell/2)^2$ originally, changing to $2m(\ell/4)^2$ finally, where ℓ is the original rope length. Then $I\omega$ and $(1/2)I\omega^2$ would give us the angular momentum and kinetic energies.

---

61. A uniform solid disk of radius R is set into rotation with an angular speed ω_i about an axis through its center. While still rotating at this speed, the disk is placed into contact with a horizontal surface and immediately released as shown in Figure P11.61. (a) What is the angular speed of the disk once pure rolling takes place? (b) Find the fractional change in kinetic energy from the moment the disk is set down until pure rolling occurs. (c) Assume that the coefficient of friction between disk and surface is μ. What is the time interval after setting the disk down before pure rolling motion begins? (d) How far does the disk travel before pure rolling begins?

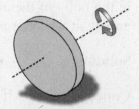

Figure P11.61

Solution

Conceptualize: At the first instant the disk will sit and spin, skidding on the table. Friction acting to the right in the picture will feed rightward momentum into the disk, so it will start moving to the right, faster and faster as mechanical energy loss slows its spinning. Eventually it will stop skidding and start pure rolling, which it can carry on forever.

Categorize: We write an equation expressing how the disk gains momentum and an equation expressing how it loses angular momentum. We combine them with $\omega = v/R$ to express the condition for pure rolling. Then we can solve for the final angular speed and the time interval required to attain it. Knowing the time along with the constant acceleration will give us the length of the skid mark.

Analyze:

(a) The normal force on the disk is mg. The constant friction force is μmg. Newton's second law for horizontal motion

$$\mu mg = ma = m(v - 0)/t \qquad a = \mu g$$

where v is the speed at any later time becomes the impulse-momentum theorem

$$\mu mgt = mv \qquad \text{or} \qquad \mu gt = v \qquad \qquad \textbf{[1]}$$

Again, Newton's second law for rotation about the center of mass is

$$-\mu mgR = I\alpha = (1/2)mR^2(\omega - \omega_i)/t$$

Here the negative sign in the first term represents how the frictional torque is opposite to the original rotational motion. It becomes the "angular-impulse-angular-momentum theorem"

$$-\mu mgRt = (1/2)mR^2\omega - (1/2)mR^2\omega_i$$

or simplified $R\omega_i - 2\mu gt = R\omega$ [2]

where ω is the speed at any later time. At the particular moment when rolling-without-slipping begins, we have $v = R\omega.$ [3]

Combining all three equations, we can eliminate v like this:

$$R\omega_i - 2v = R\omega \qquad\qquad R\omega_i - 2R\omega = R\omega \qquad\qquad \omega_i = 3\omega$$

so $\omega = \omega_i/3$ ∎

(c) While we are at it, it is convenient to do part (c) by eliminating ω from equations [1], [2], and [3], thus:

$$R\omega_i - 2\mu gt = v \qquad\qquad R\omega_i - 2\mu gt = \mu gt \qquad\qquad R\omega_i = 3\mu gt$$

Then $t = \dfrac{R\omega_i}{3\mu g}$ ∎

(b) The original kinetic energy is $K_i = \dfrac{1}{2}I\omega_i^2 = \dfrac{1}{4}MR^2\omega_i^2$

The final kinetic energy is

$$K_f = \frac{1}{2}I\omega^2 + \frac{1}{2}Mv^2 = \frac{1}{4}MR^2\left(\frac{\omega_i}{3}\right)^2 + \frac{1}{4}M\left(R\frac{\omega_i}{3}\right)^2 = \frac{3}{4}\frac{MR^2\omega_i^2}{9}$$

The fractional change in mechanical energy is then

$$\frac{K_f - K_i}{K_i} = \frac{\frac{1}{12}MR^2\omega_i^2 - \frac{1}{4}MR^2\omega_i^2}{\frac{1}{4}MR^2\omega_i^2} = -\frac{2}{3}$$ ∎

(d) The distance of travel while skidding is $\Delta x = 0 + \dfrac{1}{2}at^2$

Using results from parts (a) and (c), we find $\Delta x = \dfrac{1}{2}(\mu g)\left(\dfrac{R\omega_i}{3\mu g}\right)^2 = \dfrac{R^2\omega_i^2}{18\mu g}$ ∎

Finalize: This problem is at a goal level for excellent student accomplishment in a thorough elementary physics course. Simultaneous equations are great at keeping track of speed increasing while angular speed decreases. The missing mechanical energy turns into internal energy in the disk rim and skid mark.

12

Static Equilibrium and Elasticity

EQUATIONS AND CONCEPTS

Necessary conditions for equilibrium of an object:

The net external force acting on the object must equal zero. *This is the condition for translational equilibrium; the translational acceleration of the center of mass is zero.*

$$\sum \vec{\mathbf{F}}_{\text{ext}} = 0 \tag{12.1}$$

The net external torque on the object about any axis must be zero. *This is a statement of rotational equilibrium; the angular acceleration about any axis must be zero.*

$$\sum \vec{\tau}_{\text{ext}} = 0 \tag{12.2}$$

In the **case of coplanar forces** (e.g. all forces in the xy plane), three equations specify equilibrium. Two equations correspond to the first condition of equilibrium and the third comes from the second condition (the torque equation). *In this case, the torque vector lies along a line parallel to the z axis. All problems in this chapter fall into this category.*

$$\sum F_x = 0 \qquad \sum F_y = 0 \qquad \sum \tau_z = 0 \tag{12.3}$$

The **center of gravity** of an object is located at the center of mass when $\vec{\mathbf{g}}$ is uniform over the entire object. *In order to compute the torque due to the weight* (gravitational force), all the weight can be considered to be concentrated at a single point called the center of gravity. Equations similar to Equation 12.4 designate the y and z coordinates of the center of gravity.

$$x_{\text{CG}} = \frac{\sum_i m_i x_i}{\sum_i m_i}$$

$$x_{\text{CG}} = \frac{m_1 x_1 + m_2 x_2 + m_3 x_3 + \dots}{m_1 + m_2 + m_3 + \dots} \tag{12.4}$$

The **elastic modulus** of a material is defined as the ratio of stress to strain for that material. **Stress** is a quantity which is proportional to the force causing a deformation. **Strain** is a

$$\text{Elastic modulus} \equiv \frac{\text{stress}}{\text{strain}} \tag{12.5}$$

measure of the degree of deformation. *For a given material, there is an elastic modulus corresponding to each type of deformation:*

(i)   *resistance to change in length,*

(ii) *resistance to motion of parallel planes, and*

(iii) *resistance to change in volume.*

Young's modulus Y is a measure of the resistance of a body to *elongation* or *compression*, and is equal to the ratio of the tensile stress (the force per unit cross-sectional area) to the tensile strain (the fractional change in length).

The **elastic limit** is the maximum stress from which a substance will recover to an initial length.

The **Shear modulus** S is a measure of the resistance of a solid to internal *planes sliding past each other*. This occurs when a force is applied to one face of an object while the opposite face is held fixed. The shear modulus equals the ratio of the shear stress to the shear strain. *In Equation 12.7, Δx is the distance that the sheared face moves and h is the height of the object.*

The **Bulk modulus** B is a parameter which characterizes the resistance of an object to a *change in volume* caused by a change in pressure. It is defined as the ratio of the volume stress (the pressure) to the volume strain ($\Delta V/V_i$). *When the pressure increases (ΔP is positive) and the volume decreases (ΔV is negative). The negative sign is included in Equation 12.8 so that the value of B will be positive.*

The **compressibility** of a substance is the reciprocal of the bulk modulus.

$$Y \equiv \frac{\text{tensile stress}}{\text{tensile strain}} = \frac{F/A}{\Delta L/L_i} \qquad (12.6)$$

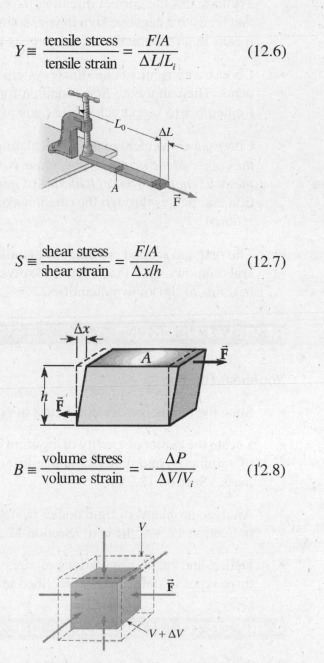

$$S \equiv \frac{\text{shear stress}}{\text{shear strain}} = \frac{F/A}{\Delta x/h} \qquad (12.7)$$

$$B \equiv \frac{\text{volume stress}}{\text{volume strain}} = -\frac{\Delta P}{\Delta V/V_i} \qquad (12.8)$$

SUGGESTIONS, SKILLS, AND STRATEGIES

The following procedure is recommended when analyzing a body in static equilibrium under the action of several external forces:

- Make a diagram of the system under consideration.

- Draw a force diagram and label all external forces acting on each object in the system. Try to guess the correct direction for each force. If you select an incorrect direction that leads to a negative sign in your solution for a force, do not be alarmed; this merely means that the direction of the force is the opposite of what you assumed.

- Choose a convenient coordinate system and resolve all forces into rectangular components. Then apply the first condition for equilibrium along each coordinate direction. Remember to keep track of the signs of the various force components.

- Choose a convenient axis for calculating the net torque on the object. *Remember that the choice of the origin for the torque equation is arbitrary; therefore, choose an origin that will simplify your calculation as much as possible.* A force that acts along a direction that passes through the origin makes a zero contribution to the torque and can be ignored.

- The first and second conditions of equilibrium give a set of linear equations with several unknowns. All that is left is to solve the simultaneous equations for the unknowns in terms of the known quantities.

REVIEW CHECKLIST

You should be able to:

- State the two necessary conditions of equilibrium for a rigid body. (Section 12.1)

- Locate the center of gravity of a system of particles or a rigid body, of sufficient degree of symmetry, and understand the difference between center of gravity and center of mass. (Section 12.2)

- Analyze problems of rigid bodies in static equilibrium using the procedures presented in Section 12.3 of the text. (Section 12.3)

- Define and make calculations of stress, strain, and elastic modulus for each of the three types of deformation described in the chapter. (Section 12.4)

ANSWER TO AN OBJECTIVE QUESTION

9.   A certain wire, 3 m long, stretches by 1.2 mm when under tension 200 N. **(i)** Does an equally thick wire 6 m long, made of the same material and under the same tension, stretch by (a) 4.8 mm, (b) 2.4 mm, (c) 1.2 mm, (d) 0.6 mm, or (e) 0.3 mm? **(ii)** A wire with twice

the diameter, 3 m long, made of the same material and under the same tension, stretches by what amount? Choose from the same possibilities (a) through (e).

Answer (i) (b) In $\Delta L = FL_i/AY$ the extension distance is directly proportional to the initial length. Doubling L_i from 3 m to 6 m, with no other changes, will make ΔL double from 1.2 mm to 2.4 mm. (ii) (e) A wire with twice the diameter will have four times the cross-sectional area. With A four times larger, $\Delta L = FL_i/AY$ will be one fourth as large, and $(1/4)(1.2 \text{ mm}) = 0.3 \text{ mm}$.

ANSWER TO A CONCEPTUAL QUESTION

7. A ladder stands on the ground, leaning against a wall. Would you feel safer climbing up the ladder if you were told that the ground is frictionless but the wall is rough or if you were told that the wall is frictionless but the ground is rough? Explain your answer.

Answer The picture shows the forces on the ladder if both the wall and floor exert friction. If the floor is perfectly smooth, it can exert no friction force to the right, to counterbalance the wall's normal force. Therefore, a ladder on a smooth floor cannot stand in equilibrium. On the other hand, a smooth wall can still exert a normal force to hold the ladder in equilibrium against horizontal motion. The counterclockwise torque of this force prevents rotation about the foot of the ladder. So you should choose a rough floor.

□   □   □   □

SOLUTIONS TO SELECTED END-OF-CHAPTER PROBLEMS

4. Consider the following distribution of objects: a 5.00-kg object with its center of gravity at (0, 0) m, a 3.00-kg object at (0, 4.00) m, and a 4.00-kg object at (3.00, 0) m. Where should a fourth object of mass 8.00 kg be placed so that the center of gravity of the four-object arrangement will be at (0, 0)?

Solution

Conceptualize: We will use a diagram to guide our calculations. Peeking at it now, we guess that the large-mass object should be in the third quadrant, about 2 m from the origin.

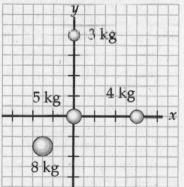

Categorize: The definition of the center of gravity as the average position of mass in the set of objects will result in equations about x and y coordinates that we can rearrange and solve to find where the last mass must be.

Analyze: From $\vec{r}_{CG} = \dfrac{\sum m_i \vec{r}_i}{\sum m_i}$ $\vec{r}_{CG}\left(\sum m_i\right) = \sum m_i \vec{r}_i$

We require the center of mass to be at the origin; this simplifies the equation,

leaving $\sum m_i x_i = 0$ and $\sum m_i y_i = 0$

To find the x-coordinate:

$[5.00 \text{ kg}][0 \text{ m}] + [3.00 \text{ kg}][0 \text{ m}] + [4.00 \text{ kg}][3.00 \text{ m}] + [8.00 \text{ kg}]x = 0$

and algebra gives $x = -1.50 \text{ m}$

Likewise, to find the y-coordinate, we solve:

$[5.00 \text{ kg}][0 \text{ m}] + [3.00 \text{ kg}][4.00 \text{ m}] + [4.00 \text{ kg}][0 \text{ m}] + [8.00 \text{ kg}]y = 0$

to find $y = -1.50 \text{ m}$

Therefore, a fourth mass of 8.00 kg should be located at

$$\vec{r}_4 = (-1.50\hat{\mathbf{i}} - 1.50\hat{\mathbf{j}}) \text{ m}$$ ∎

Finalize: The presence of the object at the origin does not enter into the calculation. Three kilograms at four meters distance has as much effect on the location of the center of gravity as four kilograms at three meters.

---

5.   Pat builds a track for his model car out of solid wood as shown in Figure P12.5. The track is 5.00 cm wide, 1.00 m high, and 3.00 m long. The runway is cut so that it forms a parabola with the equation $y = (x-3)^2/9$. Locate the horizontal coordinate of the center of gravity of this track.

Solution

Conceptualize: The center of gravity will be closer to the $x = 0$ end than to the $x = 3$ m end, because of the greater height of the $x = 0$ end. If the top edge were straight, we would have a triangle and the center of gravity would be at one third of 3 m. Because of the curvature of the top edge, the coordinate of the center of gravity will be even less than 1 m.

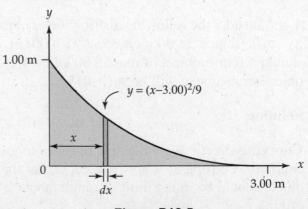

Figure P12.5

Categorize: We must do an integral, as described in Section 9.5, to find the average x coordinate of all the incremental vertical strips that form the track. In fact, we must do two integrals, to find the total mass as well as the summation of $x \, dm$.

Analyze:

Let σ represent the mass-per-face area. (It would be equal to the material's density multiplied by the constant thickness of the wood.) A vertical strip at position x, with width dx

and height $\dfrac{(x-3.00)^2}{9}$ has mass $dm = \dfrac{\sigma(x-3.00)^2\,dx}{9}$.

The total mass is

$$M = \int dm = \int_{x=0}^{3.00} \frac{\sigma(x-3)^2\,dx}{9} = \left(\frac{\sigma}{9}\right)\int_0^{3.00}\left(x^2 - 6x + 9\right)dx =$$

$$= \left(\frac{\sigma}{9}\right)\left[\frac{x^3}{3} - \frac{6x^2}{2} + 9x\right]_0^{3.00} = \sigma$$

The x-coordinate of the center of gravity is

$$x_{CG} = \frac{\int x\,dm}{M} = \frac{1}{9\sigma}\int_0^{3.00}\sigma x(x-3)^2\,dx = \frac{\sigma}{9\sigma}\int_0^{3.00}\left(x^3 - 6x^2 + 9x\right)dx$$

$$= \frac{1}{9}\left[\frac{x^4}{4} - \frac{6x^3}{3} + \frac{9x^2}{2}\right]_0^{3.00} = \frac{6.75\ \text{m}}{9.00} = 0.750\ \text{m}$$ ∎

Finalize: Our answer is less than 1 m, as predicted. The two integrals differ from each other just by one containing the extra factor x.

---

10. A 1 500-kg automobile has a wheel base (the distance between the axles) of 3.00 m. The automobile's center of mass is on the centerline at a point 1.20 m behind the front axle. Find the force exerted by the ground on each wheel.

Solution

Conceptualize: Since the center of mass lies in the front half of the car, there should be more force on the front wheels than the rear ones, and the sum of the wheel forces must equal the weight of the car.

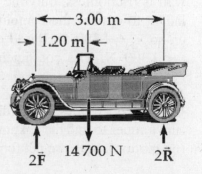

Categorize: Draw a force diagram, apply Newton's second law, and sum torques to find the unknown forces for this statics problem.

Analyze: The car's weight is

$$F_g = mg = (1\,500\ \text{kg})(9.80\ \text{m/s}^2) = 14\,700\ \text{N}$$

Call $\vec{F}$ the force of the ground on each of the front wheels and $\vec{R}$ the normal force on each of the rear wheels. If we take torques around the front axle, with counterclockwise in the picture chosen as positive, the equations are as follows:

$$\sum F_x = 0: \qquad 0 = 0$$

$$\sum F_y = 0: \qquad 2R - 14\,700\text{ N} + 2F = 0$$

$$\sum \tau = 0: \qquad +2R(3.00\text{ m}) - (14\,700\text{ N})(1.20\text{ m}) + 2F\,(0) = 0$$

The torque equation gives:

$$R = \frac{17\,640\text{ N}\cdot\text{m}}{6.00\text{ m}} = 2\,940\text{ N} = 2.94\text{ kN} \qquad \blacksquare$$

Then, from the second force equation,

$$2(2.94\text{ kN}) - 14.7\text{ kN} + 2F = 0 \qquad \text{and} \qquad F = 4.41\text{ kN} \qquad \blacksquare$$

Finalize: As expected, the front wheels experience a greater force (about 50% more) than the rear wheels. Since the friction force between the tires and road is proportional to this normal force, it makes sense that most cars today are built with front wheel drive so that the wheels under power are the ones with more traction (friction). Note that the force and torque equations are read from the same force diagram, which shows just forces, not torques.

---

13. A 15.0-m uniform ladder weighing 500 N rests against a frictionless wall. The ladder makes a 60.0° angle with the horizontal. (a) Find the horizontal and vertical forces the ground exerts on the base of the ladder when an 800-N firefighter has climbed 4.00 m along the ladder from the bottom. (b) If the ladder is just on the verge of slipping when the firefighter is 9.00 m from the bottom, what is the coefficient of static friction between ladder and ground?

Solution

Conceptualize: Refer to the force diagram as needed. Since the wall is frictionless, only the ground exerts an upward force on the ladder to oppose the combined weight of the ladder and firefighter, so $n_g = 1\,300$ N. Based on the angle of the ladder, we expect $f < 1\,300$ N. The coefficient of friction is probably somewhere between 0 and 1.

Categorize: Draw a force diagram, apply Newton's second law, and sum torques to find the unknown forces. Since this is a statics problem (no motion), both the net force and net torque are zero.

Analyze:

(a) $\quad \sum F_x = f - n_w = 0 \qquad \sum F_y = n_g - 800\text{ N} - 500\text{ N} = 0$

so that $n_g = 1\,300$ N (upwards) $\qquad \blacksquare$

Taking torques about an axis at the foot of the ladder, $\sum \tau_A = 0$:

$-(800\ \text{N})(4.00\ \text{m})\sin 30° - (500\ \text{N})(7.50\ \text{m})\sin 30° + n_w (15.0\ \text{m})\cos 30° = 0$

Solving the torque equation for n_w gives

$$n_w = \frac{[(4.00\ \text{m})(800\ \text{N}) + (7.50\ \text{m})(500\ \text{N})]\tan 30.0°}{15.0\ \text{m}} = 267.5\ \text{N}$$

Next substitute this value into the F_x equation to find:

$$f = n_w = 268\ \text{N} \text{ (the friction force is directed toward the wall)} \quad \blacksquare$$

(b) When the firefighter is 9.00 m up the ladder, the torque equation $\sum \tau_A = 0$ gives:

$-(800\ \text{N})(9.00\ \text{m})\sin 30° - (500\ \text{N})(7.50\ \text{m})\sin 30° + n_w (15.0\ \text{m})\sin 60° = 0$

Solving, $n_w = \dfrac{\left[(9.00\ \text{m})(800\ \text{N}) + (7.50\ \text{m})(500\ \text{N})\right]\tan 30.0°}{15.0\ \text{m}} = 421\ \text{N}$

Since $f = n_w = 421\ \text{N}$ and $f = f_{max} = \mu_s n_g$

$$\mu_s = \frac{f_{max}}{n_g} = \frac{421\ \text{N}}{1\ 300\ \text{N}} = 0.324 \quad \blacksquare$$

Finalize: The calculated answers seem reasonable and agree with our predictions. You should not think of a statics problem as just about adding forces and adding torques to find their sums. In Chapter 3 you added vectors to find their resultant, but in equilibrium problems we already know the total force and the total torque are zero. The unknown is something else entirely.

This problem would be more realistic if the wall were not frictionless, in which case an additional vertical force would be added. This more complicated problem could be solved if we knew at least one of the forces of friction.

---

27. A 200-kg load is hung on a wire of length 4.00 m, cross-sectional area $0.200 \times 10^{-4}\ \text{m}^2$, and Young's modulus $8.00 \times 10^{10}\ \text{N/m}^2$. What is its increase in length?

Solution

Conceptualize: Since metal wire does not stretch very much, the length will probably not change by more than 1% (<4 cm in this case) unless it is stretched beyond its elastic limit.

Categorize: Apply the Young's Modulus strain equation to find the increase in length.

Analyze: Young's Modulus is $Y = \dfrac{F/A}{\Delta L/L_i}$

The load force is $\qquad F = (200 \text{ kg})(9.80 \text{ m/s}^2) = 1\,960 \text{ N}$

So $\qquad \Delta L = \dfrac{FL_i}{AY} = \dfrac{(1\,960 \text{ N})(4.00 \text{ m})(1\,000 \text{ mm/m})}{(0.200 \times 10^{-4} \text{ m}^2)(8.00 \times 10^{10} \text{ N/m}^2)} = 4.90 \text{ mm}$ ∎

Finalize: The wire only stretched about 0.1% of its length, so this seems like a reasonable amount.

---

31. Assume that if the shear stress in steel exceeds about 4.00×10^8 N/m² the steel ruptures. Determine the shearing force necessary to (a) shear a steel bolt 1.00 cm in diameter and (b) punch a 1.00-cm-diameter hole in a steel plate 0.500 cm thick.

Solution

Conceptualize: We expect forces of thousands of newtons in both cases.

Categorize: We do not need the equation S = stress/strain. Rather, we use just the definition of stress, writing it as stress $= \sigma = F/A$, where A is the area of one of the layers sliding over each other.

Analyze:

(a)

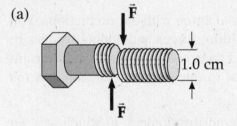

$F = \sigma A = \pi(4.00 \times 10^8 \text{ N/m}^2)(0.500 \times 10^{-2} \text{ m})^2 = 31.4 \text{ kN}$ ∎

(b) Now the area of the molecular layers sliding over each other is the curved lateral surface area of the cylinder punched out, a cylinder of radius 0.50 cm and height 0.50 cm. So,

$$F = \sigma A = \sigma(h)(2\pi r)$$

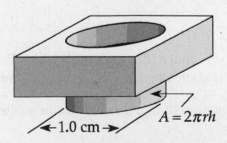

$F = 2\pi(4.00 \times 10^8 \text{ N/m}^2)(0.500 \times 10^{-2} \text{ m})(0.500 \times 10^{-2} \text{ m}) = 62.8 \text{ kN}$ ∎

Finalize: The area in stress = force/area is always the area of each one of the layers of molecules between which the stress distorts (strains) the intermolecular bonds.

---

32. When water freezes, it expands by about 9.00%. What pressure increase would occur inside your automobile engine block if the water in it froze? (The bulk modulus of ice is 2.00×10^9 N/m^2.)

Solution

Conceptualize: Don't guess about 9% of atmospheric pressure! The pressure increase will be much larger, many millions of pascals.

Categorize: We will use the definition of bulk modulus, along with some step-by-step reasoning. To find the maximum possible pressure increase, we assume the engine block does not break open, and holds the freezing water at constant volume.

Analyze: V represents the original volume; $0.090\,0V$ is the change in volume that would happen if the block cracked open. Imagine squeezing the ice, with unstressed volume $1.09V$, back down to its previous volume, so $\Delta V = -0.090\,0V$. According to the definition of the bulk modulus as given in the chapter text, we have

$$\Delta P = -\frac{B(\Delta V)}{V_i} = -\frac{\left(2.00 \times 10^9 \ \text{N/m}^2\right)(-0.090\,0V)}{1.09V} = 1.65 \times 10^8 \ \text{N/m}^2$$

∎

Finalize: Hundreds of millions of pascals is a greater pressure than the steel can sustain. You must use antifreeze to keep the coolant water from freezing.

---

37. A bridge of length 50.0 m and mass 8.00×10^4 kg is supported on a smooth pier at each end as shown in Figure P12.37. A truck of mass 3.00×10^4 kg is located 15.0 m from one end. What are the forces on the bridge at the points of support?

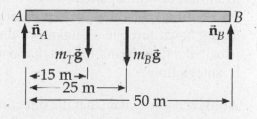

Figure P12.37

Solution

Conceptualize: If the truck were not there, each pier would exert about 400 kN to support half the weight of the bridge. With the truck present, pier B will exert slightly more force and pier A considerably more.

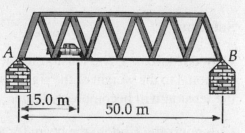

Categorize: The object can be the bridge deck. It feels the unknown forces as well as known forces. Then we use the second and first conditions for equilibrium.

Analyze: Let n_A and n_B be the normal forces at the points of support.

$$\sum F_y = 0: \quad n_A + n_B - (8.00 \times 10^4 \text{ kg})g - (3.00 \times 10^4 \text{ kg})g = 0$$

Choosing the axis at point A, we find:

$$\sum \tau = 0: \quad -(3.00 \times 10^4 \text{ kg})(15.0 \text{ m})g - (8.00 \times 10^4 \text{ kg})(25.0 \text{ m})g + n_B(50.0 \text{ m}) = 0$$

The torque equation we can solve directly to find

$$n_B = \frac{\left[(3.00 \times 10^4)15.0 + (8.00 \times 10^4)25.0\right] \text{ kg } (9.80 \text{ m/s}^2)}{50.0} = 4.80 \times 10^5 \text{ N} \quad \blacksquare$$

Then the force equation gives

$$n_A = (8.00 \times 10^4 \text{ kg} + 3.00 \times 10^4 \text{ kg})(9.80 \text{ m/s}^2) - 4.80 \times 10^5 \text{ N}$$

$$= 5.98 \times 10^5 \text{ N} \quad \blacksquare$$

Finalize: Our estimates were good. Do not get mixed up by thinking about the bridge pushing down on its supports. The little word *on* in the problem statement reminds you to consider the upward force of each pier on the bridge. Internal forces of tension and compression in the framework of the bridge do not count as external forces acting on the bridge. Internal forces are not included in the force diagram.

---

45. A uniform sign of weight F_g and width $2L$ hangs from a light, horizontal beam hinged at the wall and supported by a cable (Fig. P12.45). Determine (a) the tension in the cable and (b) the components of the reaction force exerted by the wall on the beam, in terms of F_g, d, L, and θ.

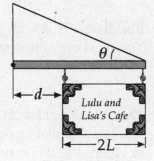

Figure P12.45

Solution

Conceptualize: The tension in the cable will be proportional to the weight of the sign. As θ approaches zero, the tension will become very large.

Categorize: Choose the beam for analysis, and draw a force diagram as shown. Note how we represent tension as an unknown-magnitude force in a known direction. By contrast, note how we represent the hinge force as an unknown force in an unknown direction. Because it is called a light beam, the gravitational force exerted on it is negligible.

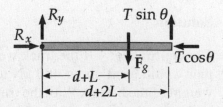

Analyze: We know that the direction of the force from the cable at the right end is along the cable, at an angle of θ above the horizontal. On the other end, we do not know magnitude or direction for the hinge force $\vec{R}$ so we show it as two unknown components.

The first condition for equilibrium gives two equations:

$$\sum F_x = 0: \qquad +R_x - T\cos\theta = 0$$

$$\sum F_y = 0: \qquad +R_y - F_g + T\sin\theta = 0$$

Taking torques about the left end, we find the second condition is

$$\sum \tau = 0: \qquad R_y(0) + R_x(0) - F_g(d + L) + (0)(T\cos\theta) + (d + 2L)(T\sin\theta) = 0$$

(a) The torque equation gives $\qquad T = \dfrac{F_g(d + L)}{(d + 2L)\sin\theta}$ ■

(b) Now from the force equations, $\qquad R_x = \dfrac{F_g(d + L)}{(d + 2L)\tan\theta}$ ■

and $\qquad\qquad\qquad\qquad R_y = F_g - \dfrac{F_g(d + L)}{d + 2L} = \dfrac{F_g L}{d + 2L}$ ■

Finalize: We confirm the proportionality of the tension to the gravitational force on the sign. And when θ approaches zero, $1/\sin\theta$ does, as predicted, grow beyond all bounds.

---

49. A 10 000-N shark is supported by a rope attached to a 4.00-m rod that can pivot at the base. (a) Calculate the tension in the cable between the rod and the wall, assuming that the cable is holding the system in the position shown in Figure P12.49. Find (b) the horizontal force and (c) the vertical force exerted on the base of the rod. Ignore the weight of the rod.

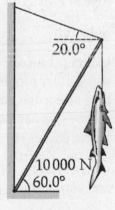

Figure P12.49

Solution

Conceptualize: Since the rod helps support the weight of the shark by exerting a vertical force, the tension in the upper portion of the cable must be less than 10 000 N. Likewise, the vertical force on the base of the rod should also be less than 10 kN.

Categorize: This is another statics problem where the sum of the forces and torques must be zero. To find the unknown forces, draw a force diagram of the rod, apply Newton's second law, and sum torques.

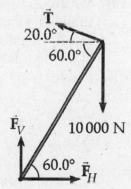

Analyze: From the diagram, the angle $\vec{T}$ makes with the rod is $\theta = 60.0° + 20.0° = 80.0°$ and the perpendicular component of $\vec{T}$ is $T\sin 80.0°$.

Summing torques around the base of the rod, and applying Newton's second law in the horizontal and vertical directions, we have

$$\sum \tau = 0: \qquad -(4.00\text{ m})(10\,000\text{ N})\cos 60° + T(4.00\text{ m})\sin(80°) = 0$$

(a)   giving $T = \dfrac{(10\ 000\ \text{N}) \cos (60.0°)}{\sin(80.0°)} = 5.08 \times 10^3\ \text{N}$ ■

$\sum F_x = 0:$ $F_H - T \cos(20.0°) = 0$

(b)   so $F_H = T \cos (20.0°) = 4.77 \times 10^3\ \text{N}$ ■

and from $\sum F_y = 0:$ $F_V + T \sin (20.0°) - 10\ 000\ \text{N} = 0$

(c)   we find $F_V = (10\ 000\ \text{N}) - T \sin (20.0°) = 8.26 \times 10^3\ \text{N}$ ■

Finalize: The forces calculated are indeed less than 10 kN as predicted. That shark sure is a big catch; he weighs about a ton!

---

51.   A uniform beam of mass m is inclined at an angle θ to the horizontal. Its upper end (point P) produces a 90° bend in a very rough rope tied to a wall, and its lower end rests on a rough floor (Fig. P12.51). Let μ_s represent the coefficient of static friction between beam and floor. Assume μ_s is less than the cotangent of θ. (a) Find an expression for the maximum mass M that can be suspended from the top before the beam slips. Determine (b) the magnitude of the reaction force at the floor and (c) the magnitude of the force exerted by the beam on the rope at P in terms of m, M, and μ_s.

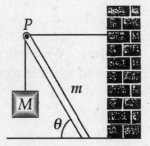

Figure P12.51

Solution

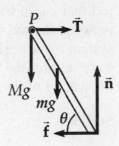

Conceptualize: The solution to this problem is not so obvious as some other problems because there are three independent variables that affect the maximum mass M. We could at least expect that more mass can be supported for higher coefficients of friction (μ_s), larger angles (θ), and a more massive beam (m).

Categorize: Draw a force diagram, apply Newton's second law, and sum torques to find the unknown forces for this statics problem.

Analyze:

(a)   We use $\sum F_x = \sum F_y = \sum \tau = 0$ and choose the axis at the point of contact with the floor to simplify the torque analysis. Since the rope is described as very rough, we will assume that it will never slip on the end of the beam. First, let us determine what friction force at the floor is necessary to put the system in equilibrium; then we can check whether that friction force can be obtained.

$\sum F_x = 0:$ $T - f = 0$

$\sum F_y = 0:$ $n - Mg - mg = 0$

$\sum \tau = 0:$ $Mg(\cos \theta)L + mg(\cos \theta) \dfrac{L}{2} - T(\sin \theta)L = 0$

Solving the torque equation, we find

$$T = (M + \tfrac{1}{2}m)g \cot \theta$$

Then the horizontal force equation implies by substitution that this same expression is equal to *f*. In order for the beam not to slip, we need $f \le \mu_s n$. Substituting for *n* and *f*, we obtain the requirement

$$\mu_s \ge \left[\frac{M + \tfrac{1}{2}m}{M + m}\right]\cot\theta$$

The factor in brackets is always less than one, so if $\mu \ge \cot\theta$ then *M* can be increased without limit. In this case there is no maximum mass!

Otherwise, if $\mu_s < \cot\theta$, on the verge of slipping the equality will apply, and solving for *M* yields

$$M = \frac{m}{2}\left[\frac{2\mu_s \sin\theta - \cos\theta}{\cos\theta - \mu_s \sin\theta}\right] \qquad \blacksquare$$

(b) At the floor we see that the normal force is in the *y* direction and the friction force is in the −*x* direction. The reaction force exerted by the floor then has magnitude

$$R = \sqrt{n^2 + (\mu_s n)^2} = g(M + m)\sqrt{1 + \mu_s^2} \qquad \blacksquare$$

(c) At point *P*, the force of the beam on the rope is in magnitude

$$F = \sqrt{T^2 + (Mg)^2} = g\sqrt{M^2 + \mu_s^2(M + m)^2} \qquad \blacksquare$$

Finalize: In our answer to part (a), notice that this result does not depend on *L*, which is reasonable since the lever arm of the beam's weight is proportional to the length of the beam.

The answer to this problem is certainly more complex than most problems. We can see that the maximum mass *M* that can be supported is proportional to *m*, but it is not clear from the solution that *M* increases proportionally to μ_s and θ as predicted. To further examine the solution to part (a), we could graph or calculate the ratio *M/m* as a function of θ for several reasonable values of μ_s, ranging from 0.5 to 1. This would show that *M* does increase with increasing μ_s and θ, as predicted. The complex requirements for stability help to explain why we don't encounter this precarious configuration very often.

---

56. A stepladder of negligible weight is constructed as shown in Figure P12.56, with *AC* = *BC* = ℓ = 4.00 m. A painter of mass *m* = 70.0 kg stands on the ladder *d* = 3.00 m from the bottom. Assuming the floor is frictionless, find (a) the tension in the horizontal bar *DE* connecting the two halves of the ladder, (b) the normal forces at *A* and *B*, and (c) the components of the reaction force at the single hinge *C* that the left half of the ladder exerts on the right half. *Suggestion*: Treat the ladder as a single object, but also each half of the ladder separately.

Solution

Conceptualize: The normal force at A will be somewhat more than half of the weight of the painter (about 700 N) and the normal force at B somewhat less than half. For the tension in the tie bar we expect about 100 N.

Categorize: Equilibrium of the whole ladder can tell us support forces but not the tension in the tie bar, because it is internal to the ladder. We must then also consider equilibrium of one side of the ladder.

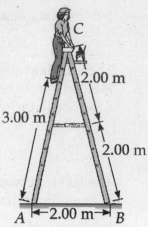

Figure P12.56

Analyze: If we think of the whole ladder, we can solve part (b).

(b) The painter is 3/4 of the way up the ladder, so the lever arm of her weight about A is $\frac{3}{4}(1.00 \text{ m}) = 0.750 \text{ m}$

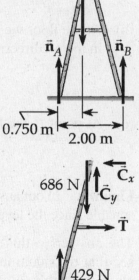

$$\sum F_x = 0: \qquad 0 = 0$$

$$\sum F_y = 0: \qquad n_A - 686 \text{ N} + n_B = 0$$

$$\sum \tau_A = 0: \qquad n_A(0) - (686 \text{ N})(0.750 \text{ m}) + n_B(2.00 \text{ m}) = 0$$

Thus, $\qquad n_B = (686 \text{ N})(0.750 \text{ m})/(2.00 \text{ m}) = 257 \text{ N}$ ∎

and $\qquad n_A = 686 \text{ N} - 257 \text{ N} = 429 \text{ N}$ ∎

Now consider the left half of the ladder. We know the direction of the bar tension, and we make guesses for the directions of the components of the hinge force. If a guess is wrong, the answer will be negative.

The side rails make an angle with the horizontal

$$\theta = \cos^{-1}(1/4) = 75.5°$$

Taking torques about the top of the ladder, we have

$$\sum \tau_C = 0: \qquad (-429 \text{ N})(1.00 \text{ m}) + T(2.00 \text{ m}) \sin 75.5° + (686 \text{ N})(0.250 \text{ m}) = 0$$

$$\sum F_x = 0: \qquad T - C_x = 0$$

$$\sum F_y = 0: \qquad 429 \text{ N} - 686 \text{ N} + C_y = 0$$

(a) From the torque equation,

$$T = \frac{(429 \text{ N})(1.00 \text{ m}) - (686 \text{ N})(0.250 \text{ m})}{2.00 \text{ m} \sin 75.5°} = \frac{257 \text{ N} \cdot \text{m}}{1.94 \text{ m}} = 133 \text{ N} \qquad ∎$$

(c) From the force equations, $\qquad C_x = T = 133 \text{ N}$

$$C_y = 686 \text{ N} - 429 \text{ N} = 257 \text{ N up}$$

This is the force that the right half exerts on the left half. The force that the left half exerts on the right half has opposite components:

133 N to the right, and 257 N down. ∎

Finalize: Our estimates were good.

---

58. Review. A wire of length L, Young's modulus Y, and cross-sectional area A is stretched elastically by an amount ΔL. By Hooke's law, the restoring force is $-k\,\Delta L$. (a) Show that $k = YA/L$. (b) Show that the work done in stretching the wire by an amount ΔL is $W = \frac{1}{2}YA(\Delta L)^2/L$.

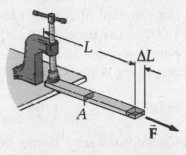

Solution

Conceptualize: A wire made of stiffer material (larger Y) should have a larger spring-stiffness constant k. If its cross-sectional area is larger there are more atomic bonds to stretch, and k should again be larger. A longer wire will stretch by the same percentage under the same tension, so it will stretch by a greater extension distance than a shorter wire. Then the longer wire has a smaller spring constant. These proportionalities agree with the relation $k = YA/L$ we are to prove.

Categorize: We will let F represent some stretching force and use algebra to combine the Hooke's-law account of the stretching with the Young's modulus account. Then integration will reveal the work done as the wire extends.

Analyze:

(a) According to Hooke's law $|\vec{F}| = k\Delta L$

Young's modulus is defined as $Y = \dfrac{F/A}{\Delta L/L}$

By substitution,

$$Y = k\frac{L}{A} \qquad \text{or} \qquad k = \frac{YA}{L} \qquad ∎$$

(b) The spring exerts force $-kx$. The outside agent stretching it exerts force $+kx$. We can determine the work done by integrating the force kx over the distance we stretch the wire.

$$W = -\int_0^{\Delta L} F\,dx = -\int_0^{\Delta L}(-kx)dx = \frac{YA}{L}\int_0^{\Delta L} x\,dx = \left[\frac{YA}{L}\left(\frac{1}{2}x^2\right)\right]_{x=0}^{x=\Delta L}$$

Therefore,

$$W = \frac{YA}{2L}\Delta L^2 \qquad ∎$$

Finalize: Our Hooke's-law theory of a spring applies much more generally than you might expect. In Chapter 15 we will study a block bouncing on a spring, and this system models a very wide class of phenomena.

---

60. (a) Estimate the force with which a karate master strikes a board, assuming the hand's speed at the moment of impact is 10.0 m/s, decreasing to 1.00 m/s during a 0.002 00-s time interval of contact between the hand and the board. The mass of his hand and arm is 1.00 kg. (b) Estimate the shear stress, assuming this force is exerted on a 1.00-cm-thick pine board that is 10.0 cm wide. (c) If the maximum shear stress a pine board can support before breaking is 3.60×10^6 N/m², will the board break?

Solution

Conceptualize: Karate masters break boards all the time.

Categorize: We will use the impulse-momentum theorem to find the force on the hand, which is equal in magnitude to the force on the board. Then the definition of shear stress will let us compute that quantity to see if it is large enough to break the board.

Analyze: The impulse-momentum theorem describes the force of the board on his hand: $Ft = mv_f - mv_i$

$$F = m\left(\frac{\Delta v}{\Delta t}\right) = (1.00 \text{ kg}) \frac{(1.00 - 10.0) \text{ m/s}}{0.002\ 00 \text{ s}} = -4\ 500 \text{ N}$$

(a)   Therefore, the force of his hand on board is 4 500 N.   ∎

(b)   That force produces a shear stress on the area that is exposed when the board snaps:

$$\text{Stress} = \frac{F}{A} = \frac{4\ 500 \text{ N}}{(10^{-1} \text{ m})\ (10^{-2} \text{ m})} = 4.50 \times 10^6 \text{ N/m}^2$$ ∎

(c)   **Yes;** this is larger than 3.60 MN/m² and suffices to break the board.   ∎

Finalize: We do not have the right combination of information to use the work-kinetic energy theorem to find the force. We could compute the acceleration of the hand and use Newton's second law to find the same 4.50 kN force, but writing the impulse-momentum theorem is more direct. The area in stress = force/area is always the area of the layers of molecules between which the stress produces distortion (strain).

---

62. Consider the rectangular cabinet of Problem 50 shown in Figure P12.50, but with a force applied $\bar{F}$ horizontally at the upper edge. (a) What is the minimum force required to start to tip the cabinet? (b) What is the minimum coefficient of static friction required for the cabinet not to slide with the application of a force of this magnitude? (c) Find the magnitude and direction of the minimum force required to tip the cabinet if the point of application can be chosen *anywhere* on the cabinet.

$w = 60$ cm

$\ell = 100$ cm

Figure P12.50

Solution

Conceptualize: To *tip* the cabinet, the applied force must cause enough torque about the front bottom edge to balance the torque caused by the cabinet's weight. If the static friction force at the floor is large enough, the cabinet will tip and not *slip*. A smaller force can be found in part (c) than that in part (a) if we choose to maximize its moment arm.

Categorize: From one force diagram we can think about torques in part (a) and forces in part (b). A new diagram will be required in (c) with the direction and application point of the applied force chosen thoughtfully.

Analyze: The weight acts at the geometric center of the uniform cabinet. Its value is given as 400 N in Problem 50.

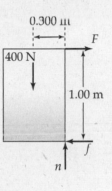

0.300 m

400 N

1.00 m

When the cabinet is just ready to tip, the whole normal force acts at the front bottom edge, shown as a corner in the diagram. Use the force diagram and sum the torques about the lower front corner of the cabinet.

$$\sum \tau = 0: \quad -F(1.00 \text{ m}) + (400 \text{ N})(0.300 \text{ m}) = 0$$

(a) yielding

$$F = \frac{(400 \text{ N})(0.300 \text{ m})}{1.00 \text{ m}} = 120 \text{ N}$$

$$\sum F_x = 0: \quad -f + 120 \text{ N} = 0, \quad \text{or} \quad f = 120 \text{ N}$$

$$\sum F_y = 0: \quad -400 \text{ N} + n = 0, \quad \text{so} \quad n = 400 \text{ N}$$

(b) Thus, $\mu_s = \dfrac{f}{n} = \dfrac{120 \text{ N}}{400 \text{ N}} = 0.300$

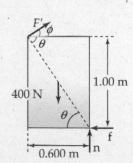

400 N

1.00 m

0.600 m

(c) Apply F' at the upper rear corner, directed so $\theta + \phi = 90.0°$ to obtain the largest possible lever arm. We have

$$\theta = \tan^{-1}\left(\frac{1.00 \text{ m}}{0.600 \text{ m}}\right) = 59.0°$$

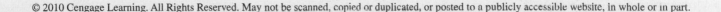

Thus, $\phi = 90.0° - 59.0° = 31.0°$.

Sum the torques about the lower front corner of the cabinet:

$$-F'\sqrt{(1.00 \text{ m})^2 + (0.600 \text{ m})^2} + (400 \text{ N})(0.300 \text{ m}) = 0$$

so $\quad F' = \dfrac{120 \text{ N} \cdot \text{m}}{1.17 \text{ m}} = 103 \text{ N}$

Therefore, the minimum force required to tip the cabinet is 103 N applied at 31.0° above the horizontal at the upper left corner. ∎

Finalize: Both 120 N and 103 N are small compared to the 400-N weight of the cabinet. Slip-or-tip problems are nice exercises in visualizing the effects of forces and torques separately, and can be nicely verified in laboratory.

13

Universal Gravitation

EQUATIONS AND CONCEPTS

Newton's law of universal gravitation states that every particle in the Universe attracts every other particle with a force that is directly proportional to the product of their masses and inversely proportional to the square of the distance between them. *This is one example of a force form referred to as an inverse square law.*

$$F_g = G\frac{m_1 m_2}{r^2} \tag{13.1}$$

$$\vec{F}_{12} = -G\frac{m_1 m_2}{r^2}\hat{r}_{12} \tag{13.3}$$

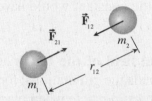

$$\vec{F}_{12} = -\vec{F}_{21} \text{ (an action-reaction pair)}$$

The **universal gravitational constant** is designated by the symbol G.

$$G = 6.673 \times 10^{-11} \text{ N} \cdot \text{m}^2/\text{kg}^2 \tag{13.2}$$

The **acceleration due to gravity** decreases with increasing altitude. In Equation 13.6, $r = (R_E + h)$ where h is measured from the Earth's surface. *At the surface of the Earth, $h = 0$ and as $r \to \infty$, the weight of an object approaches zero.*

$$g = \frac{GM_E}{r^2} = \frac{GM_E}{(R_E + h)^2} \tag{13.6}$$

Kepler's first law states that all planets move in *elliptical orbits* with the Sun at one focus (see figure at right). *This law is a direct result of the inverse square nature of the gravitational force. The semimajor axis has length **a** and the semiminor axis has length **b**. The foci F_1 and F_2 are located a distance **c** from the center of the ellipse.*

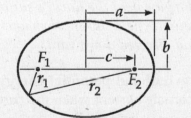

Kepler's second law states that the radius vector from the Sun to any planet sweeps out equal areas in equal times. *This result applies in the case of any central force and implies conservation of angular momentum*

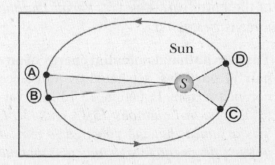

229

for an isolated system. Therefore, the speed of a planet along the elliptical path is greater when the planet is closer to the Sun.

$$\frac{dA}{dt} = \frac{L}{2M_P} \tag{13.7}$$

Kepler's third law states that the square of the orbital period of any planet is proportional to the cube of the semimajor axis of the elliptical orbit. *Kepler's third law is valid for circular and elliptical orbits and the constant of proportionality is independent of the mass of the planet orbiting the Sun.*

$$T^2 = \left(\frac{4\pi^2}{GM_S}\right)a^3 = K_S a^3 \tag{13.8}$$

$$K_S = 2.97 \times 10^{-19} \text{ s}^2/\text{m}^3$$

For an Earth satellite (e.g. the Moon orbiting the Earth), M_S is replaced by the mass of the Earth and the constant would have a different value.

$$K_E = \frac{4\pi^2}{GM_E}$$

The **gravitational field of the Earth** at a point in space equals the gravitational force experienced by a *test particle* when placed at that point divided by the mass of the test particle. *The Earth creates the gravitational field; the presence of the test particle is not necessary for the existence of the field.*

$$\vec{g} \equiv \frac{\vec{F}_g}{m} \tag{13.9}$$

The **gravitational field at a distance r from the center of the Earth** points *inward* along a radius toward the center of the Earth. The unit vector $\hat{r}$ pointing *outward* from the center of the Earth gives rise to the negative sign in Equation 13.10. *Over a small region near the Earth's surface, the gravitational field is approximately uniform and directed downward.*

$$\vec{g} = \frac{\vec{F}_g}{m} = -\frac{GM_E}{r^2}\hat{r} \tag{13.10}$$

The **gravitational potential energy** of an Earth-particle system when the particle is a distance r (where $r > R_E$) from the center of the Earth varies as $1/r$. *Recall that the force varies as $1/r^2$.*

$$U(r) = -\frac{GM_E m}{r} \tag{13.13}$$

The **gravitational potential energy of any pair of particles**, separated by a distance r, follows from Equation 13.13 and varies as $1/r$. *In Equations 13.13 and 13.14, it is assumed that the potential is zero at infinity; the negative sign indicates that the*

$$U(r) = -\frac{Gm_1 m_2}{r} \tag{13.14}$$

force is attractive. An external agent must do positive work to increase the separation of the particles.

The **total mechanical energy of a bound system** is negative for both circular orbits (of radius r) and elliptical orbits (with semimajor axis a). The kinetic energy is positive and equal to half the absolute value of the potential energy. *Here we use the convention that $U \rightarrow 0$ as $r \rightarrow \infty$.*

$$E = -\frac{G\,Mm}{2r} \quad \text{(circular orbits)} \qquad (13.18)$$

$$E = -\frac{GMm}{2a} \quad \text{(elliptical orbits)} \qquad (13.19)$$

The total energy and the total angular momentum of a gravitationally bound, two-object system are constants of the motion.

The **escape speed** is defined as the minimum speed an object must have, when projected from the Earth, in order to escape the Earth's gravitational field (that is, to just reach $r = \infty$ with zero speed). *Note that v_{esc} does not depend on the mass of the projected body.*

$$v_{esc} = \sqrt{\frac{2GM_E}{R_E}} \qquad (13.22)$$

REVIEW CHECKLIST

You should be able to:

- Calculate the net gravitational force on an object due to one or more nearby masses. (Section 13.1)

- Calculate the free-fall acceleration at given heights above the surface of a planet. (Section 13.2)

- State Kepler's three laws of planetary motion, describe the properties of an ellipse, and make calculations using the second and third laws. (Section 13.3)

- Calculate the potential energy of two or more particles (including an Earth-mass system). Calculate the work required to move a mass between two points in the Earth's gravitational field. (Section 13.5)

- Calculate the total energy of a satellite moving in a circular or elliptical orbit about a large body (e.g. planet orbiting the Sun or an Earth satellite) located at the center of motion. Note that the total energy is negative, as it must be for any closed orbit. (Section 13.6)

- Obtain the expression for v_{esc} using the principle of conservation of energy and calculate the escape speed of an object of mass m from a planet. (Section 13.6)

ANSWER TO AN OBJECTIVE QUESTION

1.   Rank the magnitudes of the following gravitational forces from largest to smallest. If two forces are equal, show their equality in your list. (a) the force exerted by a 2-kg object on a 3-kg object 1 m away. (b) the force exerted by a 2-kg object on a 9-kg object 1 m away. (c) the force exerted by a 2-kg object on a 9-kg object 2 m away. (d) the force exerted by a 9-kg object on a 2-kg object 2 m away. (e) the force exerted by a 4-kg object on another 4-kg object 2 m away.

Answer   The force is proportional to the product of the masses and inversely proportional to the square of the separation distance, so we compute the quantity $m_1 m_2/r^2$ for each case. In units of kg²/m², this quantity is (a) $2 \cdot 3/1^2 = 6$ (b) $18/1 = 18$ (c) $18/4 = 4.5$ (d) $18/4 = 4.5$ (e) $16/4 = 4$. The ranking is then $b > a > c = d > e$.

Note that the equality of the force magnitudes in situations (c) and (d) is described by Newton's third law.

☐   ☐   ☐   ☐

ANSWERS TO SELECTED CONCEPTUAL QUESTIONS

2.   Explain why it takes more fuel for a spacecraft to travel from the Earth to the Moon than for the return trip. Estimate the difference.

Answer   The masses and radii of the Earth and Moon, and the distance between the two are

$$M_E = 5.97 \times 10^{24}\,\text{kg}, R_E = 6.37 \times 10^6\,\text{m}$$

$$M_M = 7.35 \times 10^{22}\,\text{kg}, R_M = 1.74 \times 10^6\,\text{m} \quad \text{and} \quad d = 3.84 \times 10^8\,\text{m}$$

To travel between the Earth and the Moon, a distance d, a rocket engine must boost the spacecraft over the point of zero total gravitational field in between. Call x the distance of this point from Earth. To cancel, the Earth and Moon must here produce equal fields:

$$\frac{GM_E}{x^2} = \frac{GM_M}{(d-x)^2}$$

Isolating x gives $\quad \dfrac{M_E}{M_M}(d-x)^2 = x^2 \quad$ so $\quad \sqrt{\dfrac{M_E}{M_M}}(d-x) = x$

Thus, $x = \dfrac{d\sqrt{M_E}}{\sqrt{M_M}+\sqrt{M_E}} = 3.46 \times 10^8$ m and $(d-x) = \dfrac{d\sqrt{M_M}}{\sqrt{M_M}+\sqrt{M_E}} = 3.84 \times 10^7$ m

In general, we can ignore the gravitational pull of the far object when we're close to the other; our results will still be nearly correct. Thus the approximate energy difference between point x and the Earth's surface is

$$\frac{\Delta E_{E \to x}}{m} = \frac{-GM_E}{x} - \frac{-GM_E}{R_E}$$

Similarly, flying from the Moon:

$$\frac{\Delta E_{M \to x}}{m} = \frac{-GM_M}{d - x} - \frac{-GM_M}{R_M}$$

Taking the ratio of the energy terms,

$$\frac{\Delta E_{E \to x}}{\Delta E_{M \to x}} = \left(\frac{M_E}{x} - \frac{M_E}{R_E} \right) \Big/ \left(\frac{M_M}{d - x} - \frac{M_M}{R_M} \right) \approx 22.8$$

This would also be the minimum fuel ratio, if the spacecraft was driven by a method other than rocket exhaust. If rockets were used, the total fuel ratio would be much larger.

□ □ □ □

3. Why don't we put a geosynchronous weather satellite in orbit around the 45th parallel? Wouldn't such a satellite be more useful in the United States than one in orbit around the equator?

Answer While a satellite in orbit above the 45th parallel might be more useful, it isn't possible. The center of a satellite orbit must be the center of the Earth, since that is the force center for the gravitational force. If the satellite is north of the plane of the equator for part of its orbit, it must be south of the equatorial plane for the rest.

□ □ □ □

5. (a) At what position in its elliptical orbit is the speed of a planet a maximum? (b) At what position is the speed a minimum?

Answer (a) At the planet's closest approach to the Sun—its perihelion—the system's potential energy is at its minimum (most negative), so that the planet's kinetic energy and speed take their maximum values. (b) At the planet's greatest separation from the Sun—its aphelion—the gravitational energy is maximum, and the speed is minimum.

□ □ □ □

8. In his 1798 experiment, Cavendish was said to have "weighed the Earth." Explain this statement.

Answer The Earth creates a gravitational field at its surface according to $g = GM_E / R_E^2$. The factors g and R_E were known, so as soon as Cavendish measured G, he could compute the mass of the Earth.

□ □ □ □

SOLUTIONS TO SELECTED END-OF-CHAPTER PROBLEMS

5. In introductory physics laboratories, a typical Cavendish balance for measuring the gravitational constant G uses lead spheres with masses of 1.50 kg and 15.0 g whose centers are separated by about 4.50 cm. Calculate the gravitational force between these spheres, treating each as a particle located at the sphere's center.

Solution

Conceptualize: "Universal" gravitation means that the force is exerted by one ordinary-size object on another, as well as by astronomical objects. It is fun to have a bunch of students cluster up on one side of the apparatus to deflect it.

Categorize: This is a direct application of the equation expressing Newton's law of gravitation.

Analyze: $F = \dfrac{Gm_1 m_2}{r^2} = \dfrac{(6.67 \times 10^{-11} \text{ N} \cdot \text{m}^2/\text{kg})(1.50 \text{ kg})(0.015 \, 0 \text{ kg})}{(4.50 \times 10^{-2} \text{ m})^2}$

$$F = 7.41 \times 10^{-10} \text{ N} = 741 \text{ pN} \qquad \text{toward the other sphere.} \qquad \blacksquare$$

Finalize: This is the force that each sphere has exerted on it by the other. In the space literally "between the spheres," no force acts because no object is there to feel a force.

---

10. When a falling meteoroid is at a distance above the Earth's surface of 3.00 times the Earth's radius, what is its acceleration due to the Earth's gravitation?

Solution

Conceptualize: The motion of the meteoroid, or its mass, makes no difference. The gravitational field, which is the acceleration, will be smaller than the surface value we are used to because of the greater distance.

Categorize: The Earth exerts its force on the meteoroid according to $mg = GmM_E/r^2$ so we see that the gravitational acceleration, $g = GM_E/r^2$, follows an inverse-square law. At the surface, a distance of one Earth-radius (R_E) from the center, it is…

Analyze: …9.80 m/s². At an altitude $3.00R_E$ above the surface (at distance $4.00R_E$ from the center), the gravitational acceleration will be $4.00^2 = 16.0$ times smaller:

$$g = \dfrac{GM_E}{(4.00R_E)^2} = \dfrac{GM_E}{16.0R_E^2} = \dfrac{9.80 \text{ m/s}^2}{16.0} = 0.612 \text{ m/s}^2 \text{ down} \qquad \blacksquare$$

Finalize: We could have substituted the universal gravitational constant, the mass of the Earth, and the planet's radius into $GM_E/(16R_E^2)$ and obtained the same result. The proportional reasoning is simpler to calculate and to think about.

---

12. The free-fall acceleration on the surface of the Moon is about one-sixth that on the surface of the Earth. The radius of the Moon is about $0.250R_E$ (R_E = Earth's radius = 6.37×10^6 m). Find the ratio of their average densities, ρ_{Moon}/ρ_{Earth}.

Solution

Conceptualize: The Moon is made of rocky materials quite similar to the Earth, but we expect the Moon to have compressed itself less at its formation and so to have lower density.

Categorize: Making a fraction of the equations $g = \dfrac{GM}{R^2}$ expressing how Earth and Moon make surface gravitational fields will give us the answer by proportional reasoning.

Analyze: The gravitational field at the surface of the Earth or Moon is given by $g = \dfrac{GM}{R^2}$
The expression for density is

$$\rho = \frac{M}{V} = \frac{M}{\frac{4}{3}\pi R^3}$$

so

$$M = \tfrac{4}{3}\pi\rho R^3$$

and

$$g = \frac{G\frac{4}{3}\pi\rho R^3}{R^2} = G\tfrac{4}{3}\pi\rho R$$

Noting that this equation applies to both the Moon and the Earth, and dividing the two equations,

$$\frac{g_M}{g_E} = \frac{G\frac{4}{3}\pi\rho_M R_M}{G\frac{4}{3}\pi\rho_E R_E} = \frac{\rho_M R_M}{\rho_E R_E}$$

Substituting for the fractions,

$$\frac{1}{6} = \frac{\rho_M}{\rho_E}\left(\frac{1}{4}\right) \qquad \text{and} \qquad \frac{\rho_M}{\rho_E} = \frac{4}{6} = \frac{2}{3} \qquad ∎$$

Finalize: Our guess was right. With our method of proportional reasoning, we did not need to use the value of the Earth's radius given in the problem. In general, measurements of size and of exterior gravitational acceleration yield information about the masses and densities of objects smaller than moons and larger than galaxy clusters, and everything in between.

---

14. Io, a satellite of Jupiter, has an orbital period of 1.77 days and an orbital radius of 4.22×10^5 km. From these data, determine the mass of Jupiter.

Solution

Conceptualize: We estimate a hundred times the mass of the Earth.

Categorize: We use the particle under a net force and the particle in uniform circular motion models, as well as the universal law of gravitation.

Analyze: The gravitational force exerted by Jupiter on Io causes the centripetal acceleration of Io. A force diagram of the satellite would show one downward arrow.

$$\sum F_{on\ Io} = M_{Io}a: \qquad \frac{GM_J M_{Io}}{r^2} = \frac{M_{Io}v^2}{r} = \frac{M_{Io}}{r}\left(\frac{2\pi r}{T}\right)^2 = \frac{4\pi^2 r M_{Io}}{T^2}$$

Thus the mass of Io divides out and we have,

$$M_J = \frac{4\pi^2 r^3}{GT^2} = \frac{4\pi^2 \left(4.22 \times 10^8 \,\text{m}\right)^3}{\left(6.67 \times 10^{-11} \,\text{N} \cdot \text{m}^2/\text{kg}^2\right)(1.77\text{d})^2} \left(\frac{1\,\text{d}}{86\,400\,\text{s}}\right)^2 \left(\frac{\text{N} \cdot \text{s}^2}{\text{kg} \cdot \text{m}}\right) \quad \text{and}$$

$$M_J = 1.90 \times 10^{27} \,\text{kg}$$

■

Finalize: Three hundred times the mass of the Earth! Whenever an astronomer can observe a satellite in orbit, she can calculate the mass of the central object.

---

17. Plaskett's binary system consists of two stars that revolve in a circular orbit about a center of mass midway between them. This statement implies that the masses of the two stars are equal (Fig. P13.17). Assume the orbital speed of each star is $|\vec{v}| = 220$ km/s and the orbital period of each is 14.4 days. Find the mass M of each star. (For comparison, the mass of our Sun is 1.99×10^{30} kg.)

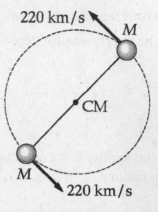

Figure P13.17

Solution

Conceptualize: From the given data, it is difficult to estimate a reasonable answer to this problem without working through the details to actually solve it. A reasonable guess might be that each star has a mass equal to or larger than our Sun because fourteen days is short compared to the periods of all the Sun's planets.

Categorize: The only force acting on each star is the central gravitational force of attraction which results in a centripetal acceleration. When we solve Newton's second law, we can find the unknown mass in terms of the variables given in the problem.

Analyze: Applying Newton's second law $\Sigma F = ma$ yields $F_g = ma_c$ for each star:

$$\frac{GMM}{(2r)^2} = \frac{Mv^2}{r} \qquad \text{so} \qquad M = \frac{4v^2 r}{G}$$

We can write r in terms of the period T by considering the time interval and the distance covered in one complete cycle. The distance traveled in one orbit is the circumference of the stars' common orbit, so $2\pi r = vT$.

By substitution

$$M = \frac{4v^2 r}{G} = \left(\frac{4v^2}{G}\right)\left(\frac{vT}{2\pi}\right) = \frac{2v^3 T}{\pi G}$$

and numerically

$$M = \frac{2\left(220 \times 10^3 \,\text{m/s}\right)^3 (14.4\,\text{d})(86\,400\,\text{s/d})}{\pi\left(6.67 \times 10^{-11} \,\text{N} \cdot \text{m}^2/\text{kg}^2\right)} = 1.26 \times 10^{32} \,\text{kg}$$

■

Finalize: The mass of each star is about 63 solar masses, much more than our initial guess! A quick check in an astronomy book reveals that stars over 8 solar masses are considered to be heavyweight stars. They are rather rare, and astronomers estimate that stable stars over about 100 solar masses cannot form. So these two stars are exceptionally massive. We know the masses of stars mainly from measurements of their orbits in binary star systems, as this problem exemplifies.

21. A synchronous satellite, which always remains above the same point on a planet's equator, is put in orbit around Jupiter to study that planet's famous red spot. Jupiter rotates once every 9.84 h. Use the data of Table 13.2 to find the altitude of the satellite above the surface of the planet.

Solution

Conceptualize: The "surface" of Jupiter that we see is not solid or liquid, but consists of the tops of atmospheric clouds. Arthur C. Clarke was the first to write about a synchronous satellite and its usefulness in communications, in a science fiction story. We expect the altitude to be more than ten times the radius of the planet.

Categorize: We will again derive a version of Kepler's third law by combining Newton's second law for a particle in uniform circular motion with Newton's law of gravitation.

Analyze: Jupiter's rotational period, in seconds, is

$$T = (9.84 \text{ h})(3\ 600 \text{ s/h}) = 35\ 424 \text{ s}$$

Jupiter's gravitational force is the force on the satellite causing its centripetal acceleration:

$$\frac{GM_sM_J}{r^2} = \frac{M_s v^2}{r} = \frac{M_s}{r}\left(\frac{2\pi r}{T}\right)^2 \quad \text{so} \quad GM_JT^2 = 4\pi^2 r^3$$

and the radius of the satellite's orbit is

$$r = \sqrt[3]{\frac{GM_JT^2}{4\pi^2}} = \left(\frac{(6.67 \times 10^{-11} \text{ N} \cdot \text{m}^2)(1.90 \times 10^{27} \text{ kg})(3.54 \times 10^4 \text{ s})^2}{\text{kg}^2 \qquad 4\pi^2}\right)^{1/3}$$

$$r = 15.9 \times 10^7 \text{ m}$$

Thus, we can calculate the altitude of the synchronous satellite to be

$$\text{Altitude} = (15.91 \times 10^7 \text{ m}) - (6.99 \times 10^7 \text{ m}) = 8.92 \times 10^7 \text{ m} \quad \blacksquare$$

Finalize: The giant planet spins so fast that the synchronous satellite is only a bit more than one radius above the cloudtops.

26. (a) Compute the vector gravitational field at a point P on the perpendicular bisector of the line joining two objects of equal mass separated by a distance $2a$ as shown in Figure P13.26. (b) Explain physically why the field should approach zero as $r \rightarrow 0$. (c) Prove mathematically that the answer to part (a) behaves in this way. (d) Explain physically why the magnitude of the field should approach $2GM/r^2$ as $r \rightarrow \infty$. (e) Prove mathematically that the answer to part (a) behaves correctly in this limit.

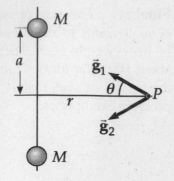

Solution

Figure P13.26

Conceptualize: Each mass by itself creates a field proportional to its mass and to the universal gravitational constant, and getting weak at large radius, so it is reasonable that both masses together do so.

Categorize: We must add the vector fields created by each mass.

Analyze:

(a)   In equation form, $\vec{g} = \vec{g}_1 + \vec{g}_2$ where

$$\vec{g}_1 = \frac{GM}{r^2 + a^2} \quad \text{to the left and upward at } \theta$$

and $\vec{g}_2 = \dfrac{GM}{r^2 + a^2}$ to the left and downward at θ

Therefore,

$$\vec{g} = \frac{GM}{r^2 + a^2} \cos\theta \left(-\hat{\mathbf{i}}\right) + \frac{GM}{r^2 + a^2} \sin\theta \left(\hat{\mathbf{j}}\right) + \frac{GM}{r^2 + a^2} \cos\theta \left(-\hat{\mathbf{i}}\right) + \frac{GM}{r^2 + a^2} \sin\theta \left(-\hat{\mathbf{j}}\right)$$

$$\vec{g} = \frac{2GM}{r^2 + a^2} \frac{r}{\sqrt{r^2 + a^2}} \left(-\hat{\mathbf{i}}\right) + 0\hat{\mathbf{j}} = \frac{-2GMr}{\left(r^2 + a^2\right)^{3/2}} \hat{\mathbf{i}}$$ ∎

(b)   At $r = 0$, the fields of the two objects are equal in magnitude and opposite in direction, to add to zero.

(c)   The expression in part (a) becomes $-2GM0a^{-3}(\hat{\mathbf{i}}) = 0$, to agree with our statement in (b).

(d)   When r is much greater than a, the fact that the two masses are separate is unimportant. They create a total field like that of a single mass $2M$. At a point on the positive x axis, this will be $G2Mr^{-2}(-\hat{\mathbf{i}})$.

(e)   When r is much greater than a, we have $r^2 + a^2 \approx r^2$, so the result from part (a) becomes $-2GMrr^{-3}(\hat{\mathbf{i}})$, to agree with the expression in (d).

Finalize: Gravitational fields add as vectors because forces add as vectors. The conceptual arguments in parts (b) and (d) are at least as important and perfectly complementary to the mathematical arguments in (c) and (e).

31. After our Sun exhausts its nuclear fuel, its ultimate fate will be to collapse to a *white dwarf* state. In this state, it would have approximately the same mass as it has now, but its radius would be equal to the radius of the Earth. Calculate (a) the average density of the white dwarf, (b) the surface free-fall acceleration, and (c) the gravitational potential energy associated with a 1.00-kg object at the surface of the white dwarf.

Solution

Conceptualize: Its density will be huge, tons per teaspoonful, and its surface gravitational field strong enough to compress atoms.

Categorize: We use the known mass of the Sun and radius of the Earth with the definition of density and the universal law of gravitation, specialized to a spherical object creating a surface field.

Analyze:

(a) The definition of density gives

$$\rho = \frac{M_S}{V} = \frac{M_S}{\frac{4}{3}\pi R_E^3} = \frac{1.99 \times 10^{30} \text{ kg}}{\frac{4}{3}\pi\left(6.37 \times 10^6 \text{ m}\right)^3} = 1.84 \times 10^9 \text{ kg/m}^3 \quad \blacksquare$$

This is on the order of one million times the density of concrete!

(b) For an object of mass m on its surface, $mg = GM_S m/R_E^2$. Thus,

$$g = \frac{GM_S}{R_E^2} = \frac{-\left(6.67 \times 10^{-11} \text{ N} \cdot \text{m}^2/\text{kg}^2\right)\left(1.99 \times 10^{30} \text{ kg}\right)}{\left(6.37 \times 10^6 \text{ m}\right)^2} = 3.27 \times 10^6 \text{ m/s}^2 \quad \blacksquare$$

This acceleration is on the order of one million times more than g_{Earth}.

(c) Relative to $U_g = 0$ at infinity, the potential energy of the kilogram at the surface of the white dwarf is

$$U_g = \frac{-GM_S m}{R_E} = \frac{-\left(6.67 \times 10^{-11} \text{ N} \cdot \text{m}^2\right)\left(1.99 \times 10^{30} \text{ kg}\right)\left(1 \text{ kg}\right)}{\text{kg}^2\left(6.37 \times 10^6 \text{ m}\right)} = -2.08 \times 10^{13} \text{ J} \quad \blacksquare$$

Such a large potential energy could yield a big gain in kinetic energy with even small changes in height. For example, dropping the 1.00-kg object from 1.00 m would result in a final velocity of 2 560 m/s.

Finalize: Light from a white dwarf is not easy to observe. The object is very hot from its formation, but its surface area is so small that its output power is small. Still more highly compressed are neutron stars and black holes, the other compact objects that dead stars can become.

---

34. A space probe is fired as a projectile from the Earth's surface with an initial speed of 2.00×10^4 m/s. What will its speed be when it is very far from the Earth? Ignore atmospheric friction and the rotation of the Earth.

Solution

Conceptualize: Gravity will always be pulling backward on the projectile to slow it down, but more and more weakly as distance increases. Thus its final speed can approach a nonzero limit.

Categorize: Its acceleration is not constant, but we can still use the energy model of an isolated system, …

Analyze: …between an initial point just after the payload is fired off at the Earth's surface, and a final point when it is coasting along far away. The only forces acting within the Earth-projectile system are gravitational forces and no force is exerted on the system from outside, so system energy is conserved:

$$K_i + U_i = K_f + U_f$$

$$\tfrac{1}{2} m_{probe} v_i^2 - \frac{GM_{Earth} m_{probe}}{r_i} = \tfrac{1}{2} m_{probe} v_f^2 - \frac{GM_E m_{probe}}{r_f}$$

The reciprocal of the final distance is negligible compared with the reciprocal of the original distance from the Earth's center.

$$\tfrac{1}{2} v_i^2 - \frac{GM_{Earth}}{R_{Earth}} = \tfrac{1}{2} v_f^2 - 0$$

We solve for the final speed, and substitute values tabulated on the endpapers:

$$v_f^2 = v_i^2 - \frac{2GM_{Earth}}{R_{Earth}}$$

$$v_f^2 = (2.00 \times 10^4 \text{ m/s})^2 - \frac{2\,(6.67 \times 10^{-11} \text{ N} \cdot \text{m}^2/\text{kg}^2)(5.97 \times 10^{24} \text{ kg})}{6.37 \times 10^6 \text{ m}}$$

$$v_f^2 = 2.75 \times 10^8 \text{ m}^2/\text{s}^2 \quad \text{and} \quad v_f = 1.66 \times 10^4 \text{ m/s} \quad \blacksquare$$

Finalize: For an object barely escaping, the Earth's gravity reduces the speed from 11.2 km/s (escape speed) to zero. The same gravitational field produces a smaller change in the speed of the object in this problem, from 20.0 km/s to 16.6 km/s. The object in this problem spends less time in close to the planet's surface where the backward acceleration is large.

---

36. A "treetop satellite" moves in a circular orbit just above the surface of a planet, assumed to offer no air resistance. Show that its orbital speed v and the escape speed from the planet are related by the expression $v_{esc} = \sqrt{2}\, v$.

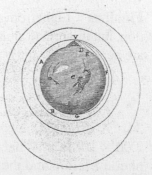

Solution

Conceptualize: We display the historic diagram of the motion of an artificial satellite, as Isaac Newton drew it in his *System of* *the World,* published in 1687. He thought of the satellite as a cannonball fired horizontally from a mountaintop, at the speed to be found in this problem.

Categorize: We use the particle in uniform circular motion model. To find the escape speed for comparison, we use the energy version of the isolated system model.

Analyze: Call M the mass of the planet and R its radius. For the orbiting treetop satellite,

$$\sum F = ma \quad \text{becomes} \quad \frac{GMm}{R^2} = \frac{mv^2}{R} \quad \text{or} \quad v = \sqrt{\frac{GM}{R}}$$

If the object is launched with escape velocity, applying conservation of energy to the object-Earth system gives

$$\tfrac{1}{2}mv^2_{esc} - \frac{GMm}{R} = 0 \quad \text{or} \quad v_{esc} = \sqrt{\frac{2GM}{R}}$$

Thus, by substitution $\quad v_{esc} = \sqrt{2}\,v$ ■

Finalize: The depth of the Earth's atmosphere is so small compared to the planet's radius, that the treetop satellite is a fair model for the Space Shuttle and other satellites in "low orbits." It is shown as VBA in Newton's diagram. Fired horizontally from a mountaintop with any speed smaller than the treetop-satellite speed v, a projectile will fall to the ground before going as much as halfway around the globe, as shown by VD, VE, and VFG in the diagram. With a speed between v and v_{esc}, the object will be in Earth orbit, moving in an ellipse of smaller or larger major axis. If it starts with a speed equal to v_{esc} it moves in a parabola, with no limit to its distance. Starting with a speed greater than v_{esc}, the projectile moves in a hyperbola, with no limit to its distance.

---

51. Two hypothetical planets of masses m_1 and m_2 and radii r_1 and r_2, respectively, are nearly at rest when they are an infinite distance apart. Because of their gravitational attraction, they head toward each other on a collision course. (a) When their center-to-center separation is d, find expressions for the speed of each planet and for their relative speed. (b) Find the kinetic energy of each planet just before they collide, taking $m_1 = 2.00 \times 10^{24}$ kg, $m_2 = 8.00 \times 10^{24}$ kg, $r_1 = 3.00 \times 10^6$ m, and $r_2 = 5.00 \times 10^6$ m. *Note:* Both energy and momentum of the isolated two-planet system are constant.

Solution

Conceptualize: The smaller-mass object will move faster. The process considered here is a time reversal of the outward flight of a projectile fired off with escape speed, so we expect $Gm_1m_2/(r_1 + r_2)$ to set the scale of the energy of each.

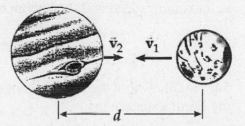

Categorize: We use both the energy version and the momentum version of the isolated system model.

Analyze:

(a)   At infinite separation, $U = 0$; and at rest, $K = 0$. Since energy of the two-planet system is conserved, we have

$$0 = \tfrac{1}{2}m_1v_1^2 + \tfrac{1}{2}m_2v_2^2 - \frac{Gm_1m_2}{d} \qquad \textbf{[1]}$$

The initial momentum of the system is zero and momentum is conserved.

As an equation,

$$0 = m_1v_1 - m_2v_2 \qquad \textbf{[2]}$$

We combine Equations [1] and [2], substituting to eliminate $v_2 = m_1v_1/m_2$, to find

$$v_1 = m_2\sqrt{\frac{2G}{d(m_1 + m_2)}} \qquad \text{and then} \qquad v_2 = m_1\sqrt{\frac{2G}{d(m_1 + m_2)}} \qquad \blacksquare$$

The relative velocity is

$$v_r = v_1 - (-v_2) = \sqrt{\frac{2G(m_1 + m_2)}{d}} \qquad \blacksquare$$

(b)   Substitute the given numerical values into the equation found for v_1 and v_2 in part (a) to find $v_1 = 1.03 \times 10^4$ m/s and $v_2 = 2.58 \times 10^3$ m/s.

Therefore,

$$K_1 = \tfrac{1}{2}m_1v_1^2 = 1.07 \times 10^{32} \text{ J} \qquad \text{and} \qquad K_2 = \tfrac{1}{2}m_2v_2^2 = 2.67 \times 10^{31} \text{ J} \qquad \blacksquare$$

Finalize: With greater speed, the smaller-mass object has more kinetic energy. As a check we can add the two kinetic energies to get 1.33×10^{33} J, to agree with

$$Gm_1m_2/(r_1 + r_2) = 6.67 \times 10^{-11}(2.00 \times 10^{24})(8.00 \times 10^{24})/(3.00 \times 10^6 + 5.00 \times 10^6)$$

$$= 1.33 \times 10^{33} \text{ J}$$

The situation in this problem is unrealistic. Two far-separated astronomical objects will essentially always be moving relative to each other fast enough that they will not collide when they fall together. Their motions will not be along the same line. They will move in parabolas, each with its focus at their center of mass, to do hairpin turns around each other at maximum speed and then zoom back out to infinite distance.

---

54.   (a) Show that the rate of change of the free-fall acceleration with vertical position near the Earth's surface is

$$\frac{dg}{dr} = -\frac{2GM_E}{R_E^3}$$

This rate of change with position is called a *gradient*. (b) Assuming h is small in comparison to the radius of the Earth, show that the difference in free-fall acceleration between two points separated by vertical distance h is

$$|\Delta g| = \frac{2GM_E h}{R_E^3}$$

(c) Evaluate this difference for $h = 6.00$ m, a typical height for a two-story building.

Solution

Conceptualize: The answer to part (a) represents the rate of decrease in the gravitational acceleration with respect to increasing altitude—you can visualize it as the amount of change in g for each meter of height increase. Then parts (b) and (c) are easier to visualize for many people. They are about the amounts by which the gravitational acceleration decreases for particular upward steps. We expect the answer for part (c) to be a millimeter per second squared or less. Previously in the course we have not had to use different values for g depending on what floor of the building we were on.

Categorize: We will use Newton's law of universal gravitation and the calculus idea of taking a derivative (also introduced by Newton). The differentiation is not with respect to time, but with respect to distance.

Analyze:

(a) The free-fall acceleration produced by the Earth is $g = \dfrac{GM_E}{r^2} = GM_E r^{-2}$, directed downward.

When our elevation changes by dy, our distance from the center of the Earth changes by that same distance, which we can represent by dr. So the rate of change of the gravitational field with vertical position is

$$\frac{dg}{dr} = GM_E(-2)r^{-3} = -2GM_E r^{-3}$$

The minus sign indicates that g decreases with increasing height.

At the Earth's surface, $\dfrac{dg}{dr} = -\dfrac{2GM_E}{R_E^3}$ ∎

(b) For small differences, we can approximate dg/dr by $\Delta g/\Delta r$, like this:

$$\frac{|\Delta g|}{\Delta r} = \frac{|\Delta g|}{h} = \frac{2GM_E}{R_E^3} \qquad \text{Thus,} \qquad |\Delta g| = \frac{2GM_E h}{R_E^3}$$ ∎

(c) On the roof of the building, g is this much smaller than on the ground below:

$$|\Delta g| = \frac{2(6.67 \times 10^{-11} \text{ N} \cdot \text{m}^2/\text{kg}^2)(5.97 \times 10^{24} \text{ kg})(6.00 \text{ m})}{(6.37 \times 10^6 \text{ m})^3} = 1.85 \times 10^{-5} \text{ m/s}^2$$ ∎

Finalize: Could you have fun and earn tips by telling tourists visiting a tall tower how much weight each loses in ascending to the top? If they take the elevator, make sure to tell them that they gain it all back again upon returning to ground level.

---

59.   The maximum distance from the Earth to the Sun (at aphelion) is 1.521×10^{11} m, and the distance of closest approach (at perihelion) is 1.471×10^{11} m. The Earth's orbital speed at perihelion is 3.027×10^4 m/s. Determine (a) the Earth's orbital speed at aphelion, and the kinetic and potential energies of the Earth-Sun system (b) at perihelion, and (c) at aphelion. (d) Is the total energy of the system constant? Explain. Ignore the effect of the Moon and other planets.

Solution

Conceptualize: The Sun-planet distance is about three percent greater at aphelion, so the speed at aphelion will be about three percent less than at perihelion, according to Kepler's second law. The amounts of energy will be huge numbers of joules, with the kinetic energy about six percent larger at perihelion than at aphelion. The gravitational energies, negative and about twice as large in magnitude as the average kinetic energy, will change by only about three percent, so that the total of kinetic and gravitational energies at both points can agree with each other.

Categorize: We use conservation of angular momentum for the Earth-Sun system to do part (a). We use the equation for kinetic energy and the general expression $-GMm/r$ for gravitational energy to do the succeeding parts.

Analyze:

(a)   The net torque exerted on the Earth is zero. Therefore, the angular momentum of the Earth is constant.

$mr_a v_a = mr_p v_p$ and

$$v_a = v_p \left(\frac{r_p}{r_a} \right) = 3.027 \times 10^4 \text{ m/s} \left(\frac{1.471}{1.521} \right) = 2.93 \times 10^4 \text{ m/s} \quad \blacksquare$$

(b)   At the maximum-speed perihelion point, the Earth's kinetic energy is

$$K_p = \tfrac{1}{2}mv_p^2 = \tfrac{1}{2}\left(5.97 \times 10^{24}\right)\left(3.027 \times 10^4\right)^2 = 2.74 \times 10^{33} \text{ J} \quad \blacksquare$$

And the gravitational energy of the Earth-Sun system is

$$U_p = -\frac{GmM}{r_p} = -\frac{\left(6.673 \times 10^{-11}\right)\left(5.97 \times 10^{24}\right)\left(1.99 \times 10^{30}\right)}{1.471 \times 10^{11}} = -5.39 \times 10^{33} \text{ J} \quad \blacksquare$$

(c)   Using the same method as in part (b), at aphelion we find $K_a = 2.56 \times 10^{33}$ J and $U_a = -5.21 \times 10^{33}$ J. $\quad \blacksquare$

(d) We add up the total mechanical energy to compare, finding that at perihelion $K_p + U_p = -2.65 \times 10^{33}$ J and at aphelion $K_a + U_a = -2.65 \times 10^{33}$ J. These values agree within the precision of our calculation. Physically, they must agree because gravity is a conservative force. ∎

Finalize: It is a highlight of an intermediate mechanics course to prove that a planet moves in a closed ellipse when the force exerted on it is described by an inverse-square law. In an elementary physics course we state the theorem, see that it corresponds to observations, and here see that it is compatible with conservation of system energy.

---

64. Two stars of masses M and m, separated by a distance d, revolve in circular orbits about their center of mass (Fig. P13.64). Show that each star has a period given by

$$T^2 = \frac{4\pi^2 d^3}{G(M+m)}$$

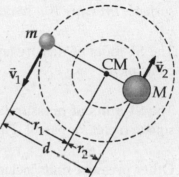

Figure P13.64

Solution

Conceptualize: The textbook derives an equation like this to represent Kepler's third law for a low-mass satellite in circular orbit around a massive central object. Here we prove that the generalization just contains the total mass of the system.

Categorize: We use the particle in uniform circular motion model and the universal law of gravitation.

Analyze: For the star of mass M and orbital radius r_2,

$$\sum F = ma \quad \text{gives} \quad \frac{GMm}{d^2} = \frac{Mv_2^2}{r_2} = \frac{M}{r_2}\left(\frac{2\pi r_2}{T}\right)^2$$

For the star of mass m,

$$\sum F = ma \quad \text{gives} \quad \frac{GMm}{d^2} = \frac{mv_1^2}{r_1} = \frac{m}{r_1}\left(\frac{2\pi r_1}{T}\right)^2$$

Clearing fractions, we then obtain the simultaneous equations

$$GmT^2 = 4\pi^2 d^2 r_2 \quad \text{and} \quad GMT^2 = 4\pi^2 d^2 r_1$$

Adding, we find $G(M+m)T^2 = 4\pi^2 d^2 (r_1 + r_2) = 4\pi^2 d^3$

so we have

$$T^2 = \frac{4\pi^2 d^3}{G(M+m)} \quad \text{the result to be proven.} \quad ∎$$

Finalize: In many binary star systems the period and stellar separation can be measured to determine the total system mass. In some visual binary star systems T, d, r_1, and r_2 can all be measured, so the mass of each component can be computed.

14

Fluid Mechanics

EQUATIONS AND CONCEPTS

Pressure is defined as the magnitude of the normal force per unit area acting on a surface. *Pressure is a scalar quantity.*

$$P \equiv \frac{F}{A} \tag{14.1}$$

The **SI units of density** are kilograms per cubic meter.

$$1 \text{ g/cm}^3 = 1\,000 \text{ kg/m}^3$$

The **SI units of pressure** are newtons per square meter, or pascal (Pa).

$$1 \text{ Pa} \equiv 1 \text{ N/m}^2 \tag{14.3}$$

Other pressure units include: atmospheres.

$$1 \text{ atm} = 1.013 \times 10^5 \text{ Pa}$$

mm of mercury (Torr).

$$1 \text{ Torr} = 133.3 \text{ Pa}$$

pounds per sq. inch.

$$1 \text{ lb/in}^2 = 6.895 \times 10^3 \text{ Pa}$$

The **absolute pressure**, P, at a depth, h, below the surface of a liquid which is open to the atmosphere is greater than atmospheric pressure, P_0, by an amount which depends on the depth below the surface. *The pressure below the surface of a liquid has the same value at all points at a given depth and does not depend on the shape of the container.* In Eq 14.4:

$$P = P_0 + \rho g h \tag{14.4}$$

P = absolute pressure
P_0 = atmospheric pressure
$\rho g h$ = *gauge pressure*

$$\text{absolute pressure} = \text{atmospheric pressure} + \text{gauge pressure}$$

Pascal's law states that a change in the pressure applied to an enclosed fluid (liquid or gas) is transmitted undiminished to every point within the fluid and to the walls of the container.

Archimedes's principle states that when an object is partially or fully immersed in a fluid, the fluid exerts an upward buoyant force on the object. The magnitude of the buoyant force equals the *weight of the fluid displaced* by the object. The weight of the displaced fluid depends on the density and volume of the displaced fluid.

$$B = \rho_{\text{fluid}} g V_{\text{disp}} \qquad (14.5)$$

The **submerged fraction of a floating object** is equal to the ratio of the density of the object to that of the fluid. *In Equation 14.6, V_{disp} is the volume of the displaced fluid, and is therefore the volume of the submerged portion of the object.*

$$\frac{V_{\text{disp}}}{V_{\text{obj}}} = \frac{\rho_{\text{obj}}}{\rho_{\text{fluid}}} \qquad (14.6)$$

The **ideal fluid model** is based on the following four assumptions:

- **Nonviscous** –internal friction between adjacent fluid layers is negligible.

- **Incompressible**—the density of the fluid is constant throughout the fluid.

- **Steady flow**—the velocity at each point in the fluid remains constant.

- **Irrotational** (nonturbulent)—there are no eddy currents within the fluid; each element of the fluid has zero angular momentum about its center of mass.

The **equation of continuity for fluids** states that the flow rate (product of area and speed of flow) of an incompressible fluid (ρ = constant) is constant at every point along a pipe.

$$A_1 v_1 = A_2 v_2 = \text{constant} \qquad (14.7)$$

Bernouilli's equation states that the sum of pressure, kinetic energy per unit volume, and potential energy per unit volume remains constant in streamline flow of an ideal fluid. *The equation is a statement of the law of conservation of mechanical energy as applied to an ideal fluid.*

$$P + \tfrac{1}{2}\rho v^2 + \rho g y = \text{constant} \qquad (14.9)$$

REVIEW CHECKLIST

You should be able to:

- Calculate pressure below the surface of a liquid and determine the total force exerted on a surface due to hydrostatic pressure. (Section 14.2)

- Convert among the several different units commonly used to express pressure.

- Determine the buoyant force on a floating or submerged object and determine the fraction of a floating object that is below the surface of a fluid. (Section 14.4)

- Calculate the density of an object or a fluid using Archimedes's principle. (Section 14.4)

- State the simplifying assumptions of an ideal fluid moving with streamline flow. (Section 14.5)

- Make calculations using the equation of continuity and Bernoulli's equation. (Sections 14.5 and 14.6)

ANSWER TO AN OBJECTIVE QUESTION

6. A solid iron sphere and a solid lead sphere of the same size are each suspended by strings and are submerged in a tank of water. (Note that the density of lead is greater than that of iron.) Which of the following statements are valid? (Choose all correct statements.) (a) The buoyant force on each is the same. (b) The buoyant force on the lead sphere is greater than the buoyant force on the iron sphere because lead has the greater density. (c) The tension in the string supporting the lead sphere is greater than the tension in the string supporting the iron sphere. (d) The buoyant force on the iron sphere is greater than the buoyant force on the lead sphere because lead displaces more water. (e) None of those statements is true.

Answer (a) and (c). Statement (a) is true, with (b), (d), and (e) being false, because the spheres are the same size. The buoyant force on each is equal in magnitude to the weight of the water that would occupy that volume if the metal were not there. Statement (c) is true. The metals are more than twice the density of water, so the tension in the strings holding up the spheres must support more than half of the weight of each. For the denser lead sphere the string must support more weight than for the iron sphere.

ANSWERS TO SELECTED CONCEPTUAL QUESTIONS

2. Two thin-walled drinking glasses having equal base areas but different shapes, with very different cross-sectional areas above the base, are filled to the same level with water. According to the expression $P = P_0 + \rho g h$, the pressure is the same at the bottom of both glasses. In view of this equality, why does one glass weigh more than the other?

Answer For the cylindrical container shown, the weight of the water is equal to the gauge pressure at the bottom multiplied by the area of the bottom. For the narrow-necked bottle, on the other hand, the weight of the fluid is much less than *PA*, the bottom pressure times the area. The water does exert the large *PA* force downward on the bottom of the bottle, but the water also exerts an upward force nearly as large on the ring-shaped horizontal area surrounding the neck. The net vector force that the water exerts on the glass is equal to the small weight of the water.

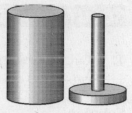

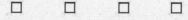

4. A fish rests on the bottom of a bucket of water while the bucket is being weighed on a scale. When the fish begins to swim around, does the scale reading change? Explain your answer.

Answer In either case, the scale is supporting the container, the water, and the fish. Therefore the weight remains the same. The reading on the scale, however, can change if the net center of mass accelerates in the vertical direction, as when the fish jumps out of the water. In that case, the scale, which registers force, will also show an additional force associated with Newton's second law $\sum F = ma$.

12. Place two cans of soft drinks, one regular and one diet, in a container of water. You will find that the diet drink floats while the regular one sinks. Use Archimedes's principle to devise an explanation.

Answer A regular soda pop is sugar syrup. Its density is higher than the density of a diet soft drink, which is nearly pure water. The low-density air inside the can has a bigger effect than the thin aluminum shell, so the can of diet soda floats.

13. The water supply for a city is often provided from reservoirs built on high ground. Water flows from the reservoir, through pipes, and into your home when you turn the tap on your faucet. Why does water flow more rapidly out of a faucet on the first floor of a building than in an apartment on a higher floor?

Answer The water supplied to the building flows through a pipe connected to the water tower. Near the ground, the water pressure is greater because the pressure increases with increasing depth beneath the surface of the water. The penthouse apartment is not so far below the water surface; hence the water flow will not be as rapid as on a lower floor.

SOLUTIONS TO SELECTED END-OF-CHAPTER PROBLEMS

1. Calculate the mass of a solid gold rectangular bar that has dimensions of 4.50 cm × 11.0 cm × 26.0 cm.

Solution

Conceptualize: We guess some kilograms. This object is around the size of a brick, and gold is far more dense than clay.

Categorize: We use the definition of density and a bit of geometry.

Analyze: The definition of density $\rho = m/V$ is often most directly useful in the form $m = \rho V$.

$V = \ell w h$ so $m = \rho V = \rho \ell w h$

Thus $m = (19.3 \times 10^3 \text{ kg/m}^3)(4.50 \text{ cm})(11.0 \text{ cm})(26.0 \text{ cm})$

$\qquad = (19.3 \times 10^3 \text{ kg/m}^3)(1\,290 \text{ cm}^3)(1 \text{ m}^3/10^6 \text{ cm}^3) = 24.8 \text{ kg}$ ∎

Finalize: Notice the million cubic centimeters in one cubic meter. The density of gold is nineteen million grams per cubic meter.

———————————————

3. A 50.0-kg woman wearing high-heeled shoes is invited into a home in which the kitchen has vinyl floor covering. The heel on each shoe is circular and has a radius of 0.500 cm. (a) If the woman balances on one heel, what pressure does she exert on the floor? (b) Should the homeowner be concerned? Explain your answer.

Solution

Conceptualize: We expect a pressure much higher than atmospheric pressure, because the force is concentrated on a small area.

Categorize: We use the definition of pressure.

Analyze: (a) The area of the circular base of the heel is

$$\pi r^2 = \pi (0.500 \text{ cm})^2 \left(\frac{1 \text{ m}^2}{10\,000 \text{ cm}^2} \right) = 7.85 \times 10^{-5} \text{ m}^2$$

The force she exerts is equal to her weight,

$$mg = (50.0 \text{ kg})(9.80 \text{ m/s}^2) = 490 \text{ N}$$

Then,

$$P = \frac{F}{A} = \frac{490 \text{ N}}{7.85 \times 10^{-5} \text{ m}^2} = 6.24 \text{ MPa}$$ ∎

(b) This pressure is about sixty atmospheres. The homeowner should be concerned, because the heel can dent or even cut the floor covering.

Finalize: The pressure can dent linoleum but not concrete. Make sure you can visualize how ten thousand square centimeters make up one square meter.

---

7. The spring of the pressure gauge shown in Figure P14.7 has a force constant of 1 250 N/m, and the piston has a diameter of 1.20 cm. As the gauge is lowered into water in a lake, what change in depth causes the piston to move in by 0.750 cm?

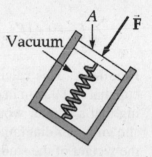

Figure P14.7

Solution

Conceptualize: We estimate a few meters. The spring is pretty stiff.

Categorize: A depth change h makes the water pressure increase by $\rho g h$.

Analyze: We assume that the cylinder contains no air to offer extra resistance to compression.

Then for the piston as a particle in equilibrium

$$F_{\text{spring}} = F_{\text{fluid}} \quad \text{or} \quad kx = \rho g h A \quad \text{and}$$

$$h = \frac{kx}{\rho g A} = \frac{(1\ 250\ \text{N/m})(0.007\ 50\ \text{m})}{\left(1\ 000\ \text{kg/m}^3\right)\left(9.80\ \text{m/s}^2\right)(0.006\ 00\ \text{m})^2 \pi} = 8.46\ \text{m} \quad \blacksquare$$

Finalize: This device can be used as an aneroid barometer for predicting weather and as an altimeter for an airplane or a skydiver, as well as a scuba diver's depth gauge.

---

9. What must be the contact area between a suction cup (completely evacuated) and a ceiling if the cup is to support the weight of an 80.0-kg student?

Solution

Conceptualize: The suction cups used by burglars seen in movies are about 10 cm in diameter, and it seems reasonable that one of these might be able to support the weight of an 80-kg student. The face area of a 10-cm cup is approximately $A = \pi r^2 \approx 3(0.05\ \text{m})^2 \approx 0.008\ \text{m}^2$

Categorize: "Suction" is not a new kind of force. Familiar forces hold the cup in equilibrium, one of which is the atmospheric pressure acting over the area of the cup. This problem is simply another application of Newton's second law, or the particle-in-equilibrium model.

Analyze: The vacuum between cup and ceiling exerts no force on either. The atmospheric pressure of the air below the cup pushes up on it with a force $(P_{atm})(A)$. If the cup barely supports the student's weight, then the normal force of the ceiling is approximately zero, and

$$\sum F_y = 0 + (P_{atm})(A) - mg = 0: \qquad A = \frac{mg}{P_{atm}} = \frac{784 \text{ N}}{1.013 \times 10^5 \text{ N/m}^2} = 7.74 \times 10^{-3} \text{m}^2 \quad \blacksquare$$

Finalize: This calculated area agrees with our prediction and corresponds to a suction cup that is 9.93 cm in diameter. Our 10-cm estimate was right on—a lucky guess, considering that a burglar would probably use at least two suction cups, not one. The suction cup shown in our diagram appears to be about 30 cm in diameter, plenty big enough to support the weight of the student.

Pressure is a scalar. In the expression for force $\vec{F} = P\vec{A}$, it is the area that can be defined as a vector, having a direction perpendicular to the surface on which the pressure acts. We will use the idea of vector area when we are thinking about electric and magnetic fields.

---

16. Blaise Pascal duplicated Torricelli's barometer using a red Bordeaux wine, of density 984 kg/m³, as the working liquid (Fig. P14.16). (a) What was the height h of the wine column for normal atmospheric pressure? (b) Would you expect the vacuum above the column to be as good as for mercury?

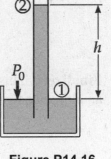

Figure P14.16

Solution

Conceptualize: Any liquid can be used for a barometer. Mercury is poisonous, but people use it because it makes a compact instrument less than a meter tall. The wine is less than one-tenth as dense, and the wine barometer will be more than ten times taller.

Categorize: A barometer is simple to think about if we assume there is a vacuum above the column, at point 2 in the diagram.

Analyze: We choose to start with Bernoulli's equation

$$P_1 + \tfrac{1}{2}\rho v_1^2 + \rho g y_1 = P_2 + \tfrac{1}{2}\rho v_2^2 + \rho g y_2$$

Take point 1 at the wine surface in the pan, where $P_1 = P_{atm}$, and point 2 at the wine surface up in the tube. Here we approximate $P_2 = 0$, although some alcohol and water will evaporate. The vacuum is not so good as with mercury. (This is the answer to part (b).) $\quad \blacksquare$

(a) Now, since the speed of the fluid at both points is zero, we can rearrange the terms to arrive at the simple equation about hydrostatic pressure varying with depth:

$$P_1 = P_2 + \rho g(y_2 - y_1)$$

$$y_2 - y_1 = \frac{P_1 - P_2}{\rho g} = \frac{1 \text{ atm} - 0}{(984 \text{ kg/m}^3)(9.80 \text{ m/s}^2)}$$

$$y_2 - y_1 = \frac{1.013 \times 10^5 \text{ N/m}^2}{9\,643 \text{ N/m}^3} = 10.5 \text{ m} \qquad \blacksquare$$

Finalize: A water barometer in a stairway of a three-story building is a nice display. Boil the water first, or else a lot of dissolved oxygen or carbon dioxide may come bubbling out. Red wine makes the fluid level easier to see.

---

23. A table-tennis ball has a diameter of 3.80 cm and average density of 0.084 0 g/cm³. What force is required to hold it completely submerged under water?

Solution

Conceptualize: According to Archimedes's Principle, the buoyant force acting on the submerged ball will be equal to the weight of the water the ball displaces. The ball has a volume of about 30 cm³, so the weight of this water is approximately:

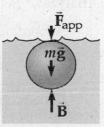

$$B = F_g = \rho V g \approx (1 \text{ g/cm}^3)(30 \text{ cm}^3)(10 \text{ m/s}^2) = 0.3 \text{ N}$$

Since the ball is much less dense than the water, its weight is almost negligible and the downward force that must be applied will be approximately equal to this buoyant force.

Categorize: Apply Newton's 2nd law to find the applied force on a particle in equilibrium.

Analyze: At equilibrium, $\sum F = 0$ or $-F_{app} - mg + B = 0$

where the buoyant force is $B = \rho_w V g$ and $\rho_w = 1\,000 \text{ kg/m}^3$

The applied force is then $F_{app} = \rho_w V g - mg$

Using $m = \rho_{ball} V$ to eliminate the unknown mass of the ball, this becomes

$$F_{app} = Vg(\rho_w - \rho_{ball}) = \tfrac{4}{3}\pi r^3 g(\rho_w - \rho_{ball})$$

$$F_{app} = \tfrac{4}{3}\pi (1.90 \times 10^{-2} \text{ m})^3(9.80 \text{ m/s}^2)(1\,000 \text{ kg/m}^3 - 84 \text{ kg/m}^3)$$

$$F_{app} = 0.258 \text{ N} \qquad \blacksquare$$

Finalize: The force is approximately what we expected. We did not need to know the depth of the ball. At greater depth, both the pressure at the top and at the bottom of the ball would be greater, but the buoyant force would be the same.

———————————————

27. A cube of wood having an edge dimension of 20.0 cm and a density of 650 kg/m³ floats on water. (a) What is the distance from the horizontal top surface of the cube to the water level? (b) What mass of lead should be placed on the cube so that the top of the cube will be just level with the water surface?

Solution

Conceptualize: We estimate that more than half of the wood is below the water line, so answer (a) should be about 8 cm. The volume of water to be displaced in (b) is (20 cm)³ = 8 000 cm³, with mass 8 000 g, so we estimate that the cube can support 3 000 g of lead.

Categorize: Knowing what each factor stands for in the buoyant force expression $\rho g V$ let's you use it both for a floating and for a submerged object, as one of the forces on a particle in equilibrium.

Analyze: Set h equal to the distance from the top of the cube to the water level.

(a) According to Archimedes's principle,

$$B = \rho_w V_{immersed} g = \left(1.00 \text{ g/cm}^3\right)\left[(20.0 \text{ cm})^2(20.0 \text{ cm} - h \text{ cm})\right]g$$

For equilibrium, the one upward force must be equal to the one downward force.

$$B = \text{weight of block} = mg = \rho_{wood} V_{wood} g = \left(0.650 \text{ g/cm}^3\right)(20.0 \text{ cm})^3 g$$

Setting these two expressions equal,

$$\left(0.650 \text{ g/cm}^3\right)(20.0 \text{ cm})^3 g = \left(1.00 \text{ g/cm}^3\right)(20.0 \text{ cm})^2(20.0 \text{ cm} - h \text{ cm})g$$

and $\quad$ 20.0 cm $- h =$ 20.0 (0.650) cm

so $\quad h =$ 20.0(1.00 cm $-$ 0.650 cm) $=$ 7.00 cm $\quad\blacksquare$

(b) Now for a net vertical force of zero on the loaded block we require

$$B = mg + Mg \qquad \text{where} \qquad M = \text{mass of lead}$$

$$\rho_{fluid} V g = \rho_{wood} V g + Mg$$

$$M = V(\rho_{fluid} - \rho_{wood})$$

$$M = (20.0 \text{ cm})^3\left(1.00 \text{ g/cm}^3 - 0.650 \text{ g/cm}^3\right)$$

$$= (20.0 \text{ cm})^3\left(0.350 \text{ g/cm}^3\right) = 2.80 \text{ kg} \qquad\blacksquare$$

Finalize: The most common mistake is to think that the density in $B = \rho g V$ is the density of the wood. A person who thinks this thinks that the buoyant object exerts a force on itself, which is nonsense.

---

29. A plastic sphere floats in water with 50.0% of its volume submerged. This same sphere floats in glycerin with 40.0% of its volume submerged. Determine the densities of (a) the glycerin and (b) the sphere.

Solution

Conceptualize: The sphere should have a density of about 0.5 g/cm³, half that of water. The sphere floats higher in the glycerin, so the glycerin should have somewhat higher density than water.

Categorize: The sphere is a particle in equilibrium. The buoyant force is not the only force on it.

Analyze: The forces on the ball are its weight $\quad F_g = mg = \rho_{\text{plastic}} V_{\text{ball}} g$

and the buoyant force of the liquid $\quad B = \rho_{\text{fluid}} V_{\text{immersed}} g$

When floating in water, $\sum F_y = 0$: $\quad -\left(\rho_{\text{plastic}}\right)\left(V_{\text{ball}}\right)g + \left(\rho_{\text{water}}\right)\left(0.500\, V_{\text{ball}}\right)g = 0$

then answer (b) is $\quad \rho_{\text{plastic}} = 0.500\, \rho_{\text{water}} = 500 \text{ kg/m}^3 \quad\blacksquare$

(a) When the ball is floating in glycerin, $\sum F_y = 0$

$$-\left(\rho_{\text{plastic}}\right)\left(V_{\text{ball}}\right)g + \left(\rho_{\text{glycerin}}\right)\left(0.400\, V_{\text{ball}}\right)g = 0$$

$$\rho_{\text{plastic}} = 0.400\, \rho_{\text{glycerin}}$$

$$\rho_{\text{glycerin}} = \frac{500 \text{ kg/m}^3}{0.400} = 1\,250 \text{ kg/m}^3 \quad\blacksquare$$

Finalize: This glycerin would sink in water. Calibrations on the side of the block could be used to gauge the density of the fluid in which it floats, by reading the one at the liquid surface—this is the idea of the hydrometer.

---

36. How many cubic meters of helium are required to lift a balloon with a 400-kg payload to a height of 8 000 m? Take $\rho_{\text{He}} = 0.179$ kg/m³. Assume the balloon maintains a constant volume and that the density of air decreases with the altitude z according to the expression $\rho_{\text{air}} = \rho_0 e^{-z/8\,000}$, where z is in meters and $\rho_0 = 1.20$ kg/m³ is the density of air at sea level.

Solution

Conceptualize: Many people would start by thinking about the volume of 400 kg of helium, which is $(400 \text{ kg})/(0.18 \text{ kg/m}^3) \approx 2\ 000 \text{ m}^3$. This is a misleading step based on thinking that the helium exerts some upward force on itself, when really the air around it exerts the buoyant force.

Categorize: The loaded balloon will be a particle in equilibrium at 8 km elevation, feeling an upward force exerted by the surrounding air.

Analyze: We assume that the mass of the balloon envelope is included in the 400 kg. We assume that the 400-kg total load is much denser than air and so has negligible volume compared to the helium. At $z = 8\ 000$ m, the density of air is

$$\rho_{\text{air}} = \rho_0 e^{-z/8\ 000} = (1.20 \text{ kg/m}^3)e^{-1} = (1.20 \text{ kg/m}^3)(0.368) = 0.441 \text{ kg/m}^3$$

Think of the balloon reaching equilibrium at this height. The weight of its payload is $Mg = (400 \text{ kg})(9.80 \text{ m/s}^2) = 3\ 920$ N. The weight of the helium in it is $mg = \rho_{\text{He}} Vg$

$$\sum F_y = 0 \qquad \text{becomes} \qquad +\rho_{\text{air}} Vg - Mg - \rho_{\text{He}} Vg = 0$$

Solving, $\qquad (\rho_{\text{air}} - \rho_{\text{He}})V = M$

and $\qquad V = \dfrac{M}{\rho_{\text{air}} - \rho_{\text{He}}} = \dfrac{400 \text{ kg}}{(0.441 - 0.179) \text{ kg/m}^3} = 1.52 \times 10^3 \text{ m}^3$ ■

Finalize: Hydrogen has about half the density of helium, but approximately the same volume of hydrogen would be required to support this load, because it is the air that does the lifting.

---

37.   A large storage tank, open at the top and filled with water, develops a small hole in its side at a point 16.0 m below the water level. The rate of flow from the leak is found to be 2.50×10^{-3} m³/min. Determine (a) the speed at which the water leaves the hole and (b) the diameter of the hole.

Solution

Conceptualize: An object falling freely 16 m from rest reaches speed $(2(9.8)16)^{1/2}$ m/s = 17.7 m/s. For ideal flow we can expect this same speed for the spurting water. A few millimeters seems reasonable for the size of a hole in the base of a dam that sprays out only a couple of liters every minute.

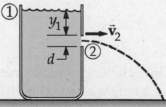

Categorize: We use Bernoulli's equation and the expression for volume flow rate contained in the equation of continuity.

Analyze: Take point 1 at the top surface and point 2 at the exiting stream. Assuming the top is open to the atmosphere, then $P_1 = P_0$. At point 2, by Newton's third law, the water must push on the air just as strongly as the air pushes on the water, so $P_2 = P_0$.

(a) Bernoulli says $P_1 + \frac{1}{2}\rho v_1^2 + \rho g y_1 = P_2 + \frac{1}{2}\rho v_2^2 + \rho g y_2$

Here $A_1 \gg A_2$, so $v_1 \ll v_2$

With the simplifications $v_1 \approx 0$ and $P_1 = P_2 = P_0$, we have

$$v_2 = \sqrt{2g(y_1 - y_2)} = \sqrt{2(9.80 \text{ m/s}^2)(16 \text{ m})} = 17.7 \text{ m/s}$$ ∎

(b) We identify the measured flow rate as $A_2 v_2 = 2.50 \times 10^{-3} \text{ m}^3/\text{min}$

$$\left(\frac{\pi d^2}{4}\right) v_2 = 2.50 \times 10^{-3} \text{ m}^3/\text{min}$$

and $d = \left(\dfrac{4\left(2.50 \times 10^{-3} \text{ m}^3/\text{min}\right)}{\pi v_2}\right)^{1/2} = \left(\dfrac{4\left(2.50 \times 10^{-3} \text{ m}^3/\text{min}\right)}{\pi (17.7 \text{ m/s})(60 \text{ s}/1 \text{ min})}\right)^{1/2}$

Thus, $d = 1.73 \times 10^{-3} \text{ m} = 1.73 \text{ mm}$ ∎

Finalize: Our estimates were good. Some people think that the pressure of the exiting stream is higher than atmospheric pressure, because it is so deep below the free water surface. Some people have heard of a "Bernoulli effect" and think that the pressure is lower than 1 atm because the water moves so fast. Really it is the surrounding air that exerts a set pressure and Newton's third law is the dynamic principle determining the water pressure.

---

51. A hypodermic syringe contains a medicine with the density of water (Figure P14.51). The barrel of the syringe has a cross-sectional area $A = 2.50 \times 10^{-5} \text{ m}^2$, and the needle has a cross-sectional area $a = 1.00 \times 10^{-8} \text{ m}^2$. In the absence of a force on the plunger, the pressure everywhere is 1.00 atm. A force $\vec{F}$ of magnitude 2.00 N acts on the plunger, making medicine squirt horizontally from the needle. Determine the speed of the medicine as it leaves the needle's tip.

Solution

Conceptualize: Ideal flow is frictionless. We might estimate a speed on the order of 10 m/s.

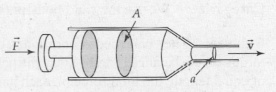

Categorize: In the barrel or reservoir, the speed is very low and the pressure higher. In the needle, the pressure is down to one atmosphere and the speed is much higher. The conditions in the two locations will be related by the equation of continuity and Bernoulli's equation.

Figure P14.51

Analyze: In the reservoir, the gauge pressure is

$$\Delta P = \frac{2.00 \text{ N}}{2.50 \times 10^{-5} \text{ m}^2} = 8.00 \times 10^4 \text{ Pa}$$

From the equation of continuity we have $A_1 v_1 = A_2 v_2$

$$\left(2.50 \times 10^{-5} \text{ m}^2\right) v_1 = \left(1.00 \times 10^{-8} \text{ m}^2\right) v_2 \quad \text{so} \quad v_1 = \left(4.00 \times 10^{-4}\right) v_2$$

Thus, v_1^2 is negligible in comparison to v_2^2.

In Bernoulli's equation $\left(P_1 - P_2\right) + \frac{1}{2}\rho v_1^2 + \rho g y_1 = \frac{1}{2}\rho v_2^2 + \rho g y_2$

the term in v_1^2 is essentially zero and the terms in y_1 and y_2 cancel each other.

Then $v_2 = \left(\dfrac{2(P_1 - P_2)}{\rho}\right)^{1/2}$

$$v_2 = \sqrt{\frac{2(8.00 \times 10^4 \text{ Pa})}{1\,000 \text{ kg/m}^3}} = 12.6 \text{ m/s} \qquad \blacksquare$$

Finalize: Our estimate was good. The units work out as $[(\text{N/m}^2)/(\text{kg/m}^3)]^{1/2} = [(\text{kg}\cdot\text{m/s}^2\cdot\text{m}^2)(\text{m}^3/\text{kg})]^{1/2} = \text{m/s}$. Note again that the pressure in the stream moving through the air must be just 1 atm, because the liquid can push on the air no more strongly and no less strongly than the air pushes on the liquid.

---

58. The true weight of an object can be measured in a vacuum, where buoyant forces are absent. A measurement in air, however, is disturbed by buoyant forces. An object of volume V is weighed in air on an equal-arm balance with the use of counterweights of density ρ. Representing the density of air as ρ_{air} and the balance reading as F_g', show that the true weight F_g is

$$F_g = F_g' + \left(V - \frac{F_g'}{\rho g}\right)\rho_{\text{air}} g$$

Solution

Conceptualize: The true weight is greater than the apparent weight because the apparent weight is reduced by the buoyant effect of the air.

Categorize: We use the rigid body in equilibrium model.

Analyze: The "balanced" condition is one in which the net torque on the balance is zero. Since the balance has lever arms of equal length, the total force on each pan is equal. Applying $\sum \tau = 0$ around the pivot gives us this in equation form:

$$F_g - B = F_g' - B'$$

where B and B' are the buoyant forces on the test mass and the counterweights, respectively. The buoyant force experienced by an object of volume V in air is

$$B = V\rho_{\text{air}} g$$

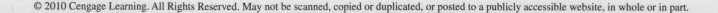

So for the test mass and for the counterweights, respectively,

$$B = V\rho_{\text{air}}g \qquad \text{and} \qquad B' = V'\rho_{\text{air}}g$$

Since the volume of the weights is not given explicitly, we must use the density equation to eliminate it:

$$V' = \frac{m'}{\rho} = \frac{m'g}{\rho g} = \frac{F'_g}{\rho g}$$

With this substitution, the buoyant force on the weights is

$$B' = \left(F'_g/\rho g\right)\rho_{\text{air}}g$$

Therefore by substitution

$$F_g = F'_g + \left(V - \frac{F'_g}{\rho g}\right)\rho_{\text{air}}g \qquad\blacksquare$$

Finalize: A pharmacist will use this equation to precisely weigh a low-density drug. For brass counterweights, the second term inside the parentheses can be negligible.

We can now answer the riddle: Which weighs more, a pound of feathers or a pound of bricks? As in the problem above, the feathers feel a greater buoyant force than the bricks, so if they "weigh" the same on a scale as a pound of bricks, then the feathers must have more mass and therefore a greater "true weight."

---

64. Review. With reference to the dam studied in Example 14.4 and shown in Figure 14.5, (a) show that the total torque exerted by the water behind the dam about a horizontal axis through O is $\frac{1}{6}\rho gwH^3$. (b) Show that the effective line of action of the total force exerted by the water is at a distance $\frac{1}{3}H$ above O.

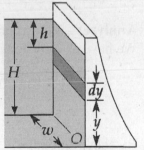

Solution

Conceptualize: A wider, higher dam holding back a denser liquid in a higher gravitational field should feel greater torque for all four reasons. It is a little remarkable that the torque is proportional to the cube of the depth.

Figure 14.5

Categorize: The pressure increases with depth, so we must do an integral to find the total force on the dam and another integral to find the torque about the base.

Analyze: The torque is calculated from the equation

$$\tau = \int d\tau = \int r \, dF \qquad \text{where the integral is over all depths.}$$

From Figure 14.5, we have

$$\tau = \int y \, dF = \int_0^H y\rho g \, (H - y)w \, dy = \frac{1}{6}\rho gwH^3 \qquad\blacksquare$$

The force on the dam is $\quad F = \int dF = \int_0^H \rho g(H-y)w\,dy = \frac{1}{2}\rho gwH^2$

If this were applied at a height y_{eff} such that the torque remains unchanged,

we would have $\quad \frac{1}{6}\rho gwH^3 = y_{eff}\left[\frac{1}{2}\rho gwH^2\right]$

so $\quad\quad\quad y_{eff} = \frac{1}{3}H$ ∎

Finalize: The proportionality of the torque to the cube of the depth means that the torque increases steeply as the liquid gets deeper, by 33% for a 10% depth increase, for example. This grimly reminds us of concrete-wall levees failing when Hurricane Katrina struck New Orleans.

---

67. In 1983, the United States began coining the one-cent piece out of copper-clad zinc rather than pure copper. The mass of the old copper penny is 3.083 g and that of the new cent is 2.517 g. The density of copper is 8.920 g/cm³ and that of zinc is 7.133 g/cm³. The new and old coins have the same volume. Calculate the percent of zinc (by volume) in the new cent.

Solution

Conceptualize: Zinc has 80.0% of the density of copper. The new pennies have 81.6% of the mass of the copper ones. We conclude that the new pennies are mostly zinc.

Categorize: We use just the definition of density, applied to the different coins, and so giving us simultaneous equations.

Analyze: Let f represent the fraction of the volume V occupied by zinc in the new coin. We have $m = \rho V$ for both coins:

$$3.083\text{ g} = (8.920\text{ g/cm}^3)V \quad\text{and}\quad 2.517\text{ g} = (7.133\text{ g/cm}^3)(fV)$$
$$+ (8.920\text{ g/cm}^3)(1-f)V$$

By substitution,

$$2.517\text{ g} = (7.133\text{ g/cm}^3)fV + 3.083\text{ g} - (8.920\text{ g/cm}^3)fV$$

$$fV = \frac{3.083\text{ g} - 2.517\text{ g}}{8.920\text{ g/cm}^3 - 7.133\text{ g/cm}^3}$$

and again substituting to eliminate the volume,

$$f = \frac{0.566\text{ g}}{1.787\text{ g/cm}^3}\left(\frac{8.920\text{ g/cm}^3}{3.083\text{ g}}\right) = 0.916\,4 = 91.6\%$$ ∎

Finalize: Our estimate was right. Algebra did the job! Never be afraid to introduce symbols for quantities you do not know.

15

Oscillatory Motion

EQUATIONS AND CONCEPTS

Hooke's law gives the force exerted by a spring that is stretched or compressed beyond the equilibrium position. The force is proportional in magnitude to the displacement from the equilibrium position. *The constant of proportionality k is the elastic constant or the spring constant; it is always positive and has a value which corresponds to the relative stiffness of the elastic medium.*

$$F_s = -kx \qquad (15.1)$$

Simple harmonic motion as exhibited by the mass-spring system shown in the figure is one form of periodic motion. *The spring force is a restoring force, always directed toward the equilibrium position.*

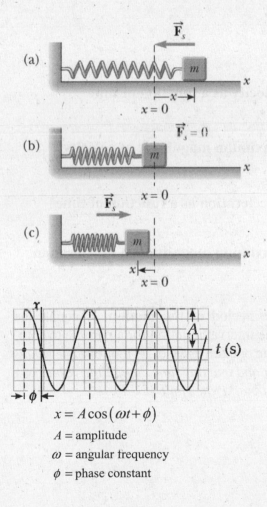

The acceleration of an object in simple harmonic motion is proportional to the displacement from equilibrium and oppositely directed. Equation 15.2 is the result of applying Newton's second law to a mass, m, where the force is given by Equation 15.1. *Note that the acceleration of a harmonic oscillator is not constant.*

A **displacement versus time graph** for an ideal oscillator undergoing simple harmonic motion along the x-axis is shown in the figure at right.

The amplitude, A, represents the maximum displacement along either positive or negative x.

The angular frequency, $\omega = 2\pi f$, has units of radians per second and is a measure of how rapidly the oscillations are occurring.

$$x = A\cos(\omega t + \phi)$$

A = amplitude
ω = angular frequency
ϕ = phase constant

261

The phase constant ϕ, is determined by the position and velocity of the oscillating particle when $t = 0$. The quantity $(\omega t + \phi)$ is called the phase.

The **mathematical representation** of a mass in simple harmonic motion along the x axis is a second order differential equation. The ratio k/m from Equation 15.2 is denoted by the factor ω^2 in Equation 15.5.

$$a_x = -\left(\frac{k}{m}\right)x \qquad (15.2)$$

$$\frac{d^2x}{dt^2} = -\omega^2 x \qquad (15.5)$$

Equations for the position, velocity, and acceleration as functions of time can be obtained from the equation of motion 15.5 above. *In these equations, ω must be expressed in units of radians per second.*

Position as a function of time

$$x(t) = A\cos(\omega t + \phi) \qquad (15.6)$$

$$\text{where } \omega = \sqrt{\frac{k}{m}} \qquad (15.9)$$

Velocity as a function of time

$$v = \frac{dx}{dt} = -\omega A\sin(\omega t + \phi) \qquad (15.15)$$

Maximum magnitude of velocity

$$v_{max} = \omega A = \sqrt{\frac{k}{m}}\,A \qquad (15.17)$$

Acceleration as a function of time

$$a = \frac{d^2x}{dt^2} = -\omega^2 A\cos(\omega t + \phi) \qquad (15.16)$$

Maximum magnitude of acceleration

$$a_{max} = \omega^2 A = \left(\frac{k}{m}\right)A \qquad (15.18)$$

The period of the motion, T, equals the time interval required for a particle to complete one full cycle of its motion. *The values of x and v at any time t equal the values of x and v at a later time $(t + T)$.*

$$T = 2\pi\sqrt{\frac{m}{k}} \qquad (15.13)$$

The **frequency of the motion** is the inverse of the period. The frequency (f) represents the number of oscillations per unit time and is measured in cycles per second or hertz (Hz). The **angular frequency** (ω) is measured in radians per second. *The frequency depends only on the values of m and k.*

$$f = \frac{1}{T}$$

$$f = \frac{1}{2\pi}\sqrt{\frac{k}{m}} \qquad (15.14)$$

$$\omega = 2\pi f$$

The **kinetic and potential energies** of a harmonic oscillator are always positive quantities; and total energy $E = K + U$.

$$E = \tfrac{1}{2}mv^2 + \tfrac{1}{2}kx^2$$

The **total mechanical energy** of a simple harmonic oscillator is a constant of the motion and is proportional to the square of the amplitude.

$$E = \tfrac{1}{2}kA^2 \qquad (15.21)$$

Velocity as a function of position can be found from the principle of conservation of energy. *The magnitude of the velocity of an object in simple harmonic motion is a maximum at x = 0; the velocity is zero when the mass is at the points of maximum displacement (x = ± A).*

$$v = \pm\sqrt{\frac{k}{m}\left(A^2 - x^2\right)} \qquad (15.22)$$

$$v = \pm\omega\sqrt{A^2 - x^2}$$

The **equation of motion for the simple pendulum** assumes a small angular displacement θ (so that $\sin\theta \approx \theta$). *This small angle approximation is valid when the angle θ is measured in radians.*

$$\frac{d^2\theta}{dt^2} = -\frac{g}{L}\theta \qquad (15.24)$$

The **period of a simple pendulum** depends only on the length of the supporting string and the local value of the acceleration due to gravity. See the equation of motion in *Suggestions, Skills and Strategies.*

$$T = 2\pi\sqrt{\frac{L}{g}} \qquad (15.26)$$

The **period of a physical pendulum** depends on the moment of inertia, I, and the distance, d, between the pivot point and the center of mass. *This result can be used to measure the moment of inertia of a flat rigid body when the location of the center of mass is known.* See the equation of motion in *Suggestions, Skills, and Strategies.*

$$T = 2\pi\sqrt{\frac{I}{mgd}} \qquad (15.28)$$

The **period of a torsion pendulum** depends on the moment of inertia of the oscillating body and the torsion constant of the supporting wire (κ). *When a wire or rod is twisted through an angle θ, the restoring torque is related to the angular displacement by the parameter κ ($\tau = -\kappa\theta$).* See the equation of motion in *Suggestions, Skills, and Strategies.*

$$T = 2\pi\sqrt{\frac{I}{\kappa}} \qquad (15.30)$$

The **position as a function of time for a damped oscillator** includes an amplitude that decreases exponentially with time. *In Equation 15.32, the damping force is assumed to be proportional to the velocity, and the damping coefficient, b, is assumed to be small.*

$$x = Ae^{-\left(\frac{b}{2m}\right)t}\cos(\omega t + \phi) \qquad (15.32)$$

The **angular frequency of a damped oscillator** of a given mass depends on the values of the restoring force and the damping coefficient. ω_0 is the natural frequency in the absence of a retarding force.

$$\omega = \sqrt{\omega_0^2 - \left(\frac{b}{2m}\right)^2}$$

where $\omega_0^2 = \dfrac{k}{m}$

Damping modes determine the manner in which the amplitude of a given damped oscillator approaches zero; damping depends on the value of the damping coefficient, b, and the natural frequency of the system, ω_0. The damping modes are called *underdamped*, *critically damped*, and *overdamped*.

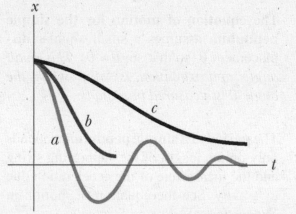

Underdamped mode: The system oscillates as the amplitude decays exponentially with time (shown as plot a in above figure).

$(b/2m) < \omega_0$

Critically damped mode: The system returns to equilibrium (zero amplitude) without oscillation (shown as plot b in above figure).

$(b/2m) = \omega_0$

Overdamped mode: The system approaches equilibrium without oscillation; and the time to reach equilibrium increases as the damping force increases (shown as plot c in figure on previous page.).

$$(b/2m) > \omega_0$$

Resonance in a forced oscillator occurs when the driving frequency matches the natural frequency of the oscillator.

$$\omega_0 = \sqrt{\frac{k}{m}}$$

SUGGESTIONS, SKILLS, AND STRATEGIES

SIMILAR STRUCTURE OF SEVERAL SYSTEMS IN SIMPLE HARMONIC MOTION

You should note the similar structure of the following equations of motion and the corresponding expressions for period, T:

Oscillating system	Equation of motion	Period
Harmonic oscillator along x axis	$\dfrac{d^2x}{dt^2} = -\left(\dfrac{k}{m}\right)x$	$T = 2\pi\sqrt{\dfrac{m}{k}}$
Simple pendulum	$\dfrac{d^2\theta}{dt^2} = -\left(\dfrac{g}{L}\right)\theta$	$T = 2\pi\sqrt{\dfrac{L}{g}}$
Physical pendulum	$\dfrac{d^2\theta}{dt^2} = -\left(\dfrac{mgd}{I}\right)\theta$	$T = 2\pi\sqrt{\dfrac{I}{mgd}}$
Torsion pendulum	$\dfrac{d^2\theta}{dt^2} = -\left(\dfrac{\kappa}{I}\right)\theta$	$T = 2\pi\sqrt{\dfrac{I}{\kappa}}$

In each case, when the equation of motion is in the form

$$\frac{d^2(\text{displacement})}{dt^2} = -C(\text{displacement})$$

the period is

$$T = 2\pi\sqrt{\frac{1}{C}}$$

COMPARISON OF SIMPLE HARMONIC MOTION AND UNIFORM CIRCULAR MOTION

Problems involving simple harmonic motion can often be solved more easily by treating simple harmonic motion along a straight line as the projection of uniform circular motion along a diameter of a reference circle. This technique is described in Section 15.4 of the text.

REVIEW CHECKLIST

You should be able to:

- Describe the general characteristics of simple harmonic motion, and the significance of the various parameters which appear in the expression for the displacement as a function of time, $x = A \cos(\omega t + \phi)$. (Sections 15.1 and 15.2)

- Start with the general expression for the displacement versus time for a simple harmonic oscillator, and obtain expressions for the velocity and acceleration as functions of time. (Section 15.2)

- Determine the frequency, period, amplitude, phase constant, and position at a specified time of a simple harmonic oscillator given an equation for $x(t)$. (Section 15.2)

- Start with an expression for $x(t)$ for an oscillator and determine the maximum speed, maximum acceleration, and the total distance traveled in a specified time interval. (Section 15.2)

- Describe the phase relations among displacement, velocity, and acceleration for simple harmonic motion. Given a curve of one of these variables versus time, sketch curves showing the time dependence of the other two. (Section 15.2)

- Calculate the mechanical energy, maximum velocity, and maximum acceleration of a mass-spring system given values for amplitude, spring constant, and mass. (Section 15.3)

- Describe the relationship between uniform circular motion and simple harmonic motion. (Section 15.4)

- Make calculations using the expressions for the period of simple, physical, and torsion pendulums. (Section 15.5)

- Describe the conditions that give rise to the different types of damped motion. (Section 15.6)

ANSWER TO AN OBJECTIVE QUESTION

12. For a simple harmonic oscillator, answer yes or no to the following questions. (a) Can the quantities position and velocity have the same sign? (b) Can velocity and acceleration have the same sign? (c) Can position and acceleration have the same sign?

Answer (a) yes (b) yes (c) no. In a simple harmonic oscillator, the velocity follows the position by 1/4 of a cycle, and the acceleration follows the position by 1/2 of a cycle. Referring to Figure 15.5 or to Active Figure 15.10 of the textbook, you can see that there are times (a) when both the position and the velocity are positive, and therefore in the same direction. (b) There also exist instants when both the velocity and the acceleration are positive, and therefore in the same direction. On the other hand, (c) when the position is positive, the acceleration is always negative, and therefore position and acceleration are always in opposite directions.

□   □   □   □

ANSWERS TO SELECTED CONCEPTUAL QUESTIONS

3. (a) If the coordinate of a particle varies as $x = -A \cos \omega t$, what is the phase constant in Equation 15.6? (b) At what position is the particle at $t = 0$?

Answer (a) Equation 15.6 says $x = A \cos(\omega t + \phi)$. Since negating a cosine function is equivalent to changing the phase by 180°, we can say that

$$\phi = 180° \left(\frac{\pi \text{ rad}}{180°} \right) = \pi \text{ rad}$$

(b) At time $t = 0$, the particle is at a position of $x = -A$.

□   □   □   □

8. Is it possible to have damped oscillations when a system is at resonance? Explain.

Answer Yes. At resonance, the amplitude of a damped oscillator will remain constant. If the system were not damped, the amplitude would increase without limit at resonance.

□   □   □   □

SOLUTIONS TO SELECTED END-OF-CHAPTER PROBLEMS

5. The position of a particle is given by the expression $x = 4.00 \cos(3.00\pi t + \pi)$, where x is in meters and t is in seconds. Determine (a) the frequency and (b) period of the motion, (c) the amplitude of the motion, (d) the phase constant, and (e) the position of the particle at $t = 0.250$ s.

Solution

Conceptualize: The given expression describes the vibration by telling the position at every instant of time as it continually changes. The frequency, period, amplitude, and phase constant are constants that describe the whole motion, …

Categorize: … and we can pick them out from the *x*-versus-*t* function. We use the particle in simple harmonic motion model.

Analyze: The particular position function $x = (4.00) \cos(3.00\pi t + \pi)$ and the general one, $x = A \cos(\omega t + \phi)$, have an especially powerful kind of equality called functional equality. They must give the same *x* value for all values of the variable *t*. This requires, then, that all parts be the same:

(a) $\omega = 3.00 \ \pi \ \text{rad/s} = (2\pi \ \text{rad/1 cycle})f$ or $f = 1.50 \ \text{Hz}$ ∎

(b) $T = \dfrac{1}{f} = 0.667 \ \text{s}$ ∎

(c) $A = 4.00 \ \text{m}$ ∎

(d) $\phi = \pi \ \text{rad}$ ∎

(e)   At $t = 0.250 \ \text{s}$, $x = (4.00 \ \text{m})\cos(1.75 \ \pi \ \text{rad}) = (4.00 \ \text{m})\cos(5.50)$

Note that the 5.50 inside the parentheses means 5.50 rad and **not** 5.50°. Instead,

$$x = (4.00 \ \text{m})\cos(5.50 \ \text{rad}) = (4.00 \ \text{m})\cos(315°) = 2.83 \ \text{m}$$ ∎

Finalize: This vibration would be impressive to watch. Imagine the top of a tree swinging back and forth through a range of 8 meters, making 15 complete vibrations in every 10 seconds.

Make sure you recognize the difference between angular frequency ω in radians per second and frequency *f* in Hz. Neither term can be used for the other physical quantity.

---

9.   A 7.00-kg object is hung from the bottom end of a vertical spring fastened to an overhead beam. The object is set into vertical oscillations having a period of 2.60 s. Find the force constant of the spring.

Solution

Conceptualize: This is a large mass in a low-frequency oscillation in laboratory terms, suggesting a fairly soft spring with *k* on the order of 10 N/m.

Categorize: We can solve directly from the expression giving the angular frequency of a block-spring system in terms of mass and spring constant.

Analyze: An object hanging from a vertical spring moves with simple harmonic motion just like an object moving without friction attached to a horizontal spring. We are given the period. It is related to the angular frequency by $\dfrac{2\pi}{\omega} = T = 2.60$ s

Solving for the angular frequency, $\quad \omega = \dfrac{2\pi \text{ rad}}{2.60 \text{ s}} = 2.42$ rad/s

Now for this system $\qquad\qquad \omega = \sqrt{\dfrac{k}{m}}, \quad$ so $\quad k = \omega^2 m = (2.42 \text{ rad/s})^2(7.00 \text{ kg})$

Thus we find the force constant: $\qquad k = 40.9 \text{ kg/s}^2 = 40.9$ N/m $\qquad\blacksquare$

Finalize: In laboratory we could test this dynamically-determined value of k with a static measurement, by measuring the force required to hold the spring extended by 10 cm. It should be 4.09 N. Note that the length of the spring never appears in calculations.

---

13. A particle moving along the x axis in simple harmonic motion starts from its equilibrium position, the origin, at $t = 0$ and moves to the right. The amplitude of its motion is 2.00 cm, and the frequency is 1.50 Hz. (a) Find an expression for the position of the particle as a function of time. Determine (b) the maximum speed of the particle and (c) the earliest time ($t > 0$) at which the particle has this speed. Find (d) the maximum positive acceleration of the particle and (e) the earliest time ($t > 0$) at which the particle has this acceleration. (f) Find the total distance traveled between $t = 0$ and $t = 1.00$ s.

Solution

Conceptualize: In its period of less than a second the particle moves 8 cm. Its minimum speed is zero, so its maximum speed must be several cm/s and its maximum acceleration a larger number of cm/s^2.

Categorize: After we determine the expression giving position as a function of time, differentiating it will give the velocity and the acceleration.

Analyze:

(a) At $t = 0$, $x = 0$ and v is positive (to the right). The sine function is zero and the cosine is positive at $\theta = 0$, so this situation corresponds to $x = A \sin\omega t$ and $v = v_i \cos \omega t$.

Since $f = 1.50$ Hz, $\qquad \omega = 2\pi f = 3\pi$ s^{-1}

Also, $\qquad\qquad\qquad A = 2.00$ cm

so that $\qquad\qquad\qquad x = (2.00)\sin(3\pi t)$ where x is in cm and t is in s.

This is equivalent to writing $x = A \cos(\omega t + \phi)$

with $\qquad A = 2.00 \text{ cm}, \omega = 3.00\pi \text{ s}^{-1} \qquad$ and $\qquad \phi = -90° = -\dfrac{\pi}{2}$

Note also that $\qquad T = \dfrac{1}{f} = 0.667 \text{ s}$

(b) The velocity is $\qquad v = \dfrac{dx}{dt} = (2.00)(3.00\ \pi)\cos(3.00\ \pi t) \text{ cm/s}$

The maximum speed is

$$v_{max} = v_i = A\omega = (2.00)(3.00\ \pi) \text{ cm/s} = 18.8 \text{ cm/s} \qquad \blacksquare$$

(c) This speed occurs at $\qquad t = 0, \qquad$ when $\qquad \cos(3.00\ \pi t) = +1$

and next at $\qquad t = \dfrac{T}{2} = 0.333 \text{ s}, \qquad$ when $\qquad \cos\left[\left(3.00\ \pi \text{s}^{-1}\right)(0.333 \text{ s})\right] = -1 \qquad \blacksquare$

(d) Again, $a = \dfrac{dv}{dt} = (-2.00 \text{ cm})\left(3.00\pi \text{ s}^{-1}\right)^2 \sin(3.00\ \pi t)$

Its maximum value is $a_{max} = A\omega^2 = (2.00 \text{ cm})\left(3.00\pi \text{ s}^{-1}\right)^2 = 178 \text{ cm/s}^2 \qquad \blacksquare$

(e) The acceleration has this positive value for the first time

at $\qquad t = \dfrac{3T}{4} = 0.500 \text{ s}$

when $\qquad a = -(2.00 \text{ cm})\left(3.00\ \pi \text{ s}^{-1}\right)^2 \sin\left[(3.00\ \pi \text{ s}^{-1})(0.500 \text{ s})\right] = 178 \text{ cm/s}^2 \qquad \blacksquare$

(f) Since $A = 2.00$ cm, the particle will travel 8.00 cm in one period, which is $T = \frac{2}{3}$s. Hence,

in $\qquad (1.00 \text{ s}) = \frac{3}{2}T$

the particle will travel $\qquad \frac{3}{2}(8.00 \text{ cm}) = 12.0 \text{ cm} \qquad \blacksquare$

Finalize: Especially if you have not recently thought about the sine and cosine functions' patterns of going positive and going negative, sketching graphs can help a lot in picking out the time when the particle has some particular speed or acceleration. At the time $t = T/4 = 0.167$ s, the particle has acceleration -178 cm/s^2, attaining maximum magnitude for the first time after $t = 0$.

---

15. A 0.500-kg object attached to a spring with a force constant of 8.00 N/m vibrates in simple harmonic motion with an amplitude of 10.0 cm. Calculate the maximum value of its (a) speed and (b) acceleration, (c) the speed and (d) the acceleration when the object is

6.00 cm from the equilibrium position, and (e) the time interval required for the object to move from $x = 0$ to $x = 8.00$ cm.

Solution

Conceptualize: This is a standard laboratory situation. The maximum speed will be a few cm/s and the maximum acceleration several cm/s². Less than one second will suffice for the object to move a few centimeters from its equilibrium point.

Categorize: We find ω first and everything follows from it. We will write an expression giving x as a function of t to find the time in part (c).

Analyze: The angular frequency is $\omega = \sqrt{\dfrac{k}{m}} = \sqrt{\dfrac{8.00 \text{ N/m}}{0.500 \text{ kg}}} = 4.00 \text{ s}^{-1}$

We can choose to start a stopwatch when the particle is passing through the origin and moving to the right. Then at all instants its position is given by $x = (10.0 \text{ cm})\sin\left[\left(4.00 \text{ s}^{-1}\right)t\right]$

This is equivalent to $x = 10.0 \cos(4.00t - \pi/2)$

(a) From the assumed equation we find that $\qquad v = \dfrac{dx}{dt} = (40.0 \text{ cm/s})\cos(4.00t)$

$$v_{\text{max}} = 40.0 \text{ cm/s} \qquad \blacksquare$$

(b) The acceleration is $\qquad a = \dfrac{dv}{dt} = -\left(160 \text{ cm/s}^2\right)\sin(4.00t)$

and its maximum value is $a_{\text{max}} = 160 \text{ cm/s}^2 \qquad \blacksquare$

(c) We solve for t in the equation giving position as a function of time, to find

$$t = \frac{1}{4.00} \sin^{-1}\left(\frac{x}{10.0}\right)$$

When $x = 6.00$ cm,

$$t = \left(\frac{1 \text{ s}}{4.00 \text{ rad}}\right)\sin^{-1}\left(\frac{6.00 \text{ cm}}{10.0 \text{ cm}}\right) = \left(\frac{1 \text{ s}}{4.00 \text{ rad}}\right)\sin^{-1}(0.600) = \frac{1 \text{ s}}{4.00 \text{ rad}}36.9° = \frac{0.644 \text{ rad} \cdot \text{s}}{4.00 \text{ rad}}$$

$t = 0.161$ s and we find that

$$v = (40.0 \text{ cm/s})\cos\left[\left(4.00 \text{ s}^{-1}\right)(0.161 \text{ s})\right] = 32.0 \text{ cm/s} \qquad \blacksquare$$

(d) The acceleration is $\quad a = -\left(160 \text{ cm/s}^2\right)\sin\left[\left(4.00 \text{ s}^{-1}\right)(0.161 \text{ s})\right] = -96.0 \text{ cm/s}^2 \qquad \blacksquare$

(e) Using $t = \frac{1}{4}\sin^{-1}\left(\frac{x}{10.0 \text{ cm}}\right)$ we similarly find that when $x = 0$, $t = 0$, and when $x = 8.00$ cm,

$t = 0.232$ s

Therefore, the interval is $\Delta t = 0.232$ s ∎

Finalize: Make sure you know that the sine and cosine functions vary just between 1 and −1. Then it is easy to pick out maximum values.

---

17. To test the resilience of its bumper during low-speed collisions, a 1 000-kg automobile is driven into a brick wall. The car's bumper behaves like a spring with a force constant of 5.00×10^6 N/m and compresses 3.16 cm as the car is brought to rest. What was the speed of the car before impact, assuming that no mechanical energy is transformed or transferred away during impact with the wall?

Solution

Conceptualize: If the bumper is only compressed 3 cm, the car is probably not permanently damaged, so v is likely less than about 5 m/s.

Categorize: Assuming no mechanical energy turns into internal energy during impact with the wall, the initial energy (kinetic) equals the final energy (elastic potential).

Analyze: Energy conservation for the car-bumper system, or the isolated-system energy model, gives $K_i = U_f$ or $\frac{1}{2}mv^2 = \frac{1}{2}kx^2$

Solving for the velocity, $v = x\sqrt{\dfrac{k}{m}} = \left(3.16 \times 10^{-2} \text{ m}\right)\sqrt{\dfrac{5.00 \times 10^6 \text{ N/m}}{1\,000 \text{ kg}}}$

Thus, $v = 2.23$ m/s ∎

Finalize: The speed is less than 5 m/s as predicted, so the answer seems reasonable. If the speed of the car were sufficient to compress the bumper beyond its elastic limit, then some of the initial kinetic energy would be lost to internal energy when the front of the car is bent up. In that case, some other procedure would have to be used to determine the car's initial speed.

---

28. A particle of mass m slides without friction inside a hemispherical bowl of radius R. Show that if the particle starts from rest with a small displacement from equilibrium, it moves in simple harmonic motion with an angular frequency equal to that of a simple pendulum of length R. That is, $\omega = \sqrt{g/R}$.

Solution

Conceptualize: The force diagram is identical to that of a simple pendulum, so it is natural that the motion be the same.

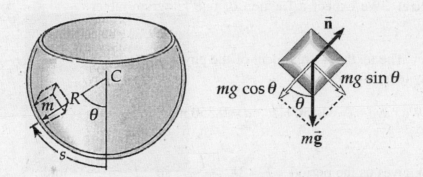

Categorize: To prove simple harmonic motion we must prove that the acceleration is a negative constant times the excursion from equilibrium.

Analyze: Locate the center of curvature C of the bowl. We can measure the excursion of the object from equilibrium by the angle θ between the radial line to C and the vertical. The distance the object moves from equilibrium is $s = R\theta$.

$\sum F_s = ma$ becomes $\quad -mg\sin\theta = m\dfrac{d^2s}{dt^2}$

For small angles $\quad \sin\theta \approx \theta$

so, by substitution $\quad -mg\theta = m\dfrac{d^2s}{dt^2} \quad$ and $\quad -mg\dfrac{s}{R} = m\dfrac{d^2s}{dt^2}$

Isolating the derivative, $\quad \dfrac{d^2s}{dt^2} = -\left(\dfrac{g}{R}\right)s$

By the form of this equation, we can see that the acceleration is proportional to the position and in the opposite direction, so we have SHM. ∎

We identify its angular frequency by comparing our equation to the general

equation $\quad \dfrac{d^2x}{dt^2} = -\omega^2 x$

Now x and s both measure position, so we identify $\quad \omega^2 = \dfrac{g}{R} \quad$ and $\quad \omega = \sqrt{\dfrac{g}{R}}$ ∎

Finalize: Rolling is different from sliding in energy content, but a ball rolling across the bottom of a shallow bowl will also move in simple harmonic motion, and so will a rocking chair.

---

29. A physical pendulum in the form of a planar object moves in simple harmonic motion with a frequency of 0.450 Hz. The pendulum has a mass of 2.20 kg, and the pivot is located

0.350 m from the center of mass. Determine the moment of inertia of the pendulum about the pivot point.

Solution

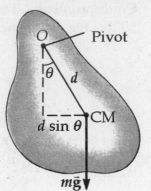

Conceptualize: We expect a fraction of one kilogram-meter-squared.

Categorize: The textbook treatment of the physical pendulum leads directly to an answer.

Analyze: We have $f = 0.450$ Hz; $d = 0.350$ m; and $m = 2.20$ kg

The textbook gives us the period $T = 2\pi \sqrt{\dfrac{I}{mgd}}$ so $T^2 = \dfrac{4\pi^2 I}{mgd}$ and

$$I = \frac{T^2 mgd}{4\pi^2} = \left(\frac{1}{f}\right)^2 \frac{mgd}{4\pi^2} = \frac{(2.20 \text{ kg})(9.80 \text{ m/s}^2)(0.350 \text{ m})}{(0.450 \text{ s}^{-1})^2 (4\pi^2)} = 0.944 \text{ kg} \cdot \text{m}^2 \quad \blacksquare$$

Finalize: Timing many cycles can be a high-precision way to measure the moment of inertia. Knowing the moment of inertia about the axis through the pivot lets you use the parallel-axis theorem to find it about any other parallel axis.

---

31. A simple pendulum has a mass of 0.250 kg and a length of 1.00 m. It is displaced through an angle of 15.0° and then released. Using the analysis model of a particle in simple harmonic motion, find (a) the maximum speed of the bob, (b) its maximum angular acceleration, and (c) the maximum restoring force on the bob. (d) **What if?** Solve parts (a) through (c) again by using analysis models introduced in earlier chapters. (e) Compare the answers.

Solution

Conceptualize: The pendulum's period will be about 2 s. Its maximum speed will be several centimeters per second. The maximum angular acceleration will be on the order of one radian per second squared, and the maximum net force on the order of one newton.

Categorize: The answers from the particle in simple harmonic motion model should agree fairly well with the results from particle-under-net-force and isolated-system-energy results.

Analyze:

(a) Since 15.0° is small enough that (in radians) $\sin\theta \approx \theta$ within 1%, it makes some sense to model the motion as simple harmonic motion. The constant angular frequency characterizing the motion is

$$\omega = \sqrt{\frac{g}{L}} = \sqrt{\frac{9.80 \text{ m/s}^2}{1.00 \text{ m}}} = 3.13 \text{ rad/s}$$

The amplitude as a distance is $A = L\theta = (1.00 \text{ m})(0.262 \text{ rad}) = 0.262 \text{ m}$

Then the maximum linear speed is

$$v_{max} = \omega A = (3.13 \text{ s}^{-1})(0.262 \text{ m}) = 0.820 \text{ m/s} \quad \blacksquare$$

(b) Similarly, $a_{max} = \omega^2 A = (3.13 \text{ s}^{-1})^2(0.262 \text{ m}) = 2.57 \text{ m/s}^2$

This implies maximum angular acceleration

$$\alpha = \frac{a}{r} = \frac{2.57 \text{ m/s}^2}{1.00 \text{ m}} = 2.57 \text{ rad/s}^2 \quad \blacksquare$$

(c) $\sum F = ma = (0.250 \text{ kg})(0.257 \text{ m/s}^2) = 0.641 \text{ N} \quad \blacksquare$

(d) We may work out slightly more precise answers by using the energy version of the isolated system model and the particle under a net force model.

At release, the pendulum has height above its equilibrium position given by $h = L - L\cos 15.0° = (1.00 \text{ m})(1 - \cos 15.0°) = 0.034 \ 1 \text{ m}$

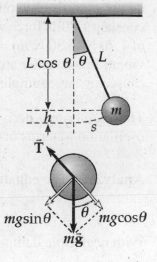

The energy of the pendulum-Earth system is conserved as the pendulum swings down:

$$(K + U)_{top} = (K + U)_{bottom}$$

$$0 + mgh = \tfrac{1}{2}mv^2_{max} + 0$$

$$v_{max} = \sqrt{2gh} = \sqrt{2(9.80 \text{ m/s}^2)(0.0341 \text{ m})} = 0.817 \text{ m/s} \quad \blacksquare$$

The restoring force at release is

$$mg \sin 15.0° = (0.250 \text{ kg})(9.80 \text{ m/s}^2)(\sin 15.0°) = 0.634 \text{ N} \quad \blacksquare$$

This produces linear acceleration

$$a = \frac{\sum F}{m} = \frac{0.634 \text{ N}}{0.250 \text{ kg}} = 2.54 \text{ m/s}^2$$

and angular acceleration $\quad \alpha = \frac{a}{r} = \frac{2.54 \text{ m/s}^2}{1.00 \text{ m}} = 2.54 \text{ rad/s}^2 \quad \blacksquare$

Finalize: (e) The answers agree to two digits. The answers computed from conservation of system energy and from Newton's second law are more precisely correct. With this amplitude the motion of the pendulum is approximately simple harmonic. ∎

---

45. Damping is negligible for a 0.150-kg object hanging from a light 6.30 N/m spring. A sinusoidal force with an amplitude of 1.70 N drives the system. At what frequency will the force make the object vibrate with an amplitude of 0.440 m?

Solution

Conceptualize: This sounds like a laboratory situation with a rather loose spring. We expect a frequency between 0.1 Hz and 10 Hz.

Categorize: Let us first dispose of a couple of possibly misleading ideas. The static extension produced by hanging the load on the spring is 0.150 kg(9.80 m/s²)/(6.30 N/m) = 0.233 m. The vibration amplitude could be less than this or more. We do not need to know the original length of the spring, but we assume that it is large enough that the coils (if it is a spiral spring) do not run into each other even when the vibration amplitude is 0.440 m. Next, note that if a constant 1.70-N force were applied, it would produce another static extension, this with a size of 1.70 N/(6.30 N/m) = 0.270 m. An oscillating force with amplitude 1.70 N could produce a vibration of smaller amplitude distance if its frequency were sufficiently high, but it can also cause any larger amplitude, such as the given 0.440 m.

The chapter text derives the equation we need, giving the distance amplitude of a forced oscillation in terms of the amplitude and frequency of the force driving it.

Analyze: The equation we use is $A = \dfrac{F_{ext}/m}{\sqrt{\left(\omega^2 - \omega_0^2\right)^2 + \left(b\omega/m\right)^2}}$

With negligible damping we take $b = 0$ and simplify to

$$A = \frac{F_{ext}/m}{\sqrt{\left(\omega^2 - \omega_0^2\right)^2}} = \frac{F_{ext}/m}{\pm\left(\omega^2 - \omega_0^2\right)} = \pm\frac{F_{ext}/m}{\omega^2 - \omega_0^2}$$

Thus, $\omega^2 = \omega_0^2 \pm \dfrac{F_{ext}/m}{A} = \dfrac{k}{m} \pm \dfrac{F_{ext}}{mA} = \dfrac{6.30 \text{ N/m}}{0.150 \text{ kg}} \pm \dfrac{1.70 \text{ N}}{\left(0.150 \text{ kg}\right)\left(0.440 \text{ m}\right)}$

$$\omega = \left(42.0/\text{s}^2 \pm 25.8/\text{s}^2\right)^{1/2}$$

This yields two answers, either $\omega = 8.23$ rad/s or $\omega = 4.03$ rad/s

Then, $f = \dfrac{\omega}{2\pi}$ gives either $f = 1.31$ Hz   or $f = 0.641$ Hz   ∎

Finalize. We might not anticipate two answers, but they both make sense because the oscillator can be driven either above or below its resonance frequency to vibrate with a particular large amplitude. Observe that the resonance angular frequency appears in the solution as $\omega_0 = (42.0/s^2)^{\frac{1}{2}}$. This problem is physically and mathematically interesting, and on another level it is a straightforward "plug and chug" problem.

---

51. A small ball of mass M is attached to the end of a uniform rod of equal mass M and length L that is pivoted at the top (Fig. P15.51). Determine the tensions in the rod (a) at the pivot and (b) at the point P when the system is stationary. (c) Calculate the period of oscillation for small displacements from equilibrium and (d) determine this period for $L = 2.00$ m.

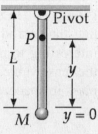

Solution

Figure P15.51

Conceptualize: The tension in the rod at the pivot is the weight of the rod plus the weight of the mass M, so at the pivot point $T = 2Mg$. The tension at point P should be slightly less since the portion of the rod between P and the pivot does not contribute to the tension.

Categorize: The tension can be found from applying Newton's Second Law. The period of this physical pendulum can be found by analyzing its moment of inertia and using Equation 15.28.

Analyze:

(a) When the pendulum is stationary, the tension at any point in the rod is simply the weight of everything below that point.

This conclusion comes from applying $\Sigma F_y = ma_y = 0$ to everything below that point. Thus, at the pivot the tension is

$$F_T = F_{g,\text{ball}} + F_{g,\text{rod}} = Mg + Mg = 2Mg$$ ∎

(b) At point P, $F_T = F_{g,\text{ball}} + F_{g,\text{rod below}P}$

$$F_T = Mg + Mg\left(\frac{y}{L}\right) = Mg\left(1 + \frac{y}{L}\right)$$ ∎

(c) For a physical pendulum where I is the moment of inertia about the pivot, $m = 2M$, and d is the distance from the pivot to the center of mass,

the period of oscillation is $T = 2\pi\sqrt{I/(mgd)}$

Relative to the pivot, $I_{\text{total}} = I_{\text{rod}} + I_{\text{ball}} = \frac{1}{3}ML^2 + ML^2 = \frac{4}{3}ML^2$

The center of mass distance is $\quad d = \dfrac{\sum m_i x_i}{\sum m_i} = \dfrac{(ML/2 + ML)}{(M+M)} = \dfrac{3L}{4}$

so we have $\quad T = 2\pi\sqrt{\dfrac{I}{mgd}} \quad$ or $\quad T = 2\pi\sqrt{\dfrac{\left(4ML^2/3\right)}{(2M)g(3L/4)}} = \dfrac{4\pi}{3}\sqrt{\dfrac{2L}{g}}$ $\quad\blacksquare$

(d) For $L = 2.00$ m, $\quad T = \dfrac{4\pi}{3}\sqrt{\dfrac{2(2.00\text{ m})}{9.80\text{ m/s}^2}} = 2.68$ s $\quad\blacksquare$

Finalize: In part (a), the tensions agree with the initial predictions. In part (b) we found that the period is slightly less (by about 6%) than a simple pendulum of length L. It is interesting to note that we were able to calculate a value for the period without knowing the mass value. This is because the period of any pendulum depends on the distribution of mass in space, and not on the **size** of the mass.

---

53. Review. A large block P attached to a light spring executes horizontal simple harmonic motion as it slides across a frictionless surface with a frequency $f = 1.50$ Hz. Block B rests on it as shown in Figure P15.53, and the coefficient of static friction between the two is $\mu_s = 0.600$. What maximum amplitude of oscillation can the system have if block B is not to slip?

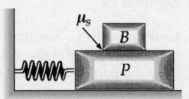

Figure P15.53

Solution

Conceptualize: If the oscillation amplitude is larger than several centimeters, static friction will be inadequate to make the upper block stick to the lower one.

Categorize: We use the particle in simple harmonic motion and particle under a net force models.

Analyze: If the block B does not slip, it undergoes simple harmonic motion with the same amplitude and frequency as those of P, and with its acceleration caused by the static friction force exerted on it by P. Think of the block when it is just ready to slip at a turning point in its motion:

$\sum F = ma \quad$ becomes $\quad f_{max} = \mu_s n = \mu_s mg = ma_{max} = mA\omega^2$

Then

$$A = \frac{\mu_s g}{\omega^2} = \frac{0.600\left(9.80\text{ m/s}^2\right)}{\left[2\pi\left(1.50\text{ s}^{-1}\right)\right]^2} = 6.62\text{ cm} \qquad\blacksquare$$

Finalize: We considered a turning point in the motion because the acceleration is greatest in magnitude there, putting friction to the most severe test.

---

55. A pendulum of length L and mass M has a spring of force constant k connected to it at a distance h below its point of suspension (Fig. P15.55). Find the frequency of vibration of the system for small values of the amplitude (small θ). Assume the vertical suspension rod of length L is rigid, but ignore its mass.

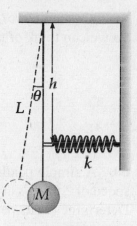

Figure P15.55

Solution

Conceptualize. The frequency of vibration should be greater than that of a simple pendulum since the spring adds an additional restoring force: $f > \dfrac{1}{2\pi}\sqrt{\dfrac{g}{L}}$

Categorize: We can find the frequency of oscillation from the angular frequency, ω, which is found in the equation for angular SHM: $d^2\theta/dt^2 = -\omega^2\theta$. The angular acceleration can be found from analyzing the torques acting on the pendulum.

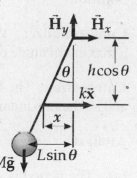

Analyze: For the pendulum (see sketch)

$$\sum \tau = I\alpha \qquad \text{and} \qquad d^2\theta/dt^2 = -\alpha$$

The negative sign appears because positive θ is measured clockwise in the picture. We take torque around the point of suspension:

$$\sum \tau = MgL \sin\theta + kxh \cos\theta = I\alpha$$

For small amplitude vibrations, we use the approximations:

$$\sin\theta \approx \theta, \quad \cos\theta \approx 1, \quad \text{and} \quad x = h\tan\theta \approx h\theta$$

Therefore, with $I = mL^2$,

$$\frac{d^2\theta}{dt^2} = -\left[\frac{MgL + kh^2}{I}\right]\theta = -\left[\frac{MgL + kh^2}{ML^2}\right]\theta$$

This is of the SHM form $\qquad \dfrac{d^2\theta}{dt^2} = -\omega^2\theta$

with angular frequency $\qquad \omega = \sqrt{\dfrac{MgL + kh^2}{ML^2}} = 2\pi f$

The frequency is $\qquad f = \dfrac{\omega}{2\pi} = \dfrac{1}{2\pi}\sqrt{\dfrac{MgL + kh^2}{ML^2}}$ ∎

Finalize: The frequency is greater than for a simple pendulum as we expected. In fact, the additional contribution inside the square root looks like the frequency of a mass on a spring scaled by h/L since the spring is connected to the rod and not directly to the mass. So we can think of the solution as

$$f^2 = \frac{1}{4\pi^2}\left(\frac{MgL + kh^2}{ML^2}\right) = f^2_{\text{pendulum}} + \frac{h^2}{L^2}f^2_{\text{spring}}$$

---

60. A simple pendulum with a length of 2.23 m and a mass of 6.74 kg is given an initial speed of 2.06 m/s at its equilibrium position. Assume it undergoes simple harmonic motion. Determine (a) its period, (b) its total energy, and (c) its maximum angular displacement.

Solution

Conceptualize: We expect a period of several seconds, an energy of several joules, and a maximum angle of a fraction of a radian.

Categorize: The length determines the period. The original speed determines the energy and so the maximum angle.

Analyze:

(a) The period is $\quad T = \dfrac{2\pi}{\omega}$

$$T = 2\pi\sqrt{\frac{L}{g}} = 2\pi\sqrt{\frac{2.23 \text{ m}}{9.80 \text{ m/s}^2}} = 3.00 \text{ s} \qquad \blacksquare$$

(b) The total energy is the energy it starts with: $\quad E = \tfrac{1}{2}mv^2_{\text{max}}$

$$E = \tfrac{1}{2}(6.74 \text{ kg})(2.06 \text{ m/s})^2 = 14.3 \text{ J} \qquad \blacksquare$$

(c) From the isolated system energy model applied to the Earth-pendulum system from its starting point to its maximum-displacement point, $0 + mgh = \tfrac{1}{2}mv^2_{\text{max}} + 0$

and $\qquad\qquad\qquad\qquad h = \dfrac{v^2_{\text{max}}}{2g} = 0.217 \text{ m}$

By geometry, $\qquad\qquad\qquad h = L - L\cos\theta_{\text{max}} = L(1 - \cos\theta_{\text{max}})$

Solving for θ_{max}, $\qquad\qquad \cos\theta_{\text{max}} = 1 - \dfrac{h}{L}$

and $\qquad\qquad \theta = \cos^{-1}\left(1 - \dfrac{0.217 \text{ m}}{2.23 \text{ m}}\right) = 25.5° \qquad \blacksquare$

Finalize: In part (c) we could alternatively write $\quad v_{\text{max}} = \omega A$

$$A = \frac{v_{\text{max}}}{\omega} = \frac{2.06 \text{ m/s}}{2.10\text{/s}} = 0.983 \text{ m} \quad \text{and} \quad \theta = \frac{A}{L} = \frac{0.983 \text{ m}}{2.23 \text{ m}} = 0.441 \text{ rad} = 25.2°$$

Our two answers are not precisely equal because the pendulum does not move with precisely simple harmonic motion.

---

63. A ball of mass m is connected to two rubber bands of length L, each under tension T as shown in Figure P15.63. The ball is displaced by a small distance y perpendicular to the length of the rubber bands. Assuming the tension does not change, show that (a) the restoring force is $-(2T/L)y$ and (b) the system exhibits simple harmonic motion with an angular frequency $\omega = \sqrt{2T/mL}$.

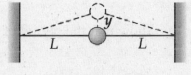

Figure P15.63

Solution

Conceptualize: It is reasonable that higher tension will raise the frequency and larger mass will lower it.

Categorize: The particle under a net force model will give us the answer to (a), and it leads directly to the answer to (b) through the particle-in-simple-harmonic-motion model.

Analyze:

(a) $\sum \vec{F} = -2T \sin \theta \, \hat{j}$ where $\theta = \tan^{-1}(y/L)$

Since for a small displacement, $\sin \theta \approx \tan \theta = \dfrac{y}{L}$

and the resultant force is $\sum \vec{F} = \left(-\dfrac{2Ty}{L} \right) \hat{j}$ ∎

(b) Since there is a restoring force that is proportional to the position, it causes the system to move with simple harmonic motion like a block-spring system. Thus, we can think of $\sum F = -\left(\dfrac{2T}{L} \right) y$ as an example of $\sum F = -kx$. By comparison, we identify the effective spring constant as $k = 2T/L$. Then the angular frequency is $\omega = \sqrt{\dfrac{k}{m}} = \sqrt{\dfrac{2T}{mL}}$ ∎

Finalize: Not everything with a stable equilibrium position will execute simple harmonic motion when disturbed and released, but a whole lot of things do when the disturbance is small.

---

72. A smaller disk of radius r and mass m is attached rigidly to the face of a second larger disk of radius R and mass M as shown in Figure P15.72. The center of the small disk is located at the edge of the large disk. The large disk is mounted at its center on a frictionless axle. The assembly is rotated through a small angle θ from its equilibrium position and released. (a) Show that the speed of the center of the small disk as it passes through the equilibrium position is

$$v = 2\left[\frac{Rg(1-\cos\theta)}{(M/m)+(r/R)^2+2}\right]^{1/2}$$

(b) Show that the period of the motion is

$$T = 2\pi\left[\frac{(M+2m)R^2+mr^2}{2mgR}\right]^{1/2}$$

Solution

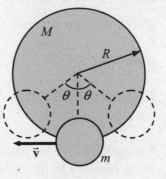

Figure P15.72

Conceptualize: This compound rigid object is a physical pendulum. Larger mass M for the big disk and larger radii R and r all tend to lower its frequency and so to lengthen its period.

Categorize: Energy conservation for the rotor-Earth system will give us the speed through equilibrium and the theory of the physical pendulum will tell us the period of the motion.

Analyze

(a) The energy version of the isolated system model, applied between the release point and the bottom of the swing, gives

$$K_{\text{top}} + U_{\text{top}} = K_{\text{bot}} + U_{\text{bot}}$$

where $\qquad K_{\text{top}} = U_{\text{bot}} = 0$

Therefore, $\qquad mgh = \tfrac{1}{2}I\omega^2$

where $\qquad h = R - R\cos\theta = R(1-\cos\theta)$

and the angular speed of the object at the equilibrium position is $\omega = v/R$. (In this solution ω does not represent the angular frequency of the motion.)

The parallel-axis theorem gives $\qquad I = \tfrac{1}{2}MR^2 + \tfrac{1}{2}mr^2 + mR^2$

Substituting, we find $\qquad mgR(1-\cos\theta) = \tfrac{1}{2}\left(\tfrac{1}{2}MR^2 + \tfrac{1}{2}mr^2 + mR^2\right)\dfrac{v^2}{R^2}$

or $\qquad mgR(1-\cos\theta) = \left(\tfrac{1}{4}M + \tfrac{1}{4}\dfrac{r^2}{R^2}m + \tfrac{1}{2}m\right)v^2$

We solve for the linear speed of the center of the small disk.

with $\qquad v^2 = \dfrac{4gR(1-\cos\theta)}{(M/m)+\left(r^2/R^2\right)+2}$

so $\qquad v = 2\sqrt{\dfrac{Rg(1-\cos\theta)}{(M/m)+(r/R)^2+2}}$ ∎

(b) A physical pendulum has period $T = 2\pi\sqrt{\dfrac{I}{M_T g d}}$

Here the total mass is $M_T = m + M$, we read the moment of inertia about the point of support from part (a), and the distance from the point of support to the center of mass is $d = \dfrac{mR + M(0)}{m+M}$

The answer assembled by substitution is

$$T = 2\pi\sqrt{\frac{\tfrac{1}{2}MR^2 + \tfrac{1}{2}mr^2 + mR^2}{mgR}} = 2\pi\sqrt{\frac{(M+2m)R^2 + mr^2}{2mgR}}$$ ∎

Finalize: This is an interesting apparatus for laboratory investigation. To display the relationship of T to r with the other variables constant, one could plot a graph of T^2 versus r^2. We read its theoretical slope as $2\pi^2/gR$.

16

Wave Motion

EQUATIONS AND CONCEPTS

The **wave function** $y(x, t)$ represents the y coordinate of an element or small segment of a medium as a wave pulse travels along the x direction. The value of y depends on two variables (position and time) and is read "y as a function of x and t". *When t has a fixed value, $y = y(x)$ defines the shape of the wave pulse at an instant in time.*

$$y(x, t) = f(x - vt) \text{ (pulse traveling right)} \quad (16.1)$$

$$y(x, t) = f(x + vt) \text{ (pulse traveling left)} \quad (16.2)$$

A **sinusoidal wave** form, $y = f(x)$, is shown in the figure at right.

Wavelength, λ, is the distance between identical points (i.e. points that are the same distance from their equilibrium positions and moving in the same direction) on adjacent crests.

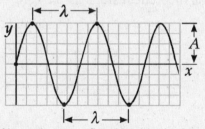

Wave shape at fixed instant in time.

Amplitude, A, is the maximum displacement from equilibrium of an element of the medium.

Period, T, is the time required for the disturbance (or pulse) to travel along the direction of propagation a distance equal to one wavelength.

The **frequency, f, of a periodic wave** is the inverse of the period. *It is the same as the frequency of the harmonic oscillation of one element of the medium through which the wave is moving.*

$$f = \frac{1}{T} \quad (16.3)$$

The **displacement of a sinusoidal wave** repeats itself when x is increased by an integral multiple of λ. *If the wave moves to the left, the quantity $(x - vt)$ is replaced with $(x + vt)$.*

$$y(x, t) = A \sin\left[\frac{2\pi}{\lambda}(x - vt)\right] \quad (16.5)$$

The **wave speed** can be expressed in terms of the wavelength and period. *A traveling wave moves a distance x = λ in a time t = T.*

$$v = \frac{\lambda}{T} \qquad (16.6)$$

The **periodic nature of y** as seen in Equation 16.7, is found by combining Equations 16.5 and 16.6.

$$y = A \sin\left[2\pi\left(\frac{x}{\lambda} - \frac{t}{T}\right)\right] \qquad (16.7)$$

Characteristic wave quantities:

    angular wave number k,

$$k \equiv \frac{2\pi}{\lambda} \qquad (16.8)$$

    angular frequency ω.

$$\omega \equiv \frac{2\pi}{T} = 2\pi f \qquad (16.9)$$

The **wave function of a sinusoidal wave** can be expressed in a more compact form in terms of the parameters defined above.

$$y = A \sin(kx - \omega t) \qquad (16.10)$$

The **general expression for a transverse wave** as stated in Equation 16.13 requires a phase constant when the transverse displacement (y) is not zero at x = 0 and t = 0.

$$y = A \sin(kx - \omega t + \phi) \qquad (16.13)$$

$\phi = $ phase constant

The **wave speed** can also be expressed in alternative forms.

$$v = \frac{\lambda}{T} \qquad (16.6)$$

$$v = \frac{\omega}{k} \qquad (16.11)$$

$$v = \lambda f \qquad (16.12)$$

The **transverse speed** v_y and **transverse acceleration** a_y of a point on a sinusoidal wave on a string are out of phase by $\pi/2$ radians. *Do not confuse transverse speed (v_y) with wave speed (v); v_y and a_y describe the motion of an element of the string that moves perpendicular to the direction of propagation of the wave.*

$$v_y = -\omega A \cos(kx - \omega t) \qquad (16.14)$$

$$a_y = -\omega^2 A \sin(kx - \omega t) \qquad (16.15)$$

Note the **maximum values of transverse speed and acceleration**.

$$v_{y,max} = \omega A \qquad (16.16)$$

$$a_{y,max} = \omega^2 A \qquad (16.17)$$

Wave speed (speed of a transverse pulse) along a stretched string depends on the tension in the string T and the linear density μ (mass

$$v = \sqrt{\frac{T}{\mu}} \qquad (16.18)$$

per unit length) of the string. *Note: The symbol T in Equation 16.18 represents the tension in the string, not the period of vibration.*

The **power** (rate of energy transfer) transmitted by a sinusodial wave on a string is proportional to the square of the angular frequency and to the square of the amplitude, where μ is the mass per unit length of the string.

$$P = \tfrac{1}{2}\mu\omega^2 A^2 v \qquad (16.21)$$

The **linear wave equation** is satisfied by any wave function having the form $y = f(x \pm vt)$. Equation 16.27 applies to various types of traveling waves (e.g. waves on strings, sound waves, electromagnetic waves).

$$\frac{\partial^2 y}{\partial x^2} = \frac{1}{v^2}\frac{\partial^2 y}{\partial t^2} \qquad (16.27)$$

REVIEW CHECKLIST

You should be able to:

- Recognize whether or not a given function is a possible description of a traveling wave and identify the direction (+ or −) in which the wave is traveling. (Section 16.1).

- Express a given sinusoidal wave function in several alternative forms involving different combinations of the wave parameters: wavelength, period, phase velocity, wave number, angular frequency, and sinusoidal frequency. (Section 16.2)

- Obtain values for the characteristic wave parameters: A, ω, k, λ, f, and ϕ if given a specific wave function for a sinusoidal wave. (Section 16.2)

- Calculate the speed of a transverse pulse traveling on a string. (Section 16.3)

- Describe the reflection of a traveling wave pulse at a fixed end and at a free end of a string. Also, describe the reflection of a pulse at a boundary between two media. (Section 16.4)

- Calculate the rate at which energy is transported by sinusoidal waves in a string. (Section 16.5)

ANSWERS TO SELECTED OBJECTIVE QUESTIONS

6. If you stretch a rubber hose and pluck it, you can observe a pulse traveling up and down the hose. **(i)** What happens to the speed of the pulse if you stretch the hose more tightly? (a) It increases. (b) It decreases. (c) It is constant. (d) It changes unpredictably. **(ii)** What happens to the speed if you fill the hose with water? Choose from the same possibilities.

Answer **(i)** (a) If you stretch the hose tighter, you increase the tension, and increase the speed of the wave. If the hose elongates at all, then you decrease the linear density, which also increases the speed of the wave. **(ii)** (b) If you fill it with water, you increase the linear density of the hose, and decrease the speed of the wave.

□ □ □ □

8.   A source vibrating at constant frequency generates a sinusoidal wave on a string under constant tension. If the power delivered to the string is doubled, by what factor does the amplitude change? (a) a factor of 4 (b) a factor of 2 (c) a factor of $\sqrt{2}$ (d) a factor of 0.707 (e) cannot be predicted

Answer   (c) Power is always proportional to the square of the amplitude. If power doubles at constant frequency, amplitude increases by $\sqrt{2}$ times, a factor of 1.41.

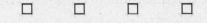

ANSWER TO A CONCEPTUAL QUESTION

2.   (a) How would you create a longitudinal wave in a stretched spring? (b) Would it be possible to create a transverse wave in a spring?

Answer   (a) A longitudinal wave can be set up in a stretched spring by compressing the coils in a small region, and releasing the compressed region. The disturbance will proceed to propagate as a longitudinal pulse. (b) It is quite possible to set up a transverse wave in a spring, simply by displacing a section of the spring in a direction perpendicular to its length and releasing it.

SOLUTIONS TO SELECTED END-OF-CHAPTER PROBLEMS

1.   At $t = 0$, a transverse pulse in a wire is described by the function

$$y = \frac{6.00}{x^2 + 3.00}$$

where x and y are in meters. If the pulse is traveling in the positive x direction with a speed of 4.50 m/s, write the function $y(x, t)$ that describes this pulse.

Solution

Conceptualize: At time zero, the wave pulse looks like a bump centered at $x = 0$. As time goes on, the wave function …

Categorize: …will be a function of t as well as x. The point about which the bump is centered will be $x_0 = 4.5t$.

Analyze: We obtain a function of the same shape by writing $y(x, t) = 6/[(x - x_0)^2 + 3]$ where the center of the pulse is at $x_0 = 4.5t$. Thus, we have

$$y(x,t) = \frac{6}{(x - 4.5t)^2 + 3} \qquad \blacksquare$$

Note that for y to stay constant as t increases, x must increase by $4.5t$, as it should to describe the wave moving at 4.5 m/s.

Finalize: In general, we can cause any waveform to move along the x axis at a velocity v_x by substituting $(x - v_x t)$ for x in the wave function $y(x)$ at $t = 0$. A wave function that depends on t through $(x + v_x t)$ describes a wave moving in the negative x direction.

---

5. The wave function for a traveling wave on a taut string is (in SI units)

$$y(x, t) = 0.350 \sin(10\pi t - 3\pi x + \pi/4)$$

(a) What are the speed and direction of travel of the wave? (b) What is the vertical position of an element of the string at $t = 0$, $x = 0.100$ m? What are (c) the wavelength and (d) the frequency of the wave? (e) What is the maximum transverse speed of an element of the string?

Solution

Conceptualize: As a function of two independent variables, the wave function represents a moving graph. The position of one bit of string at one instant is the smallest example of the information it contains. The constants amplitude, wave speed, wavelength, and frequency describe the whole wave.

Categorize: We use the traveling wave model. We compare the given equation with $y = A \sin(kx - \omega t + \phi)$

Analyze: We note that $\sin(\theta) = -\sin(-\theta) = \sin(-\theta + \pi)$

so the given wave function can be written

$$y(x, t) = (0.350) \sin(-10\pi t + 3\pi x + \pi - \pi/4)$$

Thus we find that $k = 3\pi$ rad/m and $\omega = 10\pi$ rad/s

(a) The speed and direction of the wave are both specified by the vector wave velocity:

$$\vec{v} = f\lambda \hat{i} = \frac{\omega}{k}\hat{i} = \frac{10\pi \ \text{rad/s}}{3\pi \ \text{rag/m}}\hat{i} = 3.33\hat{i} \ \text{m/s} \quad \blacksquare$$

(b) Substituting $t = 0$ and $x = 0.100$ m, we have

$$y = (0.350 \ \text{m}) \sin(-0.300\ \pi + 0.250\ \pi) = (0.350 \ \text{m}) \sin(-0.157)$$

$$= (0.350 \ \text{m})(-0.156) = -0.054\ 8 \ \text{m} = -5.48 \ \text{cm} \quad \blacksquare$$

Note that when you take the sine of a quantity with no units, the quantity is not in degrees, but in radians.

(c) The wavelength is $\quad \lambda = \dfrac{2\pi \ \text{rad}}{k} = \dfrac{2\pi \ \text{rad}}{3\pi \ \text{rad/m}} = 0.667 \ \text{m} \quad \blacksquare$

(d) and the frequency is $\quad f = \dfrac{\omega}{2\pi \ \text{rad}} = \dfrac{10\pi \ \text{rad/s}}{2\pi \ \text{rad}} = 5.00 \ \text{Hz} \quad \blacksquare$

(e) The particle speed is

$$v_y = \frac{\partial y}{\partial t} = (0.350 \ \text{m})(10\pi \ \text{rad/s}) \cos(10\pi t - 3\pi x + \pi/4)$$

The maximum occurs when the cosine factor is 1:

$$v_{y,max} = (10\pi \text{ rad/s})(0.350 \text{ m/s}) = 11.0 \text{ m/s}$$ ■

Finalize: Note the large difference between the maximum particle speed and the wave speed found in part (a). Would you write $\sin(-0.157) = -0.002\ 74$? Your teacher knows that that mistake reveals a student who has not been thinking about units through the whole course—a student who has not been thinking physically. The $\sin(-0.157)$ in step (b) refers to the sine of an angle with no units, which is the sine of an angle in radians.

---

7. A sinusoidal wave is traveling along a rope. The oscillator that generates the wave completes 40.0 vibrations in 30.0 s. A given crest of the wave travels 425 cm along the rope in 10.0 s. What is the wavelength of the wave?

Solution

Conceptualize: The given information comprises things that can be measured directly in laboratory. A high-speed photograph would reveal the wavelength...

Categorize: ...which can be computed from $v = f\lambda$.

Analyze: The frequency is

$$f = \frac{40.0 \text{ waves}}{30.0 \text{ s}} = 1.33 \text{ s}^{-1}$$

and the wave speed is

$$v = \frac{425 \text{ cm}}{10.0 \text{ s}} = 42.5 \text{ cm/s}$$

Since $v = \lambda f$, the wavelength is

$$\lambda = \frac{v}{f} = \frac{42.5 \text{ cm/s}}{1.33 \text{ s}^{-1}} = 0.319 \text{ m}$$ ■

Finalize: If we turned up the oscillator to a higher frequency, the wavelength would get shorter because the speed would be unchanged. The speed is determined just by properties of the string itself.

---

9. A wave is described by $y = 0.020\ 0 \sin(kx - \omega t)$, where $k = 2.11$ rad/m, $\omega = 3.62$ rad/s, x and y are in meters, and t is in seconds. Determine (a) the amplitude, (b) the wavelength, (c) the frequency, and (d) the speed of the wave.

Solution

Conceptualize: The wave function is a moving graph. The position of a particle of the medium, represented by y, is varying all the time and from every point to the next point at each instant. But we can pick out the parameters that characterize the whole wave and have constant values.

Categorize: We compare the given wave function with the general sinusoidal wave equation,...

Analyze: …$y = A \sin(kx - \omega t + \phi)$

(a) Its functional equality to $y = 0.020\,0 \sin(kx - \omega t)$ reveals that the amplitude is $A = 2.00$ cm   ∎

(b) The angular wave number is $k = 2.11$ rad/m   so that $\lambda = \dfrac{2\pi}{k} = 2.98$ m ∎

(c) The angular frequency is $\omega = 3.62$ rad/s   so that $f = \dfrac{\omega}{2\pi} = 0.576$ Hz   ∎

(d) The speed is $v = f\lambda = (0.576 \text{ s}^{-1})(2.98 \text{ m}) = 1.72$ m/s   ∎

Finalize: It is not important to the dynamics of the wave, but we can also identify the phase constant as $\phi = 0$. We could write the wave function to explicitly display the constant parameters as

$$y(x, t) = (2.00 \text{ cm}) \sin\left(\frac{2\pi}{2.98 \text{ m}} x - 2\pi(0.576/\text{s})t + 0\right)$$

---

19. (a) Write the expression for y as a function of x and t in SI units for a sinusoidal wave traveling along a rope in the negative x direction with the following characteristics: $A = 8.00$ cm, $\lambda = 80.0$ cm, $f = 3.00$ Hz, and $y(0, t) = 0$ at $t = 0$. (b) **What If?** Write the expression for y as a function of x and t for the wave in part (a) assuming that $y(x, 0) = 0$ at the point $x = 10.0$ cm.

Solution

Conceptualize: Think about the graph of y as a function of x at one instant as a smooth succession of identical crests and troughs, with the length in space of each cycle being 80.0 cm. The distance from the top of a crest to the bottom of a trough is 16.0 cm. Now think of the whole graph moving toward the left at 240 cm/s.

Categorize: Using the traveling wave model, we can put constants with the right values into $y = A \sin(kx + \omega t + \phi)$ to have the mathematical representation of the wave. We have the same (positive) signs for both kx and ωt so that a point of constant phase will be at a decreasing value of x as t increases—that is, so that the wave will move to the left.

Analyze: The amplitude is $A = y_{max} = 8.00$ cm $= 0.080\,0$ m

The wave number is $k = \dfrac{2\pi}{\lambda} = \dfrac{2\pi}{0.800 \text{ m}} = 2.50\,\pi\text{m}^{-1}$

The angular frequency is $\omega = 2\pi f = 2\pi(3.00 \text{ s}^{-1}) = 6.00\pi$ rad/s

(a) In $y = A \sin(kx + \omega t + \phi)$, choosing $\phi = 0$ will make it true that $y(0, 0) = 0$. Then the wave function becomes upon substitution of the constant values for this wave

$$y = (0.080\,0) \sin(2.50\pi x + 6.00\pi t)$$ ∎

(b) In general, $y = (0.080\,0) \sin(2.50\pi x + 6.00\pi t + \phi)$

If $y(x, 0) = 0$ at $x = 0.100$ m, we require $0 = (0.080\,0 \text{ m}) \sin(0.250\pi + \phi)$

so we must have the phase constant be $\qquad \phi = -0.250\pi$ rad

Therefore, the wave function for all values of the variables x and t is

$$y = (0.080\ 0)\sin(2.50\pi x + 6.00\ \pi t - 0.250\pi)$$ ■

Finalize: Instead of being a number, the answer for each part of this problem is a function y and a function of two variables at that. The sinusoidal shape of the function's graph and the values for the constant parameters that the function contains make it the wave function for the particular wave specified in the problem.

---

23. Transverse waves travel with a speed of 20.0 m/s in a string under a tension of 6.00 N. What tension is required for a wave speed of 30.0 m/s in the same string?

Solution

Conceptualize: Since v is proportional to $\sqrt{T}$, the new tension must be about twice as much as the original to achieve a 50% increase in the wave speed.

Categorize: The equation for the speed of a transverse wave on a string under tension can be used if we assume that the linear density of the string is constant. Then the ratio of the two wave speeds can be used to find the new tension.

Analyze: The two wave speeds can be written as:

$$v_1 = \sqrt{T_1/\mu} \qquad \text{and} \qquad v_2 = \sqrt{T_2/\mu}$$

Dividing the equations sets up the proportion

$$\frac{v_2}{v_1} = \sqrt{\frac{T_2}{T_1}} \qquad \text{so that} \qquad T_2 = \left(\frac{v_2}{v_1}\right)^2 T_1 = \left(\frac{30.0\ \text{m/s}}{20.0\ \text{m/s}}\right)^2 (6.00\ \text{N}) = 13.5\ \text{N}$$ ■

Finalize: The new tension is slightly more than twice the original, so the result is reasonable, agreeing with our prediction.

---

31. A steel wire of length 30.0 m and a copper wire of length 20.0 m, both with 1.00-mm diameters, are connected end to end and stretched to a tension of 150 N. During what time interval will a transverse wave travel the entire length of the two wires?

Solution

Conceptualize: This system is like a long guitar string. A hammer-tap perpendicular to the length of the wire at one end will create a pulse of wire displacement (not sound) that will travel to the far end of the compound wire, and can be used as a signal.

Categorize: Computing the speed in each material will let us find the travel time.

Analyze: The total time interval of travel is the sum of the two time intervals.

In each wire separately the travel time is $\Delta t = \dfrac{L}{v} = L\sqrt{\dfrac{\mu}{T}}$

where L is the length of each wire. Let A represent the cross-sectional area of the wire. Then the linear density can be computed from the material's mass-per-volume density as

$$\mu = \frac{m}{L} = \rho\frac{V}{L} = \rho\frac{AL}{L} = \rho A = \frac{\pi \rho d^2}{4}$$

By substitution the travel time is $\Delta t = L\sqrt{\dfrac{\pi \rho d^2}{4T}}$

For copper, $\Delta t_1 = (20.0 \text{ m})\left[\dfrac{\pi(8\,920 \text{ kg/m}^3)(0.001\,00 \text{ m})^2}{4(150 \text{ kg} \cdot \text{m/s}^2)}\right]^{1/2} = 0.137 \text{ s}$

For steel, $\Delta t_2 = (30.0 \text{ m})\left[\dfrac{\pi(7\,860 \text{ kg/m}^3)(0.001\,00 \text{ m})^2}{4(150 \text{ kg} \cdot \text{m/s}^2)}\right]^{1/2} = 0.192 \text{ s}$

The total time interval is $(0.137 \text{ s}) + (0.192 \text{ s}) = 0.329 \text{ s}$ ∎

Finalize: We could have computed numerically the cross-sectional area, then the mass of each whole wire, then the linear density, then the wave speed, and then the travel time. Proceeding in symbols saves space. It can significantly reduce time and mistakes in calculator use. And the linear density is the mass of a one-meter section, so it does not really depend on the whole length.

---

34. Sinusoidal waves 5.00 cm in amplitude are to be transmitted along a string that has a linear mass density of 4.00×10^{-2} kg/m. The source can deliver a maximum power of 300 W and the string is under a tension of 100 N. What is the highest frequency f at which the source can operate?

Solution

Conceptualize: Turning up the source frequency will increase the power carried by the wave. Perhaps on the order of a hundred hertz will be the frequency at which the energy-per-second is 300 J/s.

Categorize: We will use the expression for power carried by a wave on a string.

Analyze: The wave speed is $v = \sqrt{\dfrac{T}{\mu}} = \sqrt{\dfrac{100 \text{ N}}{4.00 \times 10^{-2} \text{ kg/m}}} = 50.0 \text{ m/s}$

From $P = \frac{1}{2}\mu\omega^2 A^2 v$, we have

$$\omega^2 = \frac{2P}{\mu A^2 v} = \frac{2(300 \text{ N} \cdot \text{m/s})}{(4.00 \times 10^{-2} \text{ kg/m})(5.00 \times 10^{-2} \text{ m})^2 (50.0 \text{ m/s})}$$

Computing, $\omega = 346.4 \text{ rad/s}$ and $f = \frac{\omega}{2\pi} = 55.1 \text{ Hz}$ ∎

Finalize: This string wave would softly broadcast sound into the surrounding air, at the frequency of the second-lowest note called A on a piano. If we tried to turn the source to a higher frequency, it might just vibrate with smaller amplitude. The power is generally proportional to the squares of both the frequency and the amplitude.

---

35. A sinusoidal wave on a string is described by the wave function

$$y = 0.15 \sin(0.80x - 50t)$$

where x and y are in meters and t is in seconds. The mass per unit length of this string is 12.0 g/m. Determine (a) the speed of the wave, (b) the wavelength, (c) the frequency, and (d) the power transmitted to the wave.

Solution

Conceptualize: The wave function is about variation of the position of points on the string in time and in space along the string. But constants in the wave function …

Categorize: …let us identify the wave number, the angular frequency, and so the quantities that the problem asks for.

Analyze: Comparing the given wave function, $y = (0.15)\sin(0.80x - 50t)$

with the general wave function, $y = A \sin(kx - \omega t)$

we have $k = 0.80 \text{ rad/m}$ and $\omega = 50 \text{ rad/s}$ and $A = 0.15 \text{ m}$

(a) The wave speed is then $v = f\lambda = \frac{\omega}{k} = \frac{50.0 \text{ rad/s}}{0.80 \text{ rad/m}} = 62.5 \text{ m/s}$ ∎

(b) The wavelength is $\lambda = \frac{2\pi}{k} = \frac{2\pi \text{ rad}}{0.80 \text{ rad/m}} = 7.85 \text{ m}$ ∎

(c) The frequency is $f = \frac{\omega}{2\pi} = \frac{50 \text{ rad/s}}{2\pi \text{ rad}} = 7.96 \text{ Hz}$ ∎

(d) The wave carries power

$$P = \frac{1}{2}\mu\omega^2 A^2 v = \frac{1}{2}(0.012\,0 \text{ kg/m})(50.0 \text{ s}^{-1})^2(0.150 \text{ m})^2(62.5 \text{ m/s}) = 21.1 \text{ W}$$ ∎

Finalize: You would have to purchase 21.1 joules every second from some energy company to run the source of this wave. The wave function is a complete specification of the wave in practical terms, as well as a moving graph telling where every particle of the medium is at every instant.

Do not confuse the angular frequency (here 50 rad/s) and the frequency (here 7.96 Hz).

---

43. Show that the wave function $y = \ln[b(x - vt)]$ is a solution to Equation 16.27, where b is a constant.

Solution

Conceptualize: We would prove that $q = 6$ satisfies the equation $2q^2 - 5q = 42$ by substituting in the value of q and observing that it gives a true result. The given function is not a realistic wave function over the whole range of x from $-\infty$ to $+\infty$, because the logarithm function diverges to infinity itself and is not defined for negative arguments. But it is a useful…

Categorize: …exercise to prove that a wave function satisfies the wave equation. We will differentiate the wave *function*, substitute the derivatives into the wave *equation* $\dfrac{\partial^2 y}{\partial x^2} = \dfrac{1}{v^2}\dfrac{\partial^2 y}{\partial t^2}$, and show that the resulting equation is true.

Analyze: The important thing to remember with partial derivatives is that **you treat all variables as constants, except the single variable of interest**. Keeping this in mind, we must apply two standard rules of differentiation to the function $y = \ln[b(x - vt)]$:

$$\frac{\partial}{\partial x}[\ln f(x)] = \frac{1}{f(x)}\frac{\partial(f(x))}{\partial x} \qquad\qquad\qquad [1]$$

$$\frac{\partial}{\partial x}\left[\frac{1}{f(x)}\right] = \frac{\partial}{\partial x}[f(x)]^{-1} = (-1)[f(x)]^{-2}\frac{\partial(f(x))}{\partial x} = -\frac{1}{[f(x)]^2}\frac{\partial(f(x))}{\partial x} \qquad [2]$$

Applying [1], $\qquad \dfrac{\partial y}{\partial x} = \left(\dfrac{1}{b(x - vt)}\right)\dfrac{\partial(bx - bvt)}{\partial x} = \left(\dfrac{1}{b(x - vt)}\right)(b) = \dfrac{1}{x - vt}$

Applying [2], $\qquad \dfrac{\partial^2 y}{\partial x^2} = -\dfrac{1}{(x - vt)^2}$

In a similar way, $\quad \dfrac{\partial y}{\partial t} = \dfrac{-v}{(x - vt)} \qquad$ and $\qquad \dfrac{\partial^2 y}{\partial t^2} = -\dfrac{v^2}{(x - vt)^2}$

From the second-order partial derivatives, we see that it is true that

$$\frac{\partial^2 y}{\partial x^2} = \frac{1}{v^2}\frac{\partial^2 y}{\partial t^2} \qquad \text{so the proposed function } \textbf{is} \text{ one solution to the wave equation.} \quad \blacksquare$$

Finalize: The wave *equation* $\dfrac{\partial^2 y}{\partial x^2} = \dfrac{1}{v^2}\dfrac{\partial^2 y}{\partial t^2}$ is analogous to Newton's laws in mechanics, specifying a rule any wave must follow to propagate along the x axis. A wave *function* corresponds to a pulse or wave train with a particular shape, given to it by its source or some external conditions. The wave function must, as in this problem, be one solution to the wave equation, which has an infinity of solutions.

---

53. Review. A block of mass M, supported by a string, rests on a frictionless incline making an angle θ with the horizontal (Fig. P16.53). The length of the string is L and its mass is $m \ll M$. Derive an expression for the time interval required for a transverse wave to travel from one end of the string to the other.

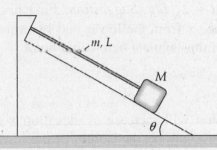

Figure P16.53

Solution

Conceptualize: A larger value of M would cause more tension in the string, a higher wave speed, and a smaller time interval for a pulse to traverse the string. A larger value of θ would have the same qualitative effect. On the other hand, if m or L were increased, the pulse would move slower or have farther to go, and the transit time would be increased.

Categorize: We will analyze the forces on the block to find the tension in the cord. Then we will represent the wave speed in terms of the quantities given, and finally use speed = distance/time to find the time interval required.

Analyze: We note the incline is frictionless and take the positive x-direction to be up the incline. Then the particle in equilibrium model applied to the block gives

$$\sum F_x = T - Mg\sin\theta = 0$$

so the tension in the string is $T = Mg\sin\theta$

The speed of transverse waves in the string is then

$$v = \sqrt{\frac{T}{\mu}} = \sqrt{\frac{Mg\sin\theta}{m/L}} = \sqrt{\frac{MgL\sin\theta}{m}}$$

The time interval for a pulse to travel the string's length is

$$\Delta t = \frac{L}{v} = L\sqrt{\frac{m}{MgL\sin\theta}} = \sqrt{\frac{mL}{Mg\sin\theta}}$$ ∎

Finalize: Our derived result agrees with the qualitative patterns of change that we identified at the Conceptualize step. We did not guess in advance that the proportionalities would be to square roots. As we study different kinds of waves, we will see that their speeds are given by square-root relationships remarkably often.

58. A rope of total mass m and length L is suspended vertically. Analysis shows that for short transverse pulses, the waves above a short distance from the free end of the rope can be represented to a good approximation by the linear wave equation discussed in Section 16.6. Show that a transverse pulse travels the length of the rope in a time interval that is given approximately by $\Delta t \approx 2\sqrt{L/g}$. *Suggestion:* First find an expression for the wave speed at any point a distance x from the lower end by considering the rope's tension as resulting from the weight of the segment below that point.

Solution

Conceptualize: The tension will increase as elevation x increases, because the upper part of the hanging rope must support the weight of the lower part. Then the wave speed increases with height.

Categorize: We will need to do an integral based on $v = dx/dt$ to find the travel time of the variable-speed pulse.

Analyze: We define $x = 0$ at the bottom of the rope and $x = L$ at the top of the rope. The tension in the rope at any point is the weight of the rope below that point. We can thus write the tension in the rope at each point x as $T = \mu x g$, where μ is the mass per unit length of the rope, which we assume uniform. The speed of the wave pulse at each point along the rope's length is therefore

$$v = \sqrt{\frac{T}{\mu}} \qquad \text{or} \qquad v = \sqrt{gx}$$

But at each point x, the wave propagates at a rate of $v = \dfrac{dx}{dt}$

So we substitute for v, and generate the differential equation

$$\frac{dx}{dt} = \sqrt{gx} \qquad \text{or} \qquad dt = \frac{dx}{\sqrt{gx}}$$

Integrating both sides from the bottom end of the rope to the top, we find

$$\Delta t = \frac{1}{\sqrt{g}} \int_0^L \frac{dx}{\sqrt{x}} = \frac{\left[2\sqrt{x}\right]_0^L}{\sqrt{g}} = 2\sqrt{\frac{L}{g}} \qquad \blacksquare$$

Finalize: What you have learned in mathematics class is another operation that can be done to both sides of an equation: integrating from one physical point to another. Here the integral extends in time and space from the situation of the pulse being at the bottom of the string to the situation with the pulse reaching the top.

---

60. Review. An aluminum wire is held between two clamps under zero tension at room temperature. Reducing the temperature, which results in a decrease in the wire's equilibrium length, increases the tension in the wire. Take the cross-sectional area of the wire to be 5.00×10^{-6} m^2, the density to be 2.70×10^3 kg/m³, and Young's modulus to be 7.00×10^{10} N/m². What strain ($\Delta L/L$) results in a transverse wave speed of 100 m/s?

Solution

Conceptualize: We expect some small fraction like 10^{-3} as an answer. The situation might remind you of fitting a hot iron rim to a wagon wheel or a shrink-fit plastic seal to a medicine bottle.

Categorize: We must review the relationship of strain to stress, and algebraically combine it with the equation for the speed of a string wave.

Analyze: The expression for the elastic modulus, $Y = \dfrac{F/A}{\Delta L/L}$

becomes an equation for strain $\qquad \dfrac{\Delta L}{L} = \dfrac{F/A}{Y} \qquad$ [1]

We substitute into the equation for the wave speed $\qquad v = \sqrt{T/\mu} = \sqrt{F/\mu}$

where tension T means the same as stretching force F.

The definition for linear density μ is related to the mass-per-volume

density of the material by $\qquad \mu = \dfrac{m}{L} = \dfrac{\rho(AL)}{L} = \rho A$

We substitute, solve, $\qquad v^2 = \dfrac{F}{\mu} = \dfrac{1}{\rho}\left(\dfrac{F}{A}\right) \qquad$ or $\qquad \left(\dfrac{F}{A}\right) = \rho v^2 \qquad$ [2]

and substitute [2] into [1], to obtain

$$\frac{\Delta L}{L} = \frac{\rho v^2}{Y} = \frac{\left(2.70 \times 10^3 \text{ kg/m}^3\right)(100 \text{ m/s})^2}{7.00 \times 10^{10} \text{ N/m}^2} = 3.86 \times 10^{-4} \qquad \blacksquare$$

Finalize: The Young's modulus equation describes stretching that does not really happen, being caused and cancelled out simultaneously by changes in tension and temperature. This wire could be used as a thermometer. The frequency at which it vibrates, after it is tapped, is determined by its constant length and the speed of the string wave. If the frequency is in the audible range, as described in the next chapter, you could hear the sound it radiates into the surrounding air. Then you could pluck it, pour liquid nitrogen onto it, pluck it again, and listen to the pitch go up as the frequency increases.

———————————

63. A rope of total mass m and length L is suspended vertically. As shown in Problem 58, a pulse travels from the bottom to the top of the rope in an approximate time interval $\Delta t = 2\sqrt{L/g}$ with a speed that varies with position x measured from the bottom of the rope as $v = \sqrt{gx}$. Assume that the linear wave equation in Section 16.6 describes waves at all locations on the rope. (a) Over what time interval does a pulse travel halfway up the rope?

Give your answer as a fraction of the quantity $2\sqrt{L/g}$. (b) A pulse starts traveling up the rope. How far has it traveled after a time interval $\sqrt{L/g}$?

Solution

Conceptualize: The wave pulse travels faster as it goes up the rope because the tension higher in the rope is greater (to support the weight of the rope below it). Therefore it should take more than half the total time Δt for the wave to travel halfway up the rope. Likewise, the pulse should travel less than halfway up the rope in time interval $\Delta t/2$.

Categorize: By using the time relationship given in the problem and making suitable substitutions, we can find the required time interval and distance.

Analyze:

(a) From the equation given, the time interval for a pulse to travel any distance, d, up from the bottom of a rope is $\Delta t_d = 2\sqrt{d/g}$. So the time for a pulse to travel a distance $L/2$ from the bottom is

$$\Delta t_{L/2} = 2\sqrt{\frac{L}{2g}} = 0.707\left(2\sqrt{\frac{L}{g}}\right) \qquad \blacksquare$$

(b) Likewise, the distance a pulse travels from the bottom of a rope in a time Δt_d is $d = g\Delta t_d^2/4$. So the distance traveled by a pulse after a time interval

$$\Delta t_d = \sqrt{\frac{L}{g}} \quad \text{is} \quad d = \frac{g(L/g)}{4} = \frac{L}{4} \qquad \blacksquare$$

Finalize: As expected, it takes the pulse more than 70% of the total time interval to cover 50% of the distance. In half the total trip time, the pulse has climbed only 1/4 of the total length. Solving the problem with proportional reasoning, we did not have to do any integrals and did not have to use the expression given in the problem for the variable pulse speed.

17

Sound Waves

EQUATIONS AND CONCEPTS

The **displacement from equilibrium** (s) of an element of a medium in which a harmonic sound wave is propagating has a sinusoidal variation in time. *The displacement is parallel to the direction of the propagation of the wave (longitudinal wave); and the maximum amplitude of the displacement is s_{max}.*

$$s(x, t) = s_{max} \cos(kx - \omega t) \qquad (17.1)$$

The **pressure variations** in a medium conducting a sound wave vary harmonically in time and are out of phase with the displacements by $\pi/2$ radians (compare Equations 17.1 and 17.2).

$$\Delta P = \Delta P_{max} \sin(kx - \omega t) \qquad (17.2)$$

The **speed of sound in a gas** depends on the bulk modulus B and equilibrium density of the medium in which the wave (a compression wave) is propagating.

$$v = \sqrt{\frac{B}{\rho}} \qquad (17.8)$$

The pressure amplitude is proportional to the displacement amplitude. *A sound wave may be considered as either a displacement wave or a pressure wave.*

$$\Delta P_{max} = \rho v \omega s_{max}$$

Intensity of a wave is defined as power per unit area perpendicular to the direction of travel of the wave.

$$I \equiv \frac{(Power)_{avg}}{A} \qquad (17.11)$$

The **intensity (power per unit area) of a periodic sound wave** is proportional to the square of the pressure amplitude. The intensity is also proportional to the square of the displacement amplitude and to the square of the angular frequency.

$$I = \frac{(\Delta P_{max})^2}{2\rho v} \qquad (17.12)$$

The **intensity I of any spherical wave** decreases as the square of the distance from the source.

$$I = \frac{(Power)_{avg}}{4\pi r^2} \quad (17.13)$$

Sound levels are measured on a logarithmic scale. β is measured in decibels (dB) and I is the corresponding intensity measured in W/m². *I_0 is a reference intensity corresponding to the threshold of hearing.*

$$\beta \equiv 10 \log\left(\frac{I}{I_0}\right) \quad (17.14)$$

$$I_0 = 10^{-12} \text{ W/m}^2$$

The **Doppler effect (apparent shift in frequency)** is observed whenever there is relative motion between a source and an observer. *Equation 17.19 is a general Doppler-shift expression; it applies to relative motion of source and observer toward or away from each other with correct use of algebraic signs.* A positive sign is used for both v_o and v_s whenever the observer or source moves toward the other. A negative sign is used for both whenever either moves away from the other. *In Equation 17.19, v_o and v_s are measured relative to the medium in which the sound travels.* See worked examples in *Suggestions, Skills, and Strategies section.*

$$f' = \left(\frac{v + v_o}{v - v_s}\right)f \quad (17.19)$$

where

v = speed of sound in a medium,

v_o = speed of the observer, and

v_s = speed of the source.

Shock waves are produced when a sound source moves through a medium with a speed which is greater than the wave speed in that medium. The shock wave front has a conical shape with a half angle which depends on the Mach number of the source, defined as the ratio v_s/v.

$$\sin\theta = \frac{v}{v_s}$$

SUGGESTIONS, SKILLS, AND STRATEGIES

SOUND INTENSITY AND THE DECIBEL SCALE

When making calculations using Equation 17.14 which defines the intensity of a sound wave on the decibel scale, the properties of logarithms must be kept clearly in mind.

In order to determine the decibel level corresponding to two sources sounded simultaneously, you must first find the intensity, I, of each source in W/m²; add these values, and then convert the resulting intensity to the decibel scale. As an illustration of this technique, determine the dB level when two sounds with intensities of $\beta_1 = 40$ dB and $\beta_2 = 45$ dB are sounded together.

Solve Equation 17.14 to find the intensity in terms of the dB level.

From: $\beta = 10 \log (I/I_0)$ find: $I = I_0 10^{\beta/10}$

Then for $\beta_1 = 40$ dB: $I_1 = (10^{-12} \text{ W/m}^2)10^4 = 1.00 \times 10^{-8} \text{ W/m}^2$

For $\beta_2 = 45$ dB: $I_2 = (10^{-12} \text{ W/m}^2)10^{4.5} = 3.16 \times 10^{-8} \text{ W/m}^2$ and

$$I_{total} = I_1 + I_2 = 4.16 \times 10^{-8} \text{ W/m}^2$$

Again using Equation 17.14 $\beta_{total} = 10 \log(I_{total}/I_0)$

So

$$\beta_{total} = 10 \log \frac{\left(4.16 \times 10^{-8} \text{ W/m}^2\right)}{\left(1.00 \times 10^{-12} \text{ W/m}^2\right)} = 10 \log(4.16 \times 10^4) = 46.2 \text{ dB}$$

The intensity level of the combined sources is 46.2 dB (not 40 dB + 45 dB = 85 dB).

DOPPLER EFFECT

Equation 17.19 is the generalized equation for the Doppler effect:

$$f' = \left(\frac{v + v_o}{v - v_s}\right) f$$

where

f = source frequency, v_o = speed of observer

f' = observed frequency, v_s = speed of source

v = speed of sound in medium

The most likely error in using Equation 17.19 is using the incorrect algebraic sign for the speed of either the observer or the source.

When the **relative motion** of source and observer is:

      either or both toward the other: **Enter v_s and v_o with + signs.**

      either or both away from the other: **Enter v_s and v_o with − signs.**

Remember in Equation 17.19 stated above, the algebraic signs (the plus sign in the numerator and the minus sign in the denominator) are part of the structure of the equation. These signs remain and correct signs, as stated above, must be entered along with the values of v_o and v_s.

Consider the following examples given a source frequency of 300 Hz and the speed of sound in air of 343 m/s.

(1) Source moving with a speed of 40 m/s toward a fixed observer. In this case, $v_o = 0$ and $v_s = +40$ m/s. Substituting into Equation 17.19,

$$f' = \left(\frac{v + v_o}{v - v_s}\right) f = \left(\frac{343 \text{ m/s} + 0}{343 \text{ m/s} - 40 \text{ m/s}}\right)(300 \text{ Hz}) = 340 \text{ Hz}$$

(2) Source moving away from a fixed observer with a speed of 40 m/s. In this case, $v_o = 0$ and $v_s = -40$ m/s; substituting into Equation 17.19,

$$f' = \left(\frac{v + v_o}{v - v_s}\right)f = \left(\frac{343 \text{ m/s} + 0}{343 \text{ m/s} - (-40 \text{ m/s})}\right)(300 \text{ Hz}) = 269 \text{ Hz}$$

(3) Observer moving with speed of 40 m/s toward a fixed source: $v_o = +40$ m/s and $v_s = 0$. Substituting into Equation 17.19,

$$f' = \left(\frac{v + v_o}{v - v_s}\right)f = \left(\frac{343 \text{ m/s} + 40 \text{ m/s}}{343 \text{ m/s} - 0}\right)(300 \text{ Hz}) = 335 \text{ Hz}$$

(4) Observer moving with speed 40 m/s away from a fixed source: $v_o = -40$ m/s and $v_s = 0$. Substituting into Equation 17.19,

$$f' = \left(\frac{v + v_o}{v - v_s}\right)f = \left(\frac{343 \text{ m/s} + (-40 \text{ m/s})}{343 \text{ m/s} - 0}\right)(300 \text{ Hz}) = 265 \text{ Hz}$$

As a check on your calculated value of f', remember:

- When the relative motion is source or observer toward the other, $f' > f$.

- When the relative motion is source or observer away from the other, $f' < f$.

REVIEW CHECKLIST

You should be able to:

- Describe the harmonic displacement and pressure variation as functions of time and position for a harmonic sound wave. Relate the displacement amplitude to the pressure amplitude for a harmonic sound wave and calculate the wave intensity from each of these parameters. (Section 17.1)

- Calculate the speed of sound in various media in terms of the appropriate elastic properties of a particular medium (including bulk modulus, Young's modulus, and the pressure-volume relationships of an ideal gas), and the corresponding inertial properties (usually mass density). (Section 17.2)

- Determine the total intensity due to one or several sound sources whose individual decibel levels are known. Calculate the decibel level due to some combination of sources whose individual intensities are known. (Section 17.3)

- Make calculations for the various situations under which the Doppler effect is observed. **Pay particular attention to the correct use of algebraic signs for the speeds of the source and the observer.** (Section 17.4)

ANSWER TO AN OBJECTIVE QUESTION

9. A point source broadcasts sound into a uniform medium. If the distance from the source is tripled, how does the intensity change? (a) It becomes one-ninth as large. (b) It becomes one-third as large. (c) It is unchanged. (d) It becomes three times larger. (e) It becomes nine times larger.

Answer (a). We suppose that a point source has no structure, and radiates sound equally in all directions (isotropically). The sound wave fronts are expanding spheres, so the area over which the sound energy spreads increases with distance according to $A = 4\pi r^2$. Thus, if the distance is tripled, the area increases by a factor of nine, and the new intensity will be one ninth of the old intensity. This answer according to the inverse-square law applies if the medium is uniform and unbounded.

For contrast, suppose that the sound is confined to move in a horizontal layer. (Thermal stratification in an ocean can have this effect on sonar "pings.") Then the area over which the sound energy is dispersed will only increase according to the circumference of an expanding circle: $A = 2\pi rh$, and so three times the distance will result in one third the intensity.

In the case of an entirely enclosed speaking tube (such as a ship's telephone), the area perpendicular to the energy flow stays the same, and increasing the distance will not change the intensity appreciably.

☐   ☐   ☐   ☐

ANSWER TO A CONCEPTUAL QUESTION

9. How can an object move with respect to an observer so that the sound from it is not shifted in frequency?

Answer For the sound from a source not to shift in frequency, the radial velocity of the source relative to the observer must be zero; that is, the source must not be moving toward or away from the observer. This can happen if the source and observer are not moving at all; if they have equal velocities relative to the medium; or, it can happen if the source moves around the observer in a circular pattern of constant radius. Even if the source accelerates along the surface of a sphere around the observer, decelerates, or stops, the frequency heard will equal the frequency emitted by the source.

☐   ☐   ☐   ☐

SOLUTIONS TO SELECTED END-OF-CHAPTER PROBLEMS

1. Write an expression that describes the pressure variation as a function of position and time for a sinusoidal sound wave in air. Assume the speed of sound is 343 m/s, $\lambda = 0.100$ m, and $\Delta P_{max} = 0.200$ Pa.

Solution

Conceptualize: The answer will be a function of the variables x and t, for a wave propagating along the x direction.

Categorize: We use the traveling wave model.

Analyze: We write the pressure variation as $\Delta P = \Delta P_{max} \sin(kx - \omega t)$

Noting that $k = \dfrac{2\pi}{\lambda}$, $k = \dfrac{2\pi \text{ rad}}{0.100 \text{ m}} = 6.28 \text{ rad/m}$

Likewise, $\omega = \dfrac{2\pi v}{\lambda}$ so $\omega = \dfrac{(2\pi \text{ rad})(343 \text{ m/s})}{0.100 \text{ m}} = 2.16 \times 10^4 \text{ rad/s}$

We now create our equation: $\Delta P = (0.200)\sin(62.8\, x - 21\,600t)$ where ΔP is in Pa, x in m, and t in s. ∎

Finalize: The wave function gives the amount by which the pressure is different from atmospheric pressure, at every point at every instant of time.

---

4.   An experimenter wishes to generate in air a sound wave that has a displacement amplitude of 5.50×10^{-6} m. The pressure amplitude is to be limited to 0.840 Pa. What is the minimum wavelength the sound wave can have?

Solution

Conceptualize: The relationship between displacement amplitude and pressure amplitude involves the frequency. The wavelength must be larger than a certain value to keep the frequency below a certain value, to keep the pressure low as specified.

Categorize: We use the relationship between ΔP_{max} and s_{max}.

Analyze: We are given $s_{max} = 5.50 \times 10^{-6}$ m   and $\Delta P_{max} = 0.840$ Pa

The pressure amplitude is

$$\Delta P_{max} = \rho v \omega_{max} s_{max} = \rho v \left(\frac{2\pi v}{\lambda_{min}} \right) s_{max} \quad \text{so} \quad \lambda_{min} = \frac{2\pi \rho v^2 s_{max}}{\Delta P_{max}}$$

$$\lambda_{min} = \frac{2\pi (1.20 \text{ kg/m}^3)(343 \text{ m/s})^2 (5.50 \times 10^{-6}\text{m})}{0.840 \text{ Pa}} = 5.81 \text{ m} \quad ∎$$

Finalize: This is the minimum wavelength allowed because a sound wave with any shorter wavelength would have a higher angular frequency ω, making $\Delta P_{max} = \rho v \omega s_{max}$ larger than the statement of the problem specifies.

---

5. Suppose you hear a clap of thunder 16.2 s after seeing the associated lightning strike. The speed of light in air is 3.00×10^8 m/s. (a) How far are you from the lightning strike? (b) Do you need to know the value of the speed of light to answer? Explain.

Solution

Conceptualize: There is a common rule of thumb that lightning is about a kilometer away for every 3 seconds of delay between the flash and thunder. Therefore, this lightning strike is about 5 km away.

Categorize: The distance can be found from the speed of sound and the elapsed time. The time interval for the light to travel to the observer will be much less than the sound delay, so the speed of light can be taken as infinite.

Analyze: (a) Assuming that the speed of sound is constant through the air between the lightning strike and the observer, we have

$$v_s = \frac{d}{\Delta t} \quad \text{or} \quad d = v_s \Delta t = (343 \text{ m/s})(16.2 \text{ s}) = 5.56 \text{ km} \qquad \blacksquare$$

(b) Let us check the validity of our assumption that the speed of light could be ignored. The time delay for the light is

$$t_{\text{light}} = \frac{d}{c_{\text{air}}} = \frac{5\,560 \text{ m}}{3.00 \times 10^8 \text{ m/s}} = 1.85 \times 10^{-5}\text{s} \quad \text{and}$$

$$\Delta t = t_{\text{sound}} - t_{\text{light}} = 16.2 \text{ s} - 1.85 \times 10^{-5} \text{ s} \approx 16.2 \text{ s}$$

Since the travel time for the light is much smaller than the uncertainty in the time of 16.2 s, t_{light} can be ignored without affecting the distance calculation. We do not need to know the actual value of the speed of light. $\qquad \blacksquare$

Finalize: Our calculated answer is consistent with our initial estimate. However, our assumption of a constant speed of sound in air is probably not valid due to local variations in air temperature during a storm. We must assume that the given speed of sound in air is an accurate **average** value for the conditions described.

---

19. Calculate the sound level (in decibels) of a sound wave that has an intensity of 4.00 μW/m^2.

Solution

Conceptualize: We expect about 60 dB.

Categorize: We use the definition of the decibel scale.

Analyze: We use the equation $\quad \beta = 10 \text{ dB} \log(I/I_0)$

where $I_0 = 10^{-12}$ W/m²

to calculate $\beta = 10 \log \left(\dfrac{4.00 \times 10^{-6} \ \text{W/m}^2}{10^{-12} \ \text{W/m}^2} \right) = 66.0$ dB ∎

Finalize: This could be the sound level of a chamber music concert, heard from a few rows back in the audience.

---

29. A family ice show is held at an enclosed arena. The skaters perform to music with level 80.0 dB. This level is too loud for your baby, who yells at 75.0 dB. (a) What total sound intensity engulfs you? (b) What is the combined sound level?

Solution

Conceptualize: Resist the temptation to say 155 dB. The decibel scale is logarithmic, so addition does not work that way.

Categorize: We must figure out the intensity of each sound, add the intensities, and then translate back to a sound level.

Analyze: From $\beta = 10 \log \left(\dfrac{I}{10^{-12} \ \text{W/m}^2} \right)$

we have $I = \left[10^{\beta/10} \right] 10^{-12}$ W/m²

(a) For your baby, $I_b = \left(10^{75.0/10} \right)\left(10^{-12} \ \text{W/m}^2 \right) = 3.16 \times 10^{-5}$ W/m²

For the music, $I_m = \left(10^{80.0/10} \right)\left(10^{-12} \ \text{W/m}^2 \right) = 10.0 \times 10^{-5}$ W/m²

The combined intensity is $I_{\text{total}} = I_m + I_b$

$I_{\text{total}} = 10.0 \times 10^{-5} \ \text{W/m}^2 + 3.16 \times 10^{-5} \ \text{W/m}^2$

$I_{\text{total}} = 13.2 \times 10^{-5} \ \text{W/m}^2$ ∎

(b) The combined sound level is then

$$\beta_{\text{total}} = 10 \log \left(\dfrac{I_{\text{total}}}{10^{-12} \ \text{W/m}^2} \right) = 10 \log \left(\dfrac{1.32 \times 10^{-4} \ \text{W/m}^2}{10^{-12} \ \text{W/m}^2} \right) = 81.2 \ \text{dB}$$ ∎

Finalize: 81.2 dB is "only a little" louder than 80 dB. The baby in this problem is the author's daughter. Fear of loud sounds is instinctive. Keep the muffler on your motor.

---

31. A firework charge is detonated many meters above the ground. At a distance of $d_1 = 500$ m from the explosion, the acoustic pressure reaches a maximum of $\Delta P_{\text{max}} = 10.0$ Pa (Fig. P17.31). Assume the speed of sound is constant at 343 m/s throughout the atmosphere over the region considered, the ground absorbs all the sound falling on it, and the air absorbs sound energy as described by the rate 7.00 dB/km. What is the sound level (in decibels) at a distance of $d_2 = 4.00 \times 10^3$ m from the explosion?

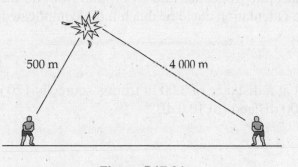

Figure P17.31

Solution

Conceptualize: At a distance of 4 km, an explosion should be audible, but probably not extremely loud. So we might expect the sound level to be somewhere between 40 and 80 dB.

Categorize: From the sound pressure data given in the problem, we can find the intensity, which is used to find the sound level in dB. The sound intensity will decrease with increased distance from the source and also because of the assumed absorption of the sound by the air.

Analyze: At a distance of 500 m from the explosion, $\Delta P_{max} = 10.0$ Pa. At this point the intensity is

$$I = \frac{\Delta P^2_{max}}{2\rho v} = \frac{\left(10.0 \text{ N/m}^2\right)^2}{2\left(1.20 \text{ kg/m}^3\right)(343 \text{ m/s})} = 0.121 \text{ W/m}^2$$

Therefore, the sound level is

$$\beta = 10 \log\left(\frac{I}{I_0}\right) = 10 \log\frac{\left(0.121 \text{ W/m}^2\right)}{\left(1.00 \times 10^{-12} \text{ W/m}^2\right)} = 111 \text{ dB}$$

From the inverse square law, we can calculate the intensity and decibel level (due to distance alone) 4 km away

$$I' = I(500 \text{ m})^2/(4\ 000 \text{ m})^2 = 1.90 \times 10^{-3} \text{ W/m}^2$$

and $$\beta = 10 \log\left(\frac{I'}{I_0}\right) = 10 \log\left(\frac{1.90 \times 10^{-3} \text{ W/m}^2}{1.00 \times 10^{-12} \text{ W/m}^2}\right) = 92.8 \text{ dB}$$

At a distance of 4 km from the explosion, absorption from the air will have decreased the sound level by an additional

$$\Delta\beta = (7.00 \text{ dB/km})(3.50 \text{ km}) = 24.5 \text{ dB}$$

So at 4 km, the sound level will be

$$\beta_f = \beta - \Delta\beta = 92.8 \text{ dB} - 24.5 \text{ dB} = 68.3 \text{ dB} \qquad\blacksquare$$

Finalize: This sound level falls within our expected range. Evidently, this explosion is rather loud (about the same as a vacuum cleaner) even at a distance of 4 km from the source.

It is interesting to note that the distance and absorption effects each reduce the sound level by about the same amount (~20 dB). If the explosion were at ground level, the sound level would be further reduced by reflection and absorption from obstacles between the source and observer, and the calculation could be much more complicated.

---

33. The sound level at a distance of 3.00 m from a source is 120 dB. At what distance is the sound level (a) 100 dB and (b) 10.0 dB?

Solution

Conceptualize: A reduction of 20 dB means reducing the intensity by a factor of 10^2, so we expect the radial distance to be 10 times larger, namely 30 m. A further reduction of 90 dB may correspond to an extra factor of $10^{4.5}$ in distance, to about $30 \times 30\ 000$ m, or about 1000 km.

Categorize: We use the definition of the decibel scale and the inverse-square law for sound intensity.

Analyze: From the definition of sound level,

$$\beta = 10 \log \left(\frac{I}{10^{-12}\ \text{W/m}^2} \right)$$

we can compute the intensities corresponding to each of the levels mentioned as $I = \left[10^{\beta/10} \right] 10^{-12}\ \text{W/m}^2$.

They are $I_{120} = 1\ \text{W/m}^2$, $I_{100} = 10^{-2}\ \text{W/m}^2$,   and $I_{10} = 10^{-11}\ \text{W/m}^2$

(a)   The power passing through any sphere around the source is $Power = 4\pi r^2 I$.

If we ignore absorption of sound by the medium, conservation of energy for the sound wave as a system requires that $r_{120}^2 I_{120} = r_{100}^2 I_{100} = r_{10}^2 I_{10}$

Then $r_{100} = r_{120} \sqrt{\dfrac{I_{120}}{I_{100}}} = (3.00\ \text{m}) \sqrt{\dfrac{1\ \text{W/m}^2}{10^{-2}\ \text{W/m}^2}} = 30.0\ \text{m}$ ∎

(b)   and $r_{10} = r_{120} \sqrt{\dfrac{I_{120}}{I_{10}}} = (3.00\ \text{m}) \sqrt{\dfrac{1\ \text{W/m}^2}{10^{-11}\ \text{W/m}^2}} = 9.49 \times 10^5\ \text{m}$ ∎

Finalize: At 949 km away, the faint 10-dB sound would not be identifiable among many other soft and loud sounds produced by other sources across the continent. And 949 km of air is such a thick screen that it may absorb a significant amount of sound power.

---

43. Standing at a crosswalk, you hear a frequency of 560 Hz from the siren of an approaching ambulance. After the ambulance passes, the observed frequency of the siren is 480 Hz. Determine the ambulance's speed from these observations.

Solution

Conceptualize: We can assume that an ambulance with its siren on is in a hurry to get somewhere, and is probably traveling between 30 and 150 km/h (~10 m/s to 50 m/s), depending on the driving conditions.

Categorize: We can use the equation for the Doppler effect to find the speed of the vehicle.

Analyze: Let v_a represent the magnitude of the velocity of the ambulance.

As it approaches you hear the frequency $$f' = \left(\frac{v}{v - v_a}\right) f = 560 \text{ Hz}$$

The negative sign appears because the source is moving toward the observer. The opposite sign with source velocity magnitude describes the ambulance moving away.

As it recedes the Doppler-shifted frequency is $$f'' = \left(\frac{v}{v + v_a}\right) f = 480 \text{ Hz}$$

Solving the second of these equations for f and substituting into the other gives

$$f' = \left(\frac{v}{v - v_a}\right)\left(\frac{v + v_a}{v}\right) f'' \qquad \text{or} \qquad f'v - f'v_a = vf'' + v_a f''$$

so the speed of the source is

$$v_a = \frac{v(f' - f'')}{f' + f''} = \frac{(343 \text{ m/s})(560 \text{ Hz} - 480 \text{ Hz})}{560 \text{ Hz} + 480 \text{ Hz}} = 26.4 \text{ m/s} \qquad \blacksquare$$

Finalize: This seems like a reasonable speed (about 85 km/h) for an ambulance, unless the street is crowded or the vehicle is traveling on an open highway. If the siren warbles instead of emitting a single tone, we could base the calculation on the highest frequencies you hear in each rising-and-falling cycle.

---

44. Review. A tuning fork vibrating at 512 Hz falls from rest and accelerates at 9.80 m/s². How far below the point of release is the tuning fork when waves of frequency 485 Hz reach the release point?

Solution

Conceptualize: We guess between 10 m and 100 m, the height of a multistory building, so that the fork can be falling at some (fairly small) fraction of the speed of sound.

Categorize: We must combine ideas of accelerated motion with the Doppler shift in frequency, and with ideas of constant-speed motion of the returning sound.

Analyze: We first determine how fast the tuning fork is falling to emit sound with apparent frequency 485 Hz. Call the magnitude of its velocity v_{fall}. The tuning fork source is moving away from the listener, so $v_s = -v_{fall}$.

Therefore, we use the equation $\quad f' = \left(\dfrac{v}{v + v_{fall}} \right) f$

Solving for v_{fall} gives $\quad \dfrac{v + v_{fall}}{v} = \dfrac{f}{f'} \quad$ and $\quad v_{fall} = v \left(\dfrac{f}{f'} - 1 \right)$

Substituting, we have $\quad v_{fall} = \left(\dfrac{512\,\text{Hz}}{485\,\text{Hz}} - 1 \right) 343\,\text{m/s} = 19.1\,\text{m/s}$

The time interval required for the tuning fork to reach this speed, from the particle under constant acceleration model, is given by

$$v_y = 0 + a_y t \quad \text{as} \quad t = v_y / a_y = (19.1\,\text{m/s})/(9.80\,\text{m/s}^2) = 1.95\,\text{s}$$

The distance that the fork has fallen is

$$\Delta y = 0 + \tfrac{1}{2}\,a_y t^2 = \tfrac{1}{2}\,(9.80\,\text{m/s}^2)(1.95\,\text{s})^2 = 18.6\,\text{m}$$

At this moment, the fork would appear to ring at 485 Hz to a stationary observer just above the fork. However, some additional time is required for the waves to reach the point of release. The fork is moving down, but the sound it radiates still travels away from its instantaneous position at 343 m/s. From the traveling wave model, the time interval it takes to return to the listener is

$$\Delta t = \Delta y / v = 18.6\,\text{m}/(343\,\text{m/s}) = 0.054\,2\,\text{s}$$

Over a total time $t + \Delta t = 1.95\,\text{s} + 0.054\,2\,\text{s} = 2.00\,\text{s}$, the fork falls a total distance $\Delta y = \tfrac{1}{2}\,a_y t^2 = \tfrac{1}{2}\,(9.80\,\text{m/s}^2)(2.00\,\text{s})^2 = 19.7\,\text{m}$ ∎

Finalize: The intensity of the sound heard at the release point would rapidly decrease as the fork falls, but a listener can learn to pay attention to just the frequency. When it is at the 19.7-m distance, the fork is moving still faster and radiating sound that will have a still lower frequency when it eventually reaches the listener. A set of pictures could help you to keep track of the sequential times involved.

---

45. A supersonic jet traveling at Mach 3.00 at an altitude of $h = 20\,000$ m is directly over a person at time $t = 0$ as shown in Figure P17.45. Assume the average speed of sound in air is 335 m/s over the path of the sound. (a) At what time will the person encounter the shock wave due to the sound emitted at $t = 0$? (b) Where will the plane be when this shock wave is heard?

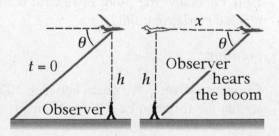

Figure P17.45

Solution

Conceptualize: The sound emitted by the plane must propagate down 20 km at about 1 km/3 s, so it may reach the person after 60 s, when the plane is far away.

Categorize: We identify the sound as a conical shock wave. The angle it makes with the plane's path will be useful to find first.

Analyze: Because the shock wave propagates at a set angle θ from the plane, we solve part (b) first.

(b) We use $\quad \sin\theta = \dfrac{v}{v_s} = \dfrac{\text{Mach 1}}{\text{Mach 3}} = \dfrac{1}{3},\quad$ and calculate $\quad \theta = 19.47°$

now in a triangle of distances, $\quad x = \dfrac{h}{\tan\theta} = \dfrac{20\,000\text{ m}}{\tan 19.47°} = 56\,570\text{ m} = 56.6\text{ km}\quad \blacksquare$

(a) It takes the plane $t = \dfrac{x}{v_s} = \dfrac{56\,570\text{ m}}{3(335\text{ m/s})} = 56.3\text{ s}$ to travel this distance $\quad \blacksquare$

Finalize: Our guess was approximately right, but the answer to (a) is not given by 20 000 m divided by 335 m/s. After computing θ, in the first picture we could find the perpendicular distance from the shock wave front to the person on the ground, as $h\cos\theta = 18.9$ km. The shock wave travels forward, perpendicular to the wave front, at the speed of sound. Then dividing 18.9 km by 335 m/s gives the answer (a) without first computing the answer to (b).

---

55. To measure her speed, a skydiver carries a buzzer emitting a steady tone at 1 800 Hz. A friend on the ground at the landing site directly below listens to the amplified sound he receives. Assume the air is calm and the speed of sound is independent of altitude. While the skydiver is falling at terminal speed, her friend on the ground receives waves of frequency 2 150 Hz. (a) What is the skydiver's speed of descent? (b) **What If?** Suppose the skydiver can hear the sound of the buzzer reflected from the ground. What frequency does she receive?

Solution

Conceptualize: Skydivers typically reach a terminal speed of about 150 mi/h (~75 m/s), so this skydiver should also fall near this rate. Since her friend receives a higher frequency as a result of the Doppler shift, the skydiver should detect a frequency with twice the Doppler shift, at approximately

$$f' = 1\,800\text{ Hz} + 2(2\,150\text{ Hz} - 1\,800\text{ Hz}) = 2\,500\text{ Hz}$$

Categorize: We can use the equation for the Doppler effect to answer both (a) and (b).

Analyze: Let $f_e = 1\,800$ Hz represent the emitted frequency; v_e the speed of the skydiver; and $f_g = 2\,150$ Hz the frequency of the wave crests reaching the ground.

(a) The skydiver source is moving toward the stationary ground, so we rearrange the equation

$$f_g = f_e\left(\frac{v}{v - v_e}\right) \quad \text{to give} \quad v_e = v\left(1 - \frac{f_e}{f_g}\right) = (343\text{ m/s})\left(1 - \frac{1\,800\text{ Hz}}{2\,150\text{ Hz}}\right) = 55.8\text{ m/s} \quad \blacksquare$$

(b) The ground now becomes a stationary source, reflecting crests with the 2 150-Hz frequency at which they reach the ground, and sending them to a moving observer, who receives them at the rate

$$f_{e2} = f_g \left(\frac{v + v_e}{v} \right) = (2\,150 \text{ Hz}) \left(\frac{343 \text{ m/s} + 55.8 \text{ m/s}}{343 \text{ m/s}} \right) = 2\,500 \text{ Hz} \qquad \blacksquare$$

Finalize: The answers appear to be consistent with our predictions, although the skydiver is falling somewhat slower than expected. The Doppler effect can be used to find the speed of many different types of moving objects, like raindrops (with Doppler radar) and cars (with police radar).

---

58. Two ships are moving along a line due east (Fig. P17.58). The trailing vessel has a speed relative to a land-based observation point of $v_1 = 64.0$ km/h, and the leading ship has a speed of $v_2 = 45.0$ km/h relative to that point. The two ships are in a region of the ocean where the current is moving uniformly due west at $v_{\text{current}} = 10.0$ km/h. The trailing ship transmits a sonar signal at a frequency of 1 200.0 Hz through the water. What frequency is monitored by the leading ship?

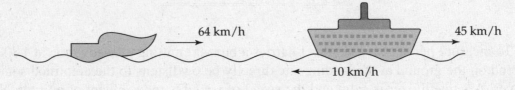

Figure P17.58

Solution

Conceptualize: The speeds of source and observer are small fractions of the speed of sound in water. The Doppler shift should be a small fraction of the radiated frequency. The gap between the ships is closing, so the frequency received will be higher than 1.20 kHz. We guess 1.25 kHz.

Categorize: Think about the speed of each ship relative to the water. From Table 17.1, sound moves at 1.533 km/s relative to the water, not relative to either ship or to the shore.

Analyze: When the observer is moving in front of and in the same direction as the source, the Doppler equation becomes

$$f' = \left(\frac{v + (-v_l)}{v - v_t} \right) f$$

where v_l and v_t are the speeds of the leading and trailing ships measured relative to the **medium** in which the sound is propagated. In this case the ocean current is opposite the direction of travel of the ships and

$$v_l = 45.0 \text{ km/h} - (-10.0 \text{ km/h}) = 55.0 \text{ km/h} = 15.3 \text{ m/s}$$

$$v_t = 64.0 \text{ km/h} - (-10.0 \text{ km/h}) = 74.0 \text{ km/h} = 20.6 \text{ m/s}$$

Therefore, $\quad f' = (1\,200.0\text{ Hz})\left(\dfrac{1\,533\text{ m/s} - 15.3\text{ m/s}}{1\,533\text{ m/s} - 20.6\text{ m/s}}\right) = 1\,204.2\text{ Hz}$ ∎

Finalize: The Doppler shift is much smaller than we guessed. Note well that sound moves much faster in water than in air. The difference in speed of the ships is only 19 km/h, and this is only 5.3 m/s.

---

61. A large meteoroid enters the Earth's atmosphere at a speed of 20.0 km/s and is not significantly slowed before entering the ocean. (a) What is the Mach angle of the shock wave from the meteoroid in the lower atmosphere? (b) If we assume the meteoroid survives the impact with the ocean surface, what is the (initial) Mach angle of the shock wave the meteoroid produces in the water?

Solution

Conceptualize: We expect an angle of a few degrees for the Mach cone of sound radiated by this object moving at orbital speed.

Categorize: The equation of the angle of a shock wave will give us the answers directly.

Analyze:

(a) The Mach angle in the air is

$$\theta_{\text{atm}} = \sin^{-1}\left(\frac{v}{v_s}\right) = \sin^{-1}\left(\frac{343\text{ m/s}}{2.00 \times 10^4\text{ m/s}}\right) = 0.983°$$ ∎

(b) At impact with the ocean,

$$\theta_{\text{ocean}} = \sin^{-1}\left(\frac{v}{v_s}\right) = \sin^{-1}\left(\frac{1\,533\text{ m/s}}{2.00 \times 10^4\text{ m/s}}\right) = 4.40°$$ ∎

Finalize: The Mach cone in the air is remarkably slim. If it happens to be headed straight down, a very dense asteroid might get through the atmosphere without slowing too much. It can never have infinite acceleration, so its speed at the bottom of the atmosphere is its initial speed in the hydrosphere.

---

62. Three metal rods are located relative to each other as shown in Figure P17.62, where $L_1 + L_2 = L_3$. The speed of sound in a rod is given by $v = \sqrt{Y/\rho}$, where Y is Young's modulus for the rod and ρ is the density. Values of density and Young's modulus for the three materials are

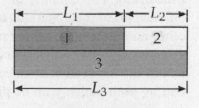

Figure P17.62

$$\rho_1 = 2.70 \times 10^3\text{ kg/m}^3, \qquad Y_1 = 7.00 \times 10^{10}\text{ N/m}^2,$$

$$\rho_2 = 11.3 \times 10^3\text{ kg/m}^3, \qquad Y_2 = 1.60 \times 10^{10}\text{ N/m}^2,$$

$$\rho_3 = 8.80 \times 10^3 \text{ kg/m}^3, \quad \text{and} \quad Y_3 = 11.0 \times 10^{10} \text{ N/m}^2.$$

If $L_3 = 1.50$ m, what must the ratio L_1/L_2 be if a sound wave is to travel the length of rods 1 and 2 in the same time interval required for the wave to travel the length of rod 3?

Solution

Conceptualize: Material 1 is stiff and low in density. It carries sound with higher speed than the others. Sound moves slowest in material 3. Therefore it will be possible to set up a tying race between sound going through 1 and 3, with sound going at a constant medium speed just through 2.

Categorize: Algebra will combine equations for travel times in the rods, to determine the required relation of lengths.

Analyze:

The time interval required for a sound pulse to travel a distance L at a speed v is given by $t = \dfrac{L}{v} = \dfrac{L}{\sqrt{Y/\rho}}$. Using this expression, we find the travel time in each rod.

$$t_1 = L_1 \sqrt{\frac{\rho_1}{Y_1}} = L_1 \sqrt{\frac{2.70 \times 10^3 \text{ kg/m}^3}{7.00 \times 10^{10} \text{ N/m}^2}} = L_1 \left(1.96 \times 10^{-4} \text{ s/m}\right)$$

$$t_2 = (1.50 - L_1) \sqrt{\frac{11.3 \times 10^3 \text{ kg/m}^3}{1.60 \times 10^{10} \text{ N/m}^2}} = 1.26 \times 10^{-3} \text{ s} - \left(8.40 \times 10^{-4} \text{ s/m}\right) L_1$$

$$t_3 = (1.50 \text{ m}) \sqrt{\frac{8.80 \times 10^3 \text{ kg/m}^3}{11.0 \times 10^{10} \text{ N/m}^2}} = 4.24 \times 10^{-4} \text{ s}$$

We require $t_1 + t_2 = t_3$, or

$$\left(1.96 \times 10^{-4} \text{ s/m}\right) L_1 + \left(1.26 \times 10^{-3} \text{ s}\right) - \left(8.40 \times 10^{-4} \text{ s/m}\right) L_1 = 4.24 \times 10^{-4} \text{ s}$$

This gives $L_1 = 1.30$ m and $L_2 = (1.50 \text{ m}) - (1.30 \text{ m}) = 0.201$ m

The ratio of lengths is $L_1/L_2 = 6.45$ ∎

Finalize: The equality of travel times could be demonstrated experimentally by tapping on one end of the apparatus to produce a sound pulse. It should arrive simultaneously at the other end, through both the pair of upper rods and the single lower rod. It would be sensed as a single pulse at that other end. Alternatively, a continuous wave could be sent into one end of the apparatus. Then each crest of the wave would arrive at the same time at the other end by both paths, with no phase difference.

63. With particular experimental methods, it is possible to produce and observe in a long, thin rod both a transverse wave whose speed depends primarily on tension in the rod and a longitudinal wave whose speed is determined by Young's modulus and the density of the material according to the expression $v = \sqrt{Y/\rho}$. The transverse wave can be modeled as a wave in a stretched string. A particular metal rod is 150 cm long and has a radius of 0.200 cm and a mass of 50.9 g. Young's modulus for the material is 6.80×10^{10} N/m². What must the tension in the rod be if the ratio of the speed of longitudinal waves to the speed of transverse waves is 8.00?

Solution

Conceptualize: The transverse wave is like the string waves we considered in Chapter 16. If this is a student-laboratory experiment, the tension might be on the order of hundreds of newtons, not likely to break the rod.

Categorize: We must mathematically (algebraically) combine equations for the speeds of longitudinal and string waves.

Analyze: The longitudinal wave and the transverse wave have respective phase speeds of

$$v_L = \sqrt{\frac{Y}{\rho}} \quad \text{where} \quad \rho = \frac{\text{mass}}{\text{volume}} = \frac{m}{\pi r^2 L}$$

and

$$v_T = \sqrt{\frac{T}{\mu}} \quad \text{where} \quad \mu = \frac{m}{L}$$

Since $v_L = 8.00\, v_T$,

$$T = \frac{\mu Y}{64.0\rho} = \frac{\pi r^2 Y}{64.0}$$

Evaluating,

$$T = \frac{\pi(0.002\,00\text{ m})^2 \left(6.80 \times 10^{10}\text{ N/m}^2\right)}{64.0} = 1.34 \times 10^4\text{ N} \quad \blacksquare$$

Finalize: The stretching of the rod is described by $Y\Delta L/L = T/\pi r^2$. Then for the strain we have just $\Delta L/L = T/\pi r^2 Y = (\pi r^2 Y/64)/\pi r^2 Y = 1/64$. This is a rather large value for strain in a metal. The experimenter should take care that the rod does not break. As we will study in Chapter 18, standing waves can be set up in the rod. If the longitudinal and transverse waves have the same frequency, the longitudinal wave will have a wavelength eight times larger. If one segment or "loop" of longitudinal vibration fits into the length of the rod, eight segments of the transverse wave will fit.

18

Superposition and Standing Waves

EQUATIONS AND CONCEPTS

A **standing wave** (an oscillation pattern with a stationary outline) can be produced in a string due to the interference of two identical sinusoidal waves traveling in opposite directions. The **amplitude** $(2A\sin kx)$ in Equation 18.1 at any point is a function of its position x along the string.

$$y = (2A\sin kx)\cos \omega t \qquad (18.1)$$

$$y = 0 \quad \text{when } x = \frac{n\lambda}{2} \quad n = 0,1,2,3,\dots$$

$$y_{\max} = 2A \quad \text{when } x = \frac{n\lambda}{4} \quad n = 1,3,5,\dots$$

Antinodes (A) are points of maximum displacement.

Nodes (N) are points of zero displacement.

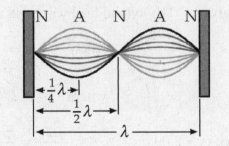

Normal modes of oscillation (a series of natural patterns of vibration) can be excited in a string of length L fixed at each end. Each mode corresponds to a quantized frequency and wavelength. *The frequencies are integral multiples of a fundamental frequency (when n = 1) and can be expressed in terms of wave speed and string length or in terms of string tension and linear mass-density.*

$$f_n = n\frac{v}{2L} \quad n = 1,2,3,\dots \qquad (18.5)$$

$$f_n = \frac{n}{2L}\sqrt{\frac{T}{\mu}} \quad n = 1,2,3,\dots \qquad (18.6)$$

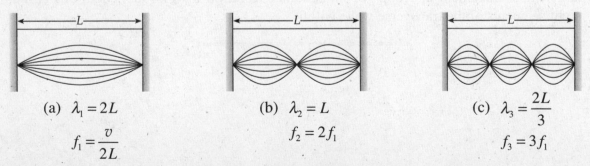

(a) $\lambda_1 = 2L$
$f_1 = \dfrac{v}{2L}$

(b) $\lambda_2 = L$
$f_2 = 2f_1$

(c) $\lambda_3 = \dfrac{2L}{3}$
$f_3 = 3f_1$

Standing wave patterns in a stretched string of length L fixed at both ends.
(a) first harmonic, (b) second harmonic, and (c) third harmonic.

In **a pipe open at both ends**, the natural frequencies of oscillation form a harmonic series that includes all integer multiples of the fundamental frequency. The wave patterns for the first three harmonics are illustrated in the figure at right.

$$f_n = n \frac{v}{2L} \quad n = 1, 2, 3, \ldots \quad (18.8)$$

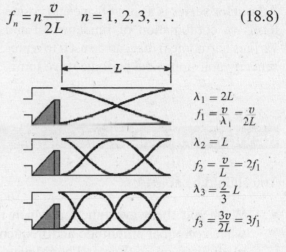

In a **pipe closed at one end**, the natural frequencies of oscillation form a harmonic series that includes only the odd integer multiples of the fundamental frequency. The wave patterns for the first three harmonics are illustrated in the figure at right.

$$f_n = n \frac{v}{4L} \quad n = 1, 3, 5, \ldots \quad (18.9)$$

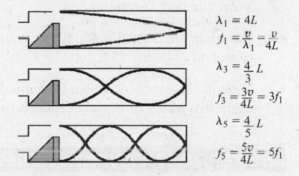

Beats are formed by the superposition of two waves of equal amplitude but having slightly different frequencies.

$$y = \left[2A \cos 2\pi \left(\frac{f_1 - f_2}{2} \right) t \right] \cos 2\pi \left(\frac{f_1 + f_2}{2} \right) t$$

$$(18.10)$$

The **amplitude (or envelope) of the resultant wave** described by Equation 18.10 above is time dependent. *There are two maxima in each period of the resultant wave; and each occurrence of maximum amplitude results in a "beat."*

$$y_{\text{envelope}} = 2A \cos 2\pi \left(\frac{f_1 - f_2}{2} \right) t \quad (18.11)$$

The **beat frequency** f_{beat} equals the absolute difference of the frequencies of the individual waves.

$$f_{\text{beat}} = |f_1 - f_2| \quad (18.12)$$

A **Fourier series** is a sum of sine and cosine terms (a combination of fundamental and various harmonics) that can be used to represent any non sinusoidal periodic wave form.

$$y(t) = \sum_n \left(A_n \sin 2\pi f_n t + B_n \cos 2\pi f_n t \right)$$

(18.13)

REVIEW CHECKLIST

You should be able to:

- Write out the wave function which represents the superposition of two sinusoidal waves of equal amplitude and frequency traveling in opposite directions in the same medium at given time and coordinate. Determine the amplitude and frequency of the resultant wave. (Section 18.1)

- Determine the angular frequency, maximum amplitude, and values of x which correspond to nodal and antinodal points of a standing wave, given an equation for the wave function. (Section 18.2)

- Calculate the normal mode frequencies for a string under tension, and for open and closed air columns. (Sections 18.3 and 18.5)

- Describe the time dependent amplitude and calculate the expected beat frequency when two waves of slightly different frequency interfere. (Section 18.7)

ANSWER TO AN OBJECTIVE QUESTION

6. Assume two identical sinusoidal waves are moving through the same medium in the same direction. Under what condition will the amplitude of the resultant wave be greater than either of the two original waves? (a) in all cases (b) only if the waves have no difference in phase (c) only if the phase difference is less than 90° (d) only if the phase difference is less than 120° (e) only if the phase difference is less than 180°.

Answer (d) Wherever the two waves are nearly enough in phase, their displacements will add to create a total displacement greater than the amplitude of either of the two original waves. When two one-dimensional waves of the same amplitude interfere, this condition is satisfied whenever the absolute value of the phase difference between the two waves is less than 120°. In the total wave with wave function

$$y = y_1 + y_2 = 2A_0 \cos\left(\frac{\phi}{2}\right) \sin\left(kx - \omega t - \frac{\phi}{2}\right)$$

the amplitude $2A_0 \cos(\phi/2)$ will be greater than A_0 if $\cos(\phi/2)$ is greater than $1/2$. This condition is satisfied if ϕ is less than 120°.

□   □   □   □

ANSWERS TO SELECTED CONCEPTUAL QUESTIONS

1. Does the phenomenon of wave interference apply only to sinusoidal waves?

Answer No. Any waves moving in the same medium interfere with each other when they disturb the medium at the same place at the same time. For example, two sawtooth-shaped pulses moving in opposite directions on a stretched string interfere when they meet each other. For waves, we can think of interference as meaning the same as superposition or addition.

□　　　□　　　□　　□

2. When two waves interfere constructively or destructively, is there any gain or loss in energy in the system of the waves? Explain.

Answer No. The energy may be transformed into other forms of energy. For example, when two pulses traveling on a stretched string in opposite directions overlap, and one is inverted, some potential energy is transferred to kinetic energy when they overlap. In fact, if they have the same shape except that one is inverted, their displacements completely cancel each other at one instant in time. In this case, all of the energy is kinetic energy of transverse motion when the resultant amplitude is zero.

□　　　□　　　□　　□

6. An airplane mechanic notices that the sound from a twin-engine aircraft rapidly varies in loudness when both engines are running. What could be causing this variation from loud to soft?

Answer Apparently the two engines are emitting sounds having frequencies that differ only slightly from each other. This results in a beat frequency, causing the variation from loud to soft, and back again. Giuseppi Tartini first described in print tuning a musical instrument by slowing down the beats until they are so infrequent that you cannot notice them. If the mechanic wishes, she might open an adjustment screw in the fuel line of the slower engine to 'eliminate' the beats.

□　　　□　　　□　　□

SOLUTIONS TO SELECTED END-OF-CHAPTER PROBLEMS

5. Two pulses traveling on the same string are described by

$$y_1 = \frac{5}{(3x - 4t)^2 + 2}, \quad \text{and} \quad y_2 = \frac{-5}{(3x + 4t - 6)^2 + 2}$$

(a) In which direction does each pulse travel? (b) At what instant do the two cancel everywhere? (c) At what point do the two pulses always cancel?

Solution

Conceptualize: Wave 1 is a roughly bell-shaped pulse of positive displacement with amplitude 2.5 units. Wave 2 has the same shape with negative displacement and, almost always, a different center point.

Categorize: Algebra on the equation $y_1 + y_2 = 0$ will reveal the time of complete destructive interference. We might not anticipate the existence of a point x of cancellation at all times, but the same equation will reveal it.

Analyze:

(a) At constant phase, $\phi = 3x - 4t$ will be constant. Then $x = \dfrac{\phi + 4t}{3}$ will change: the wave moves. As t increases in this equation, x increases, so the first wave moves to the right. ∎

  In the same way, in the second case $x = \dfrac{\phi - 4t + 6}{3}$. As t increases, x must decrease, so the second wave moves to the left. ∎

(b) We require that $y_1 + y_2 = 0$.

$$\frac{5}{(3x - 4t)^2 + 2} + \frac{-5}{(3x + 4t - 6)^2 + 2} = 0$$

  This can be written as $\qquad (3x - 4t)^2 = (3x + 4t - 6)^2$

  Solving for the positive root, $\qquad t = 0.750 \text{ s}$ ∎

(c) The negative root yields $\qquad (3x - 4t) = -(3x + 4t - 6)$

  The time terms cancel, leaving $x = 1.00$ m. At this point, the waves **always** cancel. ∎

Finalize: It may seem that algebra is more clever than can be expected. You deserve a reminder that $a^2 = 14$ has two roots. The total wave is not a standing wave, but we could call the point at $x = 1.00$ m a node of the superposition.

━━━━━━━━━

7. Two traveling sinusoidal waves are described by the wave functions

$$y_1 = 5.00 \sin\left[\pi(4.00x - 1\,200t)\right]$$
$$y_2 = 5.00 \sin\left[\pi(4.00x - 1\,200t - 0.250)\right]$$

where x, y_1, and y_2 are in meters and t is in seconds. (a) What is the amplitude of the resultant wave function $y_1 + y_2$? (b) What is the frequency of the resultant wave function?

Solution

Conceptualize: These waves have graphs like continuous sine curves. Think of water waves produced by continuously vibrating pencils on the water surface, and seen through the side wall of an aquarium.

Categorize: The pair of waves traveling in the same direction, perhaps separately created, combines to give a single traveling wave.

Analyze: We can represent the waves symbolically as

$$y_1 = A_0 \sin(kx - \omega t) \qquad \text{and} \qquad y_2 = A_0 \sin(kx - \omega t - \phi)$$

with $A_0 = 5.00$ m, $\omega = 1\,200\pi$ s^{-1} and $\phi - 0.250\pi$ rad

According to the principle of superposition, the resultant wave function has the form

$$y = y_1 + y_2 = 2A_0 \cos\left(\frac{\phi}{2}\right) \sin\left(kx - \omega t - \frac{\phi}{2}\right)$$

(a) with amplitude

$$A = 2A_0 \cos\left(\frac{\phi}{2}\right) = 2(5.00)\cos\left(\frac{\pi}{8.00}\right) = 9.24 \text{ m} \qquad \blacksquare$$

(b) and frequency

$$f = \frac{\omega}{2\pi} = \frac{1\,200\pi}{2\pi} = 600 \text{ Hz} \qquad \blacksquare$$

Finalize: When the superposed waves have the same frequency, the total wave will always have that frequency. To find the total amplitude we needed only the amplitude of each separate wave and the constant phase difference between them.

---

9. Two sinusoidal waves in a string are defined by the wave functions

$$y_1 = 2.00 \sin (20.0x - 32.0t)$$

$$y_2 = 2.00 \sin (25.0x - 40.0t)$$

where x, y_1, and y_2 are in centimeters and t is in seconds. (a) What is the phase difference between these two waves at the point $x = 5.00$ cm at $t = 2.00$ s? (b) What is the positive x value closest to the origin for which the two phases differ by $\pm\pi$ at $t = 2.00$ s? (At that location, the two waves add to zero.)

Solution

Conceptualize: Think of the two waves as moving sinusoidal graphs, with different wavelengths and frequencies. The phase difference between them is a function of the variables x and t. It controls how the waves add up at each point.

Categorize: Looking at the wave functions lets us pick out the expressions for the phase as the "angles we take the sines of"—the arguments of the sine functions. Then algebra will give particular answers.

Analyze: At any time and place, the phase shift between the waves is found by subtracting the phases of the two waves, $\Delta\psi = \phi_1 - \phi_2$.

$$\Delta\phi = (20.0 \text{ rad/cm})x - (32.0 \text{ rad/s})t - [(25.0 \text{ rad/cm})x - (40.0 \text{ rad/s})t]$$

Collecting terms, $\Delta\phi = -(5.00 \text{ rad/cm})x + (8.00 \text{ rad/s})t$

(a) At $x = 5.00$ cm   and $t = 2.00$ s, the phase difference is

$$\Delta\phi = (-5.00 \text{ rad/cm})(5.00 \text{ cm}) + (8.00 \text{ rad/s})(2.00 \text{ s})$$

$$|\Delta\phi| = 9.00 \text{ rad} = 516° = 156° \qquad\blacksquare$$

(b) The sine functions repeat whenever their arguments change by an integer number of cycles, an integer multiple of 2π radians. Then the phase shift equals $\pm\pi$ whenever $\Delta\phi = \pi + 2n\pi$, for all integer values of n.

Substituting this into the phase equation, we have

$$\pi + 2n\pi = -(5.00 \text{ rad/cm})x + (8.00 \text{ rad/s})t$$

At $t = 2.00$ s, $\pi + 2n\pi = -(5.00 \text{ rad/cm})x + (8.00 \text{ rad/s})(2.00 \text{ s})$

or $(5.00 \text{ rad/cm})x = (16.0 - \pi - 2n\pi) \text{ rad}$

The smallest positive value of x is found when $n = 2$:

$$x = \frac{(16.0 - 5\pi)\text{rad}}{5.00 \text{ rad/cm}} = 0.058\ 4 \text{ cm} \qquad\blacksquare$$

Finalize: We did not have to pay much attention to phase in previous chapters about individual waves. Now we pay attention, because phase controls how two or more waves add.

---

15. Two transverse sinusoidal waves combining in a medium are described by the wave functions

$$y_1 = 3.00 \sin\pi(x + 0.600t) \qquad y_2 = 3.00 \sin\pi(x - 0.600t)$$

where x, y_1, and y_2 are in centimeters and t is in seconds. Determine the maximum transverse position of an element of the medium at (a) $x = 0.250$ cm, (b) $x = 0.500$ cm, and (c) $x = 1.50$ cm. (d) Find the three smallest values of x corresponding to antinodes.

Solution

Conceptualize: The answers for maximum transverse position must be between 6 cm and 0. The antinodes are separated by half a wavelength, which we expect to be a couple of centimeters.

Categorize: According to the waves in interference model, we write the function $y_1 + y_2$ and start evaluating things.

Analyze: We add y_1 and y_2 using the trigonometry identity

$$\sin(\alpha + \beta) = \sin\alpha\cos\beta + \cos\alpha\sin\beta$$

We get $\quad y = y_1 + y_2 = (6.0)\sin(\pi x)\cos(0.600\ \pi t)$

Since $\cos(0) = 1$, we can find the maximum value of y by setting $t = 0$:

$$y_{max}(x) = y_1 + y_2 = (6.00)\sin(\pi x)$$

(a) At $x = 0.250$ cm, $\qquad y_{max} = (6.0)\sin(0.250\ \pi) = 4.24$ cm   ■

(b) At $x = 0.500$ cm, $\qquad y_{max} = (6.0)\sin(0.500\ \pi) = 6.00$ cm   ■

(c) At $x = 1.50$ cm, $\qquad y_{max} = |(6.0)\sin(1.50\ \pi)| = +6.00$ cm   ■

(d) The antinodes occur when $x = n\lambda/4$ for $n = 1, 3, 5, \ldots$.

But $\quad k = 2\pi/\lambda = \pi$, so $\qquad \lambda = 2.00$ cm

and $\qquad\qquad\qquad x_1 = \lambda/4 = (2.00\text{ cm})/4 = 0.500$ cm   ■

$\qquad\qquad\qquad\qquad x_2 = 3\lambda/4 = 3(2.00\text{ cm})/4 = 1.50$ cm   ■

$\qquad\qquad\qquad\qquad x_3 = 5\lambda/4 = 5(2.00\text{ cm})/4 = 2.50$ cm   ■

Finalize: Two of our answers in (d) can be read from the way the amplitude had its largest possible value in parts (b) and (c). Note again that an amplitude is defined to be always positive, as the maximum absolute value of the wave function.

---

17. Two identical loudspeakers are driven in phase by a common oscillator at 800 Hz and face each other at a distance of 1.25 m. Locate the points along the line joining the two speakers where relative minima of sound pressure amplitude would be expected.

Solution

Conceptualize: A stereophonic sound reproduction system will always play separate signals through two loudspeakers. The single tone in this problem has no musical interest and may quickly become irritating, but it sets up for sound a situation that is also important for string waves (on a guitar), light waves (in a laser), and electron waves (in a model atom).

Categorize: Two identical waves moving in opposite directions constitute a standing wave. We must find the nodes.

Analyze: The wavelength is $\quad \lambda = \dfrac{v}{f} = \dfrac{343\text{ m/s}}{800\text{ Hz}} = 0.429$ m

The two waves moving in opposite directions along the line between the two speakers will add to produce a standing wave with this distance between nodes:

$$\text{distance N to N} = \lambda/2 = 0.214 \text{ m}$$

Because the speakers vibrate in phase, air compressions from each will simultaneously reach the point halfway between the speakers, to produce an antinode of pressure here. A node of pressure will be located at this distance on either side of the midpoint:

$$\text{distance N to A} = \lambda/4 = 0.107 \text{ m}$$

Therefore nodes of sound pressure will appear at these distances from either speaker:

$$\frac{1}{2}(1.25 \text{ m}) + 0.107 \text{ m} = 0.732 \text{ m} \qquad \text{and} \qquad \frac{1}{2}(1.25 \text{ m}) - 0.107 \text{ m} = 0.518 \text{ m}$$

The standing wave contains a chain of equally-spaced nodes at distances from either speaker of

$$0.732 \text{ m} + 0.214 \text{ m} = 0.947 \text{ m}$$

$$0.947 \text{ m} + 0.214 \text{ m} = 1.16 \text{ m}$$

and also at $\qquad 0.518 \text{ m} - 0.214 \text{ m} = 0.303 \text{ m}$

$$0.303 \text{ m} - 0.214 \text{ m} = 0.089 \text{ 1 m}$$

The standing wave exists only along the line segment between the speakers. No nodes or antinodes appear at distances greater than 1.25 m or less than 0, because waves add to give a standing wave only if they are traveling in opposite directions and not in the same direction. In order, the distances from either speaker to the nodes of pressure between the speakers are 0.089 1 m, 0.303 m, 0.518 m, 0.732 m, 0.947 m, and 1.16 m. ∎

Finalize: Draw a picture of a standing wave just by writing out …N A N A N A… as appropriate for any particular case. The nodes and antinodes alternate with equal spacing. This problem is solved just by finding what fits between the speakers.

---

20. A string that is 30.0 cm long and has a mass per unit length of 9.00×10^{-3} kg/m is stretched to a tension of 20.0 N. Find (a) the fundamental frequency and (b) the next three frequencies that could cause standing-wave patterns on the string.

Solution

Conceptualize: The string described in the problem might be a laboratory model for a guitar string. We estimate that its fundamental frequency will be on the order of 100 Hz. The higher standing-wave resonance frequencies will be consecutive-integer multiples of the fundamental.

Categorize: The tension and linear density of the string can be used to find the wave speed, which can then be used along with the required wavelength to find the fundamental frequency.

Analyze: The wave speed is

$$v = \sqrt{\frac{T}{\mu}} = \sqrt{\frac{20.0 \text{ N}}{9.00 \times 10^{-3} \text{kg/m}}} = 47.1 \text{ m/s}$$

For a vibrating string of length L fixed at both ends, there are nodes at both ends. The wavelength of the fundamental is $\lambda = 2d_{NN} = 2L = 0.600$ m; and the frequency is

$$f_1 = \frac{v}{\lambda} = \frac{v}{2L} = \frac{47.1 \text{ m/s}}{0.600 \text{ m}} = 78.6 \text{ Hz}$$

After NAN, the next three vibration possibilities read NANAN, NANANAN, and NANANANAN. Each has just one more node and one more antinode than the one before. Respectively, these string waves have wavelengths of one-half, one-third, and one-quarter of 60 cm. The harmonic frequencies are

$$f_2 = 2f_1 = 157 \text{ Hz}, \quad f_3 = 3f_1 = 236 \text{ Hz}, \quad \text{and} \quad f_4 = 4f_1 = 314 \text{ Hz} \quad \blacksquare$$

Finalize: Our order-of-magnitude estimate of the fundamental frequency was right. This device is more like a bass guitar than a classical guitar. It would weakly broadcast sound into the surrounding air, but note that the sound waves are quite different from the string waves: the sound waves have longer wavelengths, move faster, and are traveling waves radiated in all directions.

---

23.   The A string on a cello vibrates in its first normal mode with a frequency of 220 Hz. The vibrating segment is 70.0 cm long and has a mass of 1.20 g. (a) Find the tension in the string. (b) Determine the frequency of vibration when the string vibrates in three segments.

Solution

Conceptualize: The tension should be less than 500 N (~100 lb) since excessive force on the four cello strings would break the neck of the instrument. If a string vibrates in three segments, there will be three antinodes (instead of one for the fundamental mode), so the frequency should be three times greater than the fundamental.

Categorize: From the string's length, we can find the wavelength. We then can use the wavelength with the fundamental frequency to find the wave speed. Finally, we can find the tension from the wave speed and the linear mass density of the string.

Analyze: When the string vibrates in the lowest frequency mode, the length of string forms a standing wave where $L = \lambda/2$ so the fundamental harmonic wavelength is

$$\lambda = 2L = 2(0.700 \text{ m}) = 1.40 \text{ m}$$

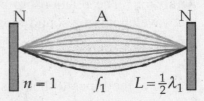

and the speed is

$$v = f\lambda = (220 \text{ s}^{-1})(1.40 \text{ m}) = 308 \text{ m/s}$$

(a) From the tension equation $v = \sqrt{\dfrac{T}{\mu}} = \sqrt{\dfrac{T}{m/L}}$

We get $T = v^2 m/L$, or $T = \dfrac{(308 \text{ m/s})^2 \left(1.20 \times 10^{-3} \text{ kg}\right)}{0.700 \text{ m}} = 163 \text{ N}$ ∎

(b) For the third harmonic, the tension, linear density, and speed are the same, but the string vibrates in three segments. Thus, that the wavelength is one third as long as in the fundamental.

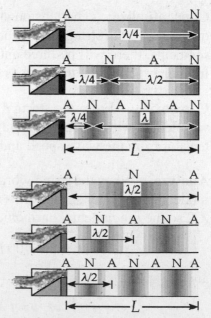

$n = 3 \qquad f_3 \qquad L = \frac{3}{2}\lambda_3$

$$\lambda_3 = \lambda_1/3$$

From the equation $v = f\lambda$, we find the frequency is three times as high.

$$f_3 = \frac{v}{\lambda_3} = 3\frac{v}{\lambda_1} = 3f_1 = 660 \text{ Hz}$$ ∎

Finalize: The tension seems reasonable, and the third harmonic is three times the fundamental frequency as expected. Related to part (b), some stringed instrument players use a technique to double the frequency of a note by playing a *natural harmonic*, in effect cutting a vibrating string in half. When the string is lightly touched at its midpoint to form a node there, the second harmonic is formed, and the resulting note is one octave higher (twice the original fundamental frequency).

---

33. Calculate the length of a pipe that has a fundamental frequency of 240 Hz assuming the pipe is (a) closed at one end and (b) open at both ends.

Solution

Conceptualize: This pipe can be thought of as like a clarinet in part (a), or like a flute in (b). We expect a length of a few decimeters in (a), and twice as much in (b).

Categorize: The pipe does not vibrate. The air within it vibrates in a standing wave.

Analyze: The relationship between the frequency and the wavelength of a sound wave is

$$v = f\lambda \qquad \text{or} \qquad \lambda = v/f$$

We draw the pipes to help us visualize the relationship between λ and L. In these diagrams lighter shading represents areas of greater vibration amplitude. The darkest shading appears at the displacement nodes.

(a) For the fundamental mode in a closed pipe, $\lambda = 4d_{NA} = 4L$

so $\quad L = \dfrac{\lambda}{4} = \dfrac{v/f}{4} \quad$ and $\quad L = \dfrac{(343 \text{ m/s})/(240 \text{ s}^{-1})}{4} = 0.357 \text{ m}$ ∎

(b) For the fundamental mode in an open pipe, $\lambda = 2L$

so $\quad L = \dfrac{\lambda}{2} = \dfrac{v/f}{2} \quad$ and $\quad L = \dfrac{(343 \text{ m/s})/(240 \text{ s}^{-1})}{2} = 0.715 \text{ m}$ ∎

Finalize: You may consider it obvious that there should be a node of displacement at a closed end of an air column. The rule that there is an antinode at an open end is equally strict. The constancy of atmospheric pressure outside forces an open end to be a node of pressure variation, which is an antinode of displacement.

---

37. An air column in a glass tube is open at one end and closed at the other by a movable piston. The air in the tube is warmed above room temperature, and a 384-Hz tuning fork is held at the open end. Resonance is heard when the piston at a distance $d_1 = 22.8$ cm from the open end and again when it is at a distance $d_2 = 68.3$ cm from the open end. (a) What speed of sound is implied by these data? (b) How far from the open end will the piston be when the next resonance is heard?

Solution

Conceptualize: The resonance is heard as amplification, when all the air in the tube vibrates in a standing-wave pattern along with the tuning fork.

$f = 384$ Hz

Warm air

Categorize: The air vibration has a node at the piston and an antinode at the top end.

Analyze: For an air column closed at one end, resonances will occur when the length of the column is equal to $\lambda/4$, $3\lambda/4$, $5\lambda/4$, and so on. Thus, the change in the length of the pipe from one resonance to the next is $d_{NN} = \lambda/2$. In this case,

$$\lambda/2 = (0.683 \text{ m} - 0.228 \text{ m}) = 0.455 \text{ m} \quad \text{and} \quad \lambda = 0.910 \text{ m}$$

(a) $v = f\lambda = (384 \text{ Hz})(0.910 \text{ m}) = 349 \text{ m/s}$ ∎

(b) $L = 0.683 \text{ m} + 0.455 \text{ m} = 1.14 \text{ m}$ ∎

Finalize: Measuring the speed of sound need not be done by setting off explosions across the valley and using a stopwatch. The speed of sound can be found quite precisely with a frequency counter and a meterstick to measure distances in standing waves.

---

40. A shower stall has dimensions 86.0 cm × 86.0 cm × 210 cm. Assume the stall acts as a pipe closed at both ends, with nodes at opposite sides. Assume singing voices range from 130 Hz to 2 000 Hz, and let the speed of sound in the hot air be 355 m/s. For people singing in this shower, which frequencies would sound the richest (because of resonance)?

Solution

Conceptualize: We expect one set of equally spaced resonance frequencies for sound moving horizontally, and another set with narrower spacing for sound moving vertically.

Categorize: Standing waves must fit between the side walls, with nodes at both surfaces, or else between floor and ceiling.

Analyze: For a closed box, the resonant air vibrations will have nodes at both sides, so the permitted wavelengths will be defined by

$$L = n d_{NN} = \frac{n\lambda}{2} = \frac{nv}{2f}, \qquad \text{with } n = 1, 2, 3, \dots$$

Rearranging, and substituting $L = 0.860$ m, the side-to-side resonant frequencies are

$$f_n = n\frac{v}{2L} = n\frac{355 \text{ m/s}}{2(0.860 \text{ m})} = n(206 \text{ Hz}), \text{ for each } n \text{ from 1 to 9} \qquad \blacksquare$$

With $L' = 2.10$ m, the top-to-bottom resonance frequencies are

$$f_n = n\frac{355 \text{ m/s}}{2(2.10 \text{ m})} = n(84.5 \text{ Hz}), \text{ for each } n \text{ from 2 to 23} \qquad \blacksquare$$

Finalize: We found the allowed values for n by checking which resonant frequencies lie in the vocal range between 130 Hz and 2 000 Hz. If the equations in this problem look to you like formulas for a pipe open at both ends or a string fixed at both ends, you should draw pictures of the vibrations instead of just thinking about formulas. The shower stall is different from any organ pipe, because it has nodes at both walls instead of antinodes. The shower stall is different from a string because the air wave is longitudinal. An equation like $d_{NN} = \lambda/2$ is true for all one-dimensional standing waves, so the best solution is based on diagrams and on this equation, with $f = v/\lambda$. A function generator connected to a loudspeaker, with your ear as detector, is all you need to do a nice experiment to observe the resonance phenomenon in any small room with hard parallel walls. Listen at different distances from one wall, and you may even discover whether your ear responds more to air displacement or to air pressure.

We have interpreted the problem statement as meaning that we should consider standing sound waves established by reflection from only two opposite walls of the stall. More realistically, a sound wave can move in any direction and can reflect from all six boundary surfaces at the same time. In article 267 of his classic book *The Theory of Sound*, J. W. S. Rayleigh demonstrates that the resonance frequencies are given by

$$f = \frac{v}{2}\left(\frac{n^2}{L^2} + \frac{o^2}{L^2} + \frac{p^2}{L'^2}\right)^{1/2} = \frac{355 \text{ m/s}}{2}\left(\frac{n^2}{(0.860 \text{ m})^2} + \frac{o^2}{(0.860 \text{ m})^2} + \frac{p^2}{(2.10 \text{ m})^2}\right)^{1/2}$$

where n, o, and p all take any integer values starting from zero. This result includes the resonances identified in our solution, such as

$$n = 0, o = 0, p = 1 : f = 84.5 \text{ Hz}$$
$$n = 0, o = 0, p = 2 : f = 169 \text{ Hz}$$
$$n = 1, o = 0, p = 0 : f = 206 \text{ Hz}$$
$$n = 0, o = 0, p = 3 : f = 254 \text{ Hz}$$

But there are many others, such as

$$n = 1, o = 0, p = 1 : f = 223 \text{ Hz}$$
$$n = 1, o = 1, p = 0 : f = 292 \text{ Hz}$$
$$n = 1, o = 1, p = 1 : f = 304 \text{ Hz}$$

---

43. Two adjacent natural frequencies of an organ pipe are determined to be 550 Hz and 650 Hz. Calculate (a) the fundamental frequency and (b) the length of this pipe.

Solution

Conceptualize: These will be fairly high harmonics of the pipe, which we expect to be a long pipe for a bass note.

Categorize: We use the wave under boundary conditions model. We are not told in advance whether the pipe is open at one end or at both ends, so we must think about both possibilities.

Analyze: Because harmonic frequencies are given by $f_1 n$ for open pipes, and $f_1(2n - 1)$ for closed pipes, the difference between adjacent harmonics is constant in both cases. Therefore, we can find each harmonic below 650 Hz by subtracting $\Delta f_{\text{Harmonic}} = (650 \text{ Hz} - 550 \text{ Hz}) = 100 \text{ Hz}$ from the previous value.

The harmonic frequencies are the set

650 Hz, 550 Hz, 450 Hz, 350 Hz, 250 Hz, 150 Hz, and 50.0 Hz

(a) The fundamental frequency, then, is 50.0 Hz. ∎

(b) The wavelength of the fundamental vibration can be calculated from the speed of sound as

$$\lambda = \frac{v}{f} = \frac{343 \text{ m/s}}{50.0 \text{ Hz}} = 6.86 \text{ m}$$

Because the step size Δf is twice the fundamental frequency, we know the pipe is closed, with an antinode at the open end, and a node at the closed end. The wavelength in this situation is four times the pipe length, so

$$L = d_{\text{NA}} = \frac{\lambda}{4} = 1.72 \text{ m}$$ ∎

Finalize: The frequency 50 Hz is a G in a low register. The harmonics measured in the problem are the 11th and the 13th, which can be perfectly observable as resonances using a function generator, a small loudspeaker, and your ear as detector. For some ranks of organ pipes, higher harmonics like these can be important contributors to a brassy or ringing sound quality.

---

47. An aluminum rod 1.60 m long is held at its center. It is stroked with a rosin-coated cloth to set up a longitudinal vibration. The speed of sound in a thin rod of aluminum is 5 100 m/s. (a) What is the fundamental frequency of the waves established in the rod? (b) What harmonics are set up in the rod held in this manner? (c) **What If?** What would be the fundamental frequency if the rod were copper, in which the speed of sound is 3 560 m/s?

Solution

Conceptualize: Standing waves are important because any wave confined to a restricted region of space will be reflected back onto itself by the boundaries of the region. Then the traveling waves moving in opposite directions constitute a standing wave. We guess that the rod can sing at a few hundred hertz and at integer-multiple higher harmonics. The frequency will be proportionately lower with copper.

Categorize: We must identify where nodes and antinodes are, use the fact that antinodes are separated by half a wavelength, and use $v = f\lambda$.

Analyze:

(a) Since the central clamp establishes a node at the center, the fundamental mode of vibration will be ANA. Thus the rod length is

$$L = d_{AA} = \lambda/2$$

Our first harmonic frequency is

$$f_1 = \frac{v}{\lambda_1} = \frac{v}{2L} = \frac{5\ 100\ \text{m/s}}{3.20\ \text{m}} = 1.59\ \text{kHz} \quad \blacksquare$$

(b) Since the rod is free at each end, the ends will be antinodes. The next vibration state will not have just one more node and one more antinode, reading ANANA, because the rod must have a node at its center. The second vibration state is ANANANA, as graphed in the diagram, with a wavelength and frequency of

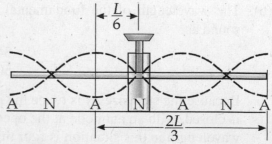

$$\lambda = \frac{2L}{3} \quad \text{and} \quad f = \frac{v}{\lambda} = \frac{3v}{2L} = 3f_1$$

Since $f = 2f_1$ was bypassed as having an antinode at the center rather than the node required, we know that we get only odd harmonics.

That is, $\quad f = \dfrac{nv}{2L} = n(1.59\ \text{kHz}) \qquad$ for $n = 1, 3, 5, \ldots$ ∎

(c) For a copper rod, the density is higher, so the speed of sound is lower, and the fundamental frequency is lower.

$$f_1 = \frac{v}{2L} = \frac{3\,560\ \text{m/s}}{3.20\ \text{m}} = 1.11\ \text{kHz}$$ ∎

Finalize: The sound is not at just a few hundred hertz, but is a squeak or a squeal at over a thousand hertz. Only a few higher harmonics are in the audible range. Sound really moves fast in materials that are stiff against compression. For a thin rod, it is Young's modulus that determines the speed of a longitudinal wave. The dashed lines in the diagram are a graph of the vibration amplitude at different positions along the rod. The vibration is longitudinal, not transverse, so the dashed lines are nothing like a photograph of the vibrating rod.

---

49. In certain ranges of a piano keyboard, more than one string is tuned to the same note to provide extra loudness. For example, the note at 110 Hz has two strings at this frequency. If one string slips from its normal tension of 600 N to 540 N, what beat frequency is heard when the hammer strikes the two strings simultaneously?

Solution

Conceptualize: Directly noticeable beat frequencies are usually only a few Hertz, so we should not expect a frequency much greater than this.

Categorize: As in previous problems, the two wave speed equations can be used together to find the frequency of vibration that corresponds to a certain tension. The beat frequency is then just the difference in the two resulting frequencies from the two strings with different tensions.

Analyze: Combining the velocity and the tension equations $v = f\lambda$ and $v = \sqrt{T/\mu}$, we find that the frequency is

$$f = \sqrt{\frac{T}{\mu \lambda^2}}$$

Since μ and λ are constant, we can apply that equation to both frequencies, and then divide the two equations to get the proportion

$$\frac{f_1}{f_2} = \sqrt{\frac{T_1}{T_2}}$$

With $f_1 = 110$ Hz, $T_1 = 600$ N, and $T_2 = 540$ N we have

$$f_2 = (110\ \text{Hz})\sqrt{\frac{540\ \text{N}}{600\ \text{N}}} = 104.4\ \text{Hz}$$

The beat frequency is

$$f_b = |f_1 - f_2| = 110 \text{ Hz} - 104.4 \text{ Hz} = 5.64 \text{ Hz} \qquad \blacksquare$$

Finalize: As expected, the beat frequency is only a few cycles per second. This result from the interference of the two sound waves with slightly different frequencies is a tone with a single pitch (low A) with tremolo (not vibrato). It varies in amplitude over time, similar to the sound made by rapidly saying "wa-wa-wa…"

The beat frequency above is written with three significant figures on our standard assumption that the original data is known precisely enough to warrant them. This assumption implies that the original frequency is known somewhat more precisely than to the three significant digits quoted in "110 Hz." For example, if the original frequency of the strings were 109.6 Hz, the beat frequency would be 5.62 Hz.

---

51. Review. A student holds a tuning fork oscillating at 256 Hz. He walks toward a wall at a constant speed of 1.33 m/s. (a) What beat frequency does he observe between the tuning fork and its echo? (b) How fast must he walk away from the wall to observe a beat frequency of 5.00 Hz?

Solution

Conceptualize: An electronic burglar detector may listen for beats between the signal it broadcasts and the Doppler-shifted reflection from a moving surface. The experiment described here is more fun. The student hears beats at a rate he controls through his walking speed.

Categorize: We must think of a double Doppler shift, for the wave reaching the wall from the moving tuning fork and the wave reaching the moving auditor.

Analyze: Let v_e represent the magnitude of the experimenter's velocity. The student broadcasts sound as a moving source. The frequency of the wave received at the wall is

$f' = f_{\text{radiated}} \dfrac{v}{v - v_e}$. The wall reflects sound at this same frequency, which the student hears

as a moving observer, with an extra Doppler shift to $f_{\text{receivedback}} = f' \dfrac{v + v_e}{v} = f_{\text{radiated}} \dfrac{v + v_e}{v - v_e}$.

The beat frequency is $\qquad f_b = |f_{\text{receivedback}} - f_{\text{radiated}}|$

Substituting gives $\qquad f_b = f_{\text{radiated}} \dfrac{2v_e}{v - v_e}$ when approaching the wall.

(a) $f_b = (256 \text{ Hz}) \dfrac{2(1.33 \text{ m/s})}{343 \text{ m/s} - 1.33 \text{ m/s}} = 1.99 \text{ Hz}$ $\qquad \blacksquare$

(b) When moving away from wall, v_e appears with opposite signs. Solving for v_e gives

$$v_e = f_b \frac{v}{2f_{\text{radiated}} - f_b} = (5.00 \text{ Hz}) \frac{343 \text{ m/s}}{2(256 \text{ Hz}) - 5.00 \text{ Hz}} = 3.38 \text{ m/s} \qquad \blacksquare$$

Finalize: Indoors, reflections from other walls may complicate the observation. Controlling with your own motion the frequency of beats that you hear may work well for children whooping while they play on swings in a playground just outside a school building with a large, high, flat vertical wall. The higher they swing, the faster the beats they hear at the lowest point in each cycle. The pianist Victor Borge would play in rhythm with a bejeweled latecomer flouncing down the aisle in the concert hall, for a remarkably similar effect.

---

65. Two train whistles have identical frequencies of 180 Hz. When one train is at rest in the station and the other is moving nearby, a commuter standing on the station platform hears beats with a frequency of 2.00 beats/s when the whistles operate together. What are the two possible speeds and directions the moving train can have?

Solution

Conceptualize: A Doppler shift in the frequency received from the moving train is the source of the beats.

Categorize: We consider velocities of approach and of recession separately in the Doppler equation, after we…

Analyze: …observe from our beat equation $f_b = |f_1 - f_2| = |f - f'|$ that the moving train must have an apparent frequency of

$$\text{either } f' = 182 \text{ Hz} \qquad \text{or else} \qquad f' = 178 \text{ Hz}$$

We let v_t represent the magnitude of the train's velocity. If the train is moving away from the station, the apparent frequency is 178 Hz, lower, as described by

$$f' = \frac{v}{v + v_t} f$$

and the train is **moving away** at

$$v_t = v\left(\frac{f}{f'} - 1\right) = (343 \text{ m/s})\left(\frac{180 \text{ Hz}}{178 \text{ Hz}} - 1\right) = 3.85 \text{ m/s} \qquad \blacksquare$$

If the train is pulling into the station, then the apparent frequency is 182 Hz. Again from the Doppler shift,

$$f' = f\frac{v}{v - v_s}$$

The train is **approaching** at

$$v_s = v\left(1 - \frac{f}{f'}\right) = (343 \text{ m/s})\left(1 - \frac{180 \text{ Hz}}{182 \text{ Hz}}\right)$$

$$v_s = 3.77 \text{ m/s} \qquad \blacksquare$$

Finalize: The Doppler effect does not measure distance, but it can be a remarkably sensitive measure of speed. With light, radar waves, or other electromagnetic waves, the Doppler equation is a bit different from the equation for sound. For electromagnetic waves the fractional shift in frequency is the same whichever is moving, source or observer.

---

66. Two wires are welded together end to end. The wires are made of the same material, but the diameter of one is twice that of the other. They are subjected to a tension of 4.60 N. The thin wire has a length of 40.0 cm and a linear mass density of 2.00 g/m. The combination is fixed at both ends and vibrated in such a way that two antinodes are present, with the node between them being right at the weld. (a) What is the frequency of vibration? (b) What is the length of the thick wire?

Solution

Conceptualize: The mass-per-volume density, the tension, and the frequency must be the same for the two wires. The linear density, wave speed, wavelength, and node-to-node distance are different.

Categorize: We know enough about the thin wire to find the frequency. The thick wire will have a predictably higher linear density, which will tell us the node-to-node distance for it.

Analyze:

(a) Since the first node is at the weld, the wavelength in the thin wire is

$$\lambda = 2d_{NN} = 2L = 80.0 \text{ cm}$$

The frequency and tension are the same in both sections, so

$$f = \frac{v}{\lambda} = \frac{1}{2L}\sqrt{\frac{T}{\mu}} = \frac{1}{2(0.400 \text{ m})}\sqrt{\frac{4.60 \text{ N}}{0.002\ 00 \text{ kg/m}}} = 59.9 \text{ Hz} \qquad \blacksquare$$

(b) Since the thick wire is twice the diameter, it will have four times the cross-sectional area, and a linear density μ' that is four times that of the thin wire. $\mu' = 4(2.00 \text{ g/m}) = 0.008\ 00 \text{ kg/m}$. Then L' varies accordingly.

$$L' = d'_{NN} = \frac{\lambda'}{2} = \frac{v'}{2f} = \frac{1}{2f}\sqrt{\frac{T}{\mu'}} = \frac{1}{2(59.9 \text{ Hz})}\sqrt{\frac{4.60 \text{ N}}{0.008\ 00 \text{ kg/m}}} = 20.0 \text{ cm} \qquad \blacksquare$$

Finalize: Note that the thick wire is half the length of the thin wire. We could have reasoned the answer by noting that the wave speed on the thick wire is half as large, so the wavelength should be half as large for the same frequency.

---

68. A standing wave is set up in a string of variable length and tension by a vibrator of variable frequency. Both ends of the string are fixed. When the vibrator has a frequency f, in a string of length L and under tension T, n antinodes are set up in the string. (a) If the length of the string is doubled, by what factor should the frequency be changed so that the same number of antinodes is produced? (b) If the frequency and length are held constant, what tension will produce $n + 1$ antinodes? (c) If the frequency is tripled and the length of the string is halved, by what factor should the tension be changed so that twice as many antinodes are produced?

Solution

Conceptualize: In (a), we expect a lower frequency to go with a longer wavelength. In (b), lower tension should go with lower wave speed for shorter wavelength at constant frequency. In (c), we will just have to divide it out.

Categorize: We will combine the equations

$$v = f\lambda, \qquad v = \sqrt{T/\mu} \qquad \text{and} \qquad \lambda_n = 2d_{NN} = 2L/n$$

where n is the number of segments or antinodes.

Analyze:

(a) By substitution we find that $\qquad f_n = \dfrac{n}{2L}\sqrt{\dfrac{T}{\mu}}$ [1]

Keeping n, T, and μ constant, we can create two equations.

$$f_n L = \frac{n}{2}\sqrt{\frac{T}{\mu}} \qquad \text{and} \qquad f_n' L' = \frac{n}{2}\sqrt{\frac{T}{\mu}}$$

Dividing the equations gives $\qquad \dfrac{f_n}{f_n'} = \dfrac{L'}{L}$

If $\qquad L' = 2L, \qquad$ then $\qquad f_n' = \dfrac{1}{2}f_n$

Therefore, in order to double the length but keep the same number of antinodes, the frequency should be halved. ∎

(b) From the same equation [1], we can hold L and f_n constant to get

$$\frac{n'}{n} = \sqrt{\frac{T}{T'}}$$

From this relation, we see that the tension must be decreased to

$$T' = T\left(\frac{n}{n+1}\right)^2 \qquad \text{to produce } n+1 \text{ antinodes} \qquad ∎$$

(c) This time, we rearrange equation [1] to produce

$$\frac{2f_n L}{n} = \sqrt{\frac{T}{\mu}} \qquad \text{and} \qquad \frac{2f_n' L'}{n'} = \sqrt{\frac{T'}{\mu}}$$

Then dividing gives

$$\frac{T'}{T} = \left(\frac{f_n'}{f_n} \cdot \frac{n}{n'} \cdot \frac{L'}{L}\right)^2 = \left(\frac{3f_n}{f_n}\right)^2 \left(\frac{n}{2n}\right)^2 \cdot \left(\frac{L/2}{L}\right)^2 = \frac{9}{16} \qquad ∎$$

Finalize: William Noble and Thomas Pigott, scholars at Oxford University, determined these proportionalities experimentally in 1677. We obtain them from dividing an equation describing one case by the equation describing another case in terms of the same parameters.

19

Temperature

EQUATIONS AND CONCEPTS

The **Celsius temperature** T_C is related to the absolute temperature T (in kelvins, K) according to Equation 19.1. *The size of a degree (unit change in temperature) on the Celsius scale equals the size of a degree on the kelvin scale.*

$$T_C = T - 273.15 \qquad (19.1)$$

$$0°C = 273.15 \text{ K}$$

Conversion between Fahrenheit and Celsius temperature scales is based on the freezing point temperature and boiling point temperature of water set for each scale.

$$T_F = \tfrac{9}{5}T_C + 32°F \qquad (19.2)$$

$$T_C = \tfrac{5}{9}\left(T_F - 32°F\right)$$

Freezing point = $0°C = 32°F$

Boiling point = $100°C = 212°F$

Thermal expansion (or contraction) due to an increase in temperature, results in changes in length, surface area, and volume of solids and a change in the volume of liquids. The fractional change is proportional to the change in temperature and to a parameter (coefficient of expansion) which is characteristic of a particular material.

Length expansion (solids)

$$\Delta L = \alpha L_i \Delta T \qquad (19.4)$$

$$L = L_i(1 + \alpha \Delta T)$$

Area expansion (solids)

$$\Delta A = \gamma A_i \Delta T$$

$$A = A_i(1 + \gamma \Delta T)$$

α = coefficient of linear expansion

Volume expansion (solids and liquids)

$\gamma \approx 2\alpha$ = coefficient of area expansion

$$\Delta V = \beta V_i \Delta T \qquad (19.6)$$

$\beta \approx 3\alpha$ = coefficient of linear expansion

$$V = V_i(1 + \beta \Delta T)$$

The **number of moles** in a sample of any substance (element or compound) equals the ratio of the mass of the sample to the molar mass characteristic of that particular substance. *One mole of any substance is that mass which contains Avogadro's number (N_A) of molecules.*

$$n = \frac{m}{M} \qquad (19.7)$$

$$N_A = 6.02 \times 10^{23} \text{ particles/mole}$$

The equation of state of an ideal gas is shown in two forms: expressed in terms of the number of moles (n) and the universal gas constant (R) as stated in Equation 19.8 or expressed in terms of the number of molecules (N) and Boltzmann's constant (k_B) as stated in Equation 19.10.

$$PV = nRT \qquad (19.8)$$

n = number of moles

$$R = 8.314 \text{ J/mol} \cdot \text{K} \qquad (19.9)$$

$$R = 0.082\,06 \text{ L} \cdot \text{atm/mol} \cdot \text{K}$$

$$PV = Nk_B T \qquad (19.10)$$

N = number of molecules

$$k_B = 1.38 \times 10^{-23} \text{ J/K} \qquad (19.11)$$

REVIEW CHECKLIST

You should be able to:

- Describe the operation of the constant-volume gas thermometer and the way in which it was used to define the absolute gas temperature scale. (Section 19.3)

- Convert among the various temperature scales: between Celsius and Kelvin, between Fahrenheit and Kelvin, and between Celsius and Fahrenheit. (Section 19.3)

- Use the thermal expansion coefficients for length, area, and volume to calculate changes in these quantities associated with changes in temperature. (Section 19.4)

- Make calculations using the ideal gas equation, expressed either in terms of n and R or in terms of N and k_B. (Section 19.5)

ANSWER TO AN OBJECTIVE QUESTION

10. A rubber balloon is filled with 1 L of air at 1 atm and 300 K and is then put into a cryogenic refrigerator at 100 K. The rubber remains flexible as it cools. **(i)** What happens to the volume of the balloon? (a) It decreases to $\frac{1}{3}$ L. (b) It decreases to $1/\sqrt{3}$ L. (c) It is constant. (d) It increases to $\sqrt{3}$ L. (e) It increases to 3 L. **(ii)** What happens to the pressure of the air in the balloon? (a) It decreases to $\frac{1}{3}$ atm. (b) It decreases to $1/\sqrt{3}$ atm. (c) It is constant. (d) It increases to $\sqrt{3}$ atm. (e) It increases to 3 atm.

Answer (i) (a) and **(ii)** (c). The tiny quantities of carbon dioxide and water in the air solidify on the way down to 100 K, but the oxygen, nitrogen, and argon remain gases, and it is reasonable to assume that their densities are low enough to permit use of the ideal gas equation of state. To qualify as a balloon, the container cannot have a fast leak, so in $PV = nRT$ we take the quantity of gas n, as well as the universal constant R, as constant in the contraction process. A refrigerator is not airtight, so the pressure inside the box stays equal to the constant atmospheric pressure outside. The flexible wall of the balloon moves to equalize the pressure of the air inside with the pressure outside. The rubber is in mechanical equilibrium throughout the process, stationary at the start and finish and moving with negligible acceleration in between, while the air inside is coming to thermal equilibrium. An ordinary person "blowing up" a balloon does not stretch the rubber enough to make the interior pressure significantly different from atmospheric pressure; as the air cools, the rubber will go slack and the interior pressure will become exactly equal to atmospheric pressure. Then the process for the sample within is isobaric and the answer to question **(ii)** is (c). In the equation of state the only quantities that change are V and T. When T falls to one-third of its initial value, $V = nRT/P$ falls to 1/3 L. The answer to **(i)** is (a).

□   □   □   □

ANSWERS TO SELECTED CONCEPTUAL QUESTIONS

2.   A piece of copper is dropped into a beaker of water. (a) If the water's temperature rises, what happens to the temperature of the copper? (b) Under what conditions are the water and copper in thermal equilibrium?

Answer   (a) If the water's temperature increases, that means that energy is being transferred by heat to the water. This can happen either if the copper changes phase from liquid to solid, or if the temperature of the copper is above that of the water, and falling.

In this case, the copper is referred to as a "piece" of copper, so it is already in the solid phase; therefore, its temperature must be falling. (b) When the temperature of the copper reaches that of the water, the copper and water will reach equilibrium, and the subsequent net energy transfer by heat between the two will be zero.

□   □   □   □

8.   When the metal ring and metal sphere in Figure CQ19.8 are both at room temperature, the sphere can barely be passed through the ring. (a) After the sphere is warmed in a flame, it cannot be passed through the ring. Explain. (b) **What if?** What if the ring is warmed and the sphere is left at room temperature? Does the sphere pass through the ring?

Figure CQ19.8

Answer   (a) The hot sphere has expanded, so it no longer fits through the ring. (b) When the ring is warmed, the cool sphere fits through more easily.

Suppose a cool sphere is put through a cool ring and then warmed so that it does not come back out. With the sphere still hot, you can separate the sphere and ring by warming the ring as shown in the drawing here. This more surprising result occurs because the thermal expansion of the ring is not like the inflation of a blood-pressure cuff. Rather, it is like a photographic enlargement; every linear dimension, including the hole diameter, increases by the same factor. The reason for this is that neighboring atoms everywhere, including those around the inner circumference, push away from each other. The only way that the atoms can accommodate the greater distances is for the circumference—and corresponding diameter—to grow. This property was once used to fit metal rims to wooden wagon and horse-buggy wheels.

☐   ☐   ☐   ☐

SOLUTIONS TO SELECTED END-OF-CHAPTER PROBLEMS

5.   Liquid nitrogen has a boiling point of −195.81°C at atmospheric pressure. Express this temperature (a) in degrees Fahrenheit and (b) in kelvins.

Solution

Conceptualize: Estimate minus several hundred degrees Fahrenheit and positive eighty-something kelvins.

Categorize: We can puzzle through the conversion to Fahrenheit. Expressing the temperature on the absolute scale will be easy.

Analyze:

(a)   By Equation 19.2, $T_F = \frac{9}{5}T_C + 32°F = \frac{9}{5}(-195.81) + 32 = -320°F$ ∎

(b)   Applying Equation 19.1, $273.15\ \text{K} - 195.81\ \text{K} = 77.3\ \text{K}$ ∎

Finalize: A convenient way to remember Equations 19.1 and 19.2 is to remember the freezing and boiling points of water, in each form:

$$T_{\text{freeze}} = 32.0°F = 0°C = 273.15\ \text{K}$$

$$T_{\text{boil}} = 212°F = 100°C$$

To convert from Fahrenheit to Celsius, subtract 32 (the freezing point), and then adjust the scale by the liquid range of water.

$$\text{Scale factor} = \frac{(100-0)°C}{(212-32)°F} = \frac{5°C}{9°F}$$

A kelvin is the same size change as a degree Celsius, but the kelvin scale takes its zero point at **absolute zero**, instead of the freezing point of water. Therefore, to convert from kelvin to Celsius, subtract 273.15 K.

---

6.   In a student experiment, a constant-volume gas thermometer is calibrated in dry ice (−78.5°C) and in boiling ethyl alcohol (78.0°C). The separate pressures are 0.900 atm and

1.635 atm. (a) What value of absolute zero in degrees Celsius does the calibration yield? What pressures would be found at the (b) freezing and (c) boiling points of water? Hint: Use the linear relationship $P = A + BT$, where A and B are constants.

Solution

Conceptualize: Visualize a straight-line graph. The temperature at which the extrapolated pressure drops to zero will be far below $-100°C$. The pressure will be around 1.2 atm at 0°C and 1.8 atm at 100°C.

Categorize: We need not actually draw the graph. We can find the equation of the graph line and solve all parts of the problem from it.

Analyze: To represent a linear graph, the pressure is related to the temperature as $P = A + BT$, where A and B are constants.

To find A and B, we use the given data:

$$0.900 \text{ atm} = A + (-78.5°C)B \quad \text{and} \quad 1.635 \text{ atm} = A + (78.0°C)B$$

Solving these simultaneously, we find

$$0.735 \text{ atm} = (156.5°C)B \quad \text{so} \quad B = 4.696 \times 10^{-3} \text{ atm/°C} \quad \text{and} \quad A = 1.269 \text{ atm}$$

Therefore, for all temperatures

$$P = 1.269 \text{ atm} + (4.696 \times 10^{-3} \text{ atm/°C})T$$

(a) At absolute zero, $P = 0 = 1.269 \text{ atm} + (4.696 \times 10^{-3} \text{ atm/°C})T$

which gives $T = -270°C$ ∎

(b) At the freezing point of water, $P = 1.269 \text{ atm} + 0 = 1.27 \text{ atm}$ ∎

(c) And at the boiling point,

$$P = 1.269 \text{ atm} + (4.696 \times 10^{-3} \text{ atm/°C})(100°C) = 1.74 \text{ atm}$$ ∎

Finalize: Our estimates were good. In a typical student experiment the extrapolated value of absolute zero would have high uncertainty.

---

7. A copper telephone wire has essentially no sag between poles 35.0 m apart on a winter day when the temperature is $-20.0°C$. How much longer is the wire on a summer day when the temperature is $35.0°C$?

Solution

Conceptualize: Normally, we do not notice a change in the length of the telephone wires. Thus, we might expect the wire to expand by less than a decimeter.

Categorize: The change in length can be found from the linear expansion of copper wire (we will assume that the insulation around the copper wire can stretch more easily than the wire itself). From Table 19.1, the coefficient of linear expansion for copper is 17×10^{-6} $(°C)^{-1}$.

Analyze: The change in length between cold and hot conditions is

$$\Delta L = \alpha L_i \Delta T = \left[17 \times 10^{-6}(°C)^{-1}\right](35.0 \text{ m})[35.0°C - (-20.0°C)]$$

$$\Delta L = 3.27 \times 10^{-2} \text{ m} = 3.27 \text{ cm} \qquad \blacksquare$$

Finalize: This expansion agrees with our expectation that the change in length is less than 10 centimeters. From ΔL we can find that if the wire sags, its midpoint can be displaced downward by 0.757 m on the hot summer day. This also seems reasonable based on everyday observations.

---

9. The active element of a certain laser is made of a glass rod 30.0 cm long and 1.50 cm in diameter. Assume the average coefficient of linear expansion of the glass is equal to $9.00 \times 10^{-6} (°C)^{-1}$. If the temperature of the rod increases by 65.0°C, what is the increase in (a) its length, (b) its diameter, and (c) its volume?

Solution

Conceptualize: We expect a change on the order of a part in a thousand. That would be a fraction of a millimeter for the length, a much smaller distance for the diameter, and maybe some cubic millimeters for the volume.

Categorize: It's all based on the definition of the coefficient of expansion.

Analyze:

(a) $\Delta L = \alpha L_i \Delta T = (9.00 \times 10^{-6} °C^{-1})(0.300 \text{ m})(65.0°C) = 1.76 \times 10^{-4} \text{ m}$ $\qquad \blacksquare$

(b) The diameter is a linear dimension, so the same equation applies:

$\Delta D = \alpha D_i \Delta T = (9.00 \times 10^{-6} °C^{-1})(0.015 \text{ 0 m})(65.0°C) = 8.78 \times 10^{-6} \text{ m}$ $\qquad \blacksquare$

(c) The original volume is

$V = \pi r^2 L = \dfrac{\pi}{4}(0.015 \text{ 0 m})^2(0.300 \text{ m}) = 5.30 \times 10^{-5} \text{ m}^3$

Using the volumetric coefficient of expansion β,

$\Delta V = \beta V_i \Delta T \approx 3\alpha V \Delta T$

$\Delta V \approx 3(9.00 \times 10^{-6} °C^{-1})(5.30 \times 10^{-5} \text{ m}^3)(65.0°C) = 93.0 \times 10^{-9} \text{ m}^3$ $\qquad \blacksquare$

Finalize: The above calculation of ΔV ignores ΔL^2 and ΔL^3 terms. Let us calculate the change in volume more precisely, and compare the answer with the approximate solution above.

The volume will increase by a factor of

$$\frac{\Delta V}{V_i} = \left(1 + \frac{\Delta D}{D_i}\right)^2 \left(1 + \frac{\Delta L}{L_i}\right) - 1$$

$$\frac{\Delta V}{V_i} = \left(1 + \frac{8.78 \times 10^{-6}\,\text{m}}{0.015\ 0\,\text{m}}\right)^2 \left(1 + \frac{1.76 \times 10^{-4}\,\text{m}}{0.300\,\text{m}}\right) - 1 = 1.76 \times 10^{-3}$$

The amount of volume change is $\quad \Delta V = \dfrac{\Delta V}{V_i} V_i$

$$\Delta V = (1.76 \times 10^{-3})(5.30 \times 10^{-5}\,\text{m}^3) = 93.1 \times 10^{-9}\,\text{m}^3 = 93.1\,\text{mm}^3 \qquad \blacksquare$$

The answer is virtually identical; the approximation $\beta \approx 3\alpha$ is a good one.

---

13. A volumetric flask made of Pyrex is calibrated at 20.0°C. It is filled to the 100-mL mark with 35.0°C acetone. After the flask is filled, the acetone cools and the flask warms, so that the combination of acetone and flask reaches a uniform temperature of 32.0°C. The combination is then cooled back to 20.0°C. (a) What is the volume of the acetone when it cools to 20.0°C? (b) At the temperature of 32.0°C, does the level of acetone lie above or below the 100-mL mark on the flask? Explain.

Solution

Conceptualize: The volume of the acetone is 100 mL at 35.0°C, so it will be slightly less at 32.0°C and still less at 20.0°C. But we guess that the error will be less than 2 mL.

Categorize: We use the definition of the volume expansion coefficient. To answer part (b), we will think about both the glass and the liquid changing in volume.

Analyze:

(a) The original volume of the acetone we take as precisely 100 mL. After it is finally cooled to 20.0°C, its volume is

$$V_f = V_i + \Delta V = V_i(1 + \beta \Delta T)$$

$$V_f = 100\ \text{mL}[1 + 1.50 \times 10^{-4}(^\circ\text{C})^{-1}(-15.0^\circ\text{C})] = 99.8\ \text{mL} \qquad \blacksquare$$

(b) At 32.0°C, the acetone has contracted a bit from its original volume and the glass has expanded its interior volume. Both effects work together to place the surface of the liquid below the fiduciary mark on the flask. $\qquad \blacksquare$

Finalize: The effect of the temperature change was smaller than we guessed in advance, but big enough to affect an experiment designed for three-digit precision. Just to show its simplicity, let us find the ratio of the amounts of volume change that the liquid and the glass undergo with equal temperature changes, such as here from 32°C to 20°C. The expansion of the acetone is described by $\Delta V_{acetone} = (\beta V_i \Delta T)_{acetone}$

and that of the glass by $\Delta V_{flask} = (\beta V_i \Delta T)_{pyrex} = (3\alpha V_i \Delta T)_{pyrex}$

for the same V_i and ΔT, we have

$$\frac{\Delta V_{acetone}}{\Delta V_{flask}} = \frac{\beta_{acetone}}{\beta_{flask}} = \frac{1.50 \times 10^{-4}}{3(3.20 \times 10^{-6})} = \frac{1}{6.40 \times 10^{-2}}$$

The volume change of flask is about 6% of the change in the volume of the acetone.

---

23. An auditorium has dimensions $10.0 \text{ m} \times 20.0 \text{ m} \times 30.0 \text{ m}$. How many molecules of air fill the auditorium at 20.0°C and a pressure of 101 kPa (1.00 atm)?

Solution

Conceptualize: The given room conditions are close to Standard Temperature and Pressure (STP is 0°C and 101.3 kPa), so we can use the estimate that one mole of an ideal gas at STP occupies a volume of about 22 L. The volume of the auditorium is $6 \times 10^3 \text{ m}^3$ and $1 \text{ m}^3 = 1\,000$ L, so we can estimate the number of molecules to be:

$$N \approx (6\,000 \text{ m}^3)\left(\frac{10^3 \text{ L}}{1 \text{ m}^3}\right)\left(\frac{1 \text{ mol}}{22 \text{ L}}\right)\left(\frac{6 \times 10^{23} \text{ molecules}}{1 \text{ mol}}\right)$$

$$\approx 1.6 \times 10^{29} \text{ molecules of air}$$

Categorize: The number of molecules can be found more precisely by applying the equation of state of an ideal gas…

Analyze: …which is $PV = nRT$. We choose to calculate first the number of moles to find N.

$$n = \frac{PV}{RT} = \frac{\left(1.01 \times 10^5 \text{ N/m}^2\right)[(10.0 \text{ m})(20.0 \text{ m})(30.0 \text{ m})]}{(8.314 \text{ J/mol}\cdot\text{K})(293 \text{ K})} = 2.49 \times 10^5 \text{ mol}$$

$$N = nN_A = (2.49 \times 10^5 \text{ mol})\left(6.02 \times 10^{23} \frac{\text{molecules}}{\text{mol}}\right) = 1.50 \times 10^{29} \text{ molecules} \qquad \blacksquare$$

Finalize: This result agrees quite well with our initial estimate. The number of molecules is somewhat smaller than it would be at 0°C.

---

26. The pressure gauge on a tank registers the gauge pressure, which is the difference between the interior pressure and exterior pressure. When the tank is full of oxygen (O_2), it contains 12.0 kg of the gas at a gauge pressure of 40.0 atm. Determine the mass of oxygen that has been withdrawn from the tank when the pressure reading is 25.0 atm. Assume that the temperature of the tank remains constant.

Solution

Conceptualize: We estimate a bit less than half of the 12 kg, or about 5 kg.

Categorize: We will use the ideal gas law and relate the mass to the number of moles.

Analyze: The equation of state for gases far from liquefaction is $PV = nRT$

At constant volume and temperature, $\dfrac{P}{n} = $ constant or

$$\frac{P_1}{n_1} = \frac{P_2}{n_2} \quad \text{and} \quad n_2 = \left(\frac{P_2}{P_1}\right) n_1$$

Further, n is proportional to m, so

$$m_2 = \left(\frac{P_2}{P_1}\right) m_1 = \left(\frac{26.0 \text{ atm}}{41.0 \text{ atm}}\right)(12.0 \text{ kg}) = 7.61 \text{ kg}$$

The mass removed is $|\Delta m| = 12.0 \text{ kg} - 7.61 \text{ kg} = 4.39 \text{ kg}$ ∎

Finalize: Our estimate was pretty good. Most of the mass of a high-pressure cylinder of gas is the mass of the metal cylinder.

---

29. An automobile tire is inflated with air originally at 10.0°C and normal atmospheric pressure. During the process, the air is compressed to 28.0% of its original volume and the temperature is increased to 40.0°C. (a) What is the tire pressure? (b) After the car is driven at high speed, the tire's air temperature rises to 85.0°C and the tire's interior volume increases by 2.00%. What is the new tire pressure (absolute)?

Solution

Conceptualize: In (a), both temperature increase and volume decrease contribute to increasing the pressure. In part (b) the temperature increase will dominate the volume increase to produce still higher pressure of some hundreds of kilopascals.

Categorize: The ideal-gas equation of state describes each condition of the air. We can set up ratios to describe the changes.

Analyze:

(a) Taking $PV = nRT$ in the initial (i) and final (f) states, and dividing, we have

$$P_i V_i = nRT_i \quad \text{and} \quad P_f V_f = nRT_f \quad \text{yielding} \quad \frac{P_f V_f}{P_i V_i} = \frac{T_f}{T_i}$$

So $P_f = P_i \dfrac{V_i T_f}{V_f T_i} = \left(1.013 \times 10^5 \text{ Pa}\right)\left(\dfrac{V_i}{0.280 V_i}\right)\left(\dfrac{273 \text{ K} + 40.0 \text{ K}}{273 \text{ K} + 10.0 \text{ K}}\right)$

$\qquad\qquad = 4.00 \times 10^5 \text{ Pa}$ ∎

(b) Introducing the hot (h) state, we have $\quad \dfrac{P_h V_h}{P_f V_f} = \dfrac{T_h}{T_f}$

$\qquad P_h = P_f \left(\dfrac{V_f}{V_h}\right)\left(\dfrac{T_h}{T_f}\right) = \left(4.00 \times 10^5 \text{ Pa}\right)\left(\dfrac{V_f (358 \text{ K})}{1.02 V_f (313 \text{ K})}\right) = 4.49 \times 10^5 \text{ Pa}$ ∎

Finalize: Those pressures are a bit higher than usual for a car tire. An absolute pressure of 400 kPa means a gauge pressure of about 300 kPa or three atmospheres. Around two atmospheres is more typical.

---

31. Review. The mass of a hot-air balloon and its cargo (not including the air inside) is 200 kg. The air outside is at 10.0°C and 101 kPa. The volume of the balloon is 400 m³. To what temperature must the air in the balloon be warmed before the balloon will lift off? (Air density at 10.0°C is 1.244 kg/m³.)

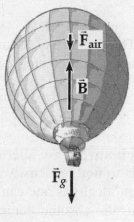

Solution

Conceptualize: The air inside the balloon must be significantly hotter than the outside air in order for the balloon to feel a net upward force, but the temperature must also be less than the melting point of the nylon used for the balloon's envelope. (Rip-stop nylon melts around 200°C.) Otherwise the results could be disastrous!

Categorize: We model the balloon with load and interior hot air as a particle in equilibrium, when it is just ready to lift off. The density of the air inside the balloon must be sufficiently low so that the buoyant force is equal to the weight of the balloon, its cargo, and the air inside. The temperature of the air required to achieve this density can be found from the equation of state of an ideal gas.

Analyze: The buoyant force equals the weight of the air at 10.0°C displaced by the balloon:

$\qquad B = m_{\text{air}}\, g = \rho_{\text{air}} V g$

$\qquad\qquad B = \left(1.244 \text{ kg/m}^3\right)\left(400 \text{ m}^3\right)\left(9.80 \text{ m/s}^2\right) = 4\,880 \text{ N}$

The weight of the balloon and its cargo is

$\qquad\qquad F_g = mg = \left(200 \text{ kg}\right)\left(9.80 \text{ m/s}^2\right) = 1\,960 \text{ N}$

Since $B > F_g$, the balloon has a chance of lifting off as long as the weight of the air inside the balloon is less than the difference in these forces:

$$F_{air} < B - F_g = 4\,880 \text{ N} - 1\,960 \text{ N} = 2\,920 \text{ N}$$

The mass of this air is

$$m_{air} = \frac{F_{air}}{g} = \frac{2\,920 \text{ N}}{9.80 \text{ m/s}^2} = 298 \text{ kg}$$

To find the required temperature of this air from $PV = nRT$, we must find the corresponding number of moles of air. Dry air is approximately 20% O_2, and 80% N_2. Using data from a periodic table, we can calculate the mass of Avogadro's number of molecules in air to be approximately

$$M = (0.80)(28.0 \text{ g/mol}) + (0.20)(32.0 \text{ g/mol}) = 28.8 \text{ g/mol}$$

so the number of moles is

$$n = \frac{m}{M} = (298 \text{ kg})\left(\frac{1\,000 \text{ g/kg}}{28.8 \text{ g/mol}}\right) = 1.03 \times 10^4 \text{ mol}$$

The pressure of this air is the ambient pressure; from $PV = nRT$, we can now find the minimum temperature required for lift off:

$$T = \frac{PV}{nR} = \frac{(1.01 \times 10^5 \text{ N/m}^2)(400 \text{ m}^3)}{(1.03 \times 10^4 \text{ mol})(8.314 \text{ J/mol} \cdot \text{K})} = 470 \text{ K} = 197°\text{C} \qquad \blacksquare$$

Finalize: An alternative approach, for the last few steps to find the hot-air temperature, is to find its density as 298 kg/400 m³ = 0.744 kg/m³. Next, from the equation of state of an ideal gas one can prove that at constant pressure the density is inversely proportional to the absolute temperature. Then the high temperature must be 283 K(1.244/0.744) = 473 K = 200°C, substantially in agreement with our first result. The average temperature of the air inside the balloon required for lift off appears to be close to the melting point of the nylon fabric, so this seems like a dangerous situation! A larger balloon would be better suited for the given load on the balloon. (Many sport balloons have volumes of about 3 000 m³ or 10 times larger than the one in this problem.)

---

41. A mercury thermometer is constructed as shown in Figure P19.41. The Pyrex glass capillary tube has a diameter of 0.004 00 cm, and the bulb has a diameter of 0.250 cm. Find the change in height of the mercury column that occurs with a temperature change of 30.0°C.

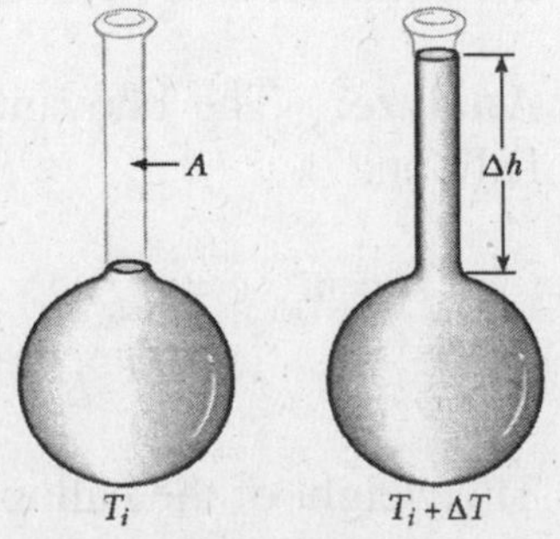

Solution

Conceptualize: For an easy-to-read thermometer, the column should rise by a few centimeters.

Figure P19.41

Categorize: We use the definition of the coefficient of expansion.

Analyze: The volume of the liquid increases as $\Delta V_l = V_i \beta \Delta T$ where V_i is the original interior volume of the bulb. Expansion of the glass makes the volume of the bulb increase as $\Delta V_g = 3\alpha V_i \Delta T$. Therefore, the overflow into the capillary is $V_c = V_i \Delta T(\beta - 3\alpha)$. For mercury $\beta = 1.82 \times 10^{-4}\,°C^{-1}$ and for Pyrex glass $\alpha = 3.20 \times 10^{-6}\,°C^{-1}$. In the capillary the mercury overflow fills a cylinder according to $V_c = A\Delta h$.

Then we have

$$\Delta h = \frac{V_i}{A}(\beta - 3\alpha)\Delta T = \frac{\frac{4}{3}\pi R_{bulb}^3}{\pi R_{cap}^2}(\beta - 3\alpha)\Delta T$$

$$= \frac{4(0.125\ cm)^3}{3(0.002\ 00\ cm)^2}[1.82 \times 10^{-4} - 3(3.20 \times 10^{-6})](°C)^{-1}(30.0°C)$$

$$= 3.37\ cm \qquad\blacksquare$$

Finalize: This is a practical thermometer. If we assumed that all of the glass undergoes the temperature change, not just the bulb, then we would include an extra factor of $[1 + 2(3.2 \times 10^{-6})(30)]$ in the bottom of the fraction that we used to compute Δh. It would account for expansion of the area of the capillary. But it does not affect the answer to three significant digits.

---

45. A liquid has a density ρ. (a) Show that the fractional change in density for a change in temperature ΔT is $\Delta\rho/\rho = -\beta\,\Delta T$. (b) What does the negative sign signify? (c) Fresh water has a maximum density of $1.000\ 0\ g/cm^3$ at $4.0°C$. At $10.0°C$, its density is $0.999\ 7\ g/cm^3$. What is β for water over this temperature interval? (d) At $0°C$, the density of water is $0.999\ 9\ g/cm^3$. What is the value for β over the temperature range $0°C$ to $4.00°C$?

Solution

Conceptualize: Mass does not change with temperature, but density does because of thermal expansion in volume.

Categorize: We will use the definitions of density and of expansion coefficient.

Analyze: We start with the two equations

$$\rho = \frac{m}{V} \qquad \text{and} \qquad \frac{\Delta V}{V} = \beta\Delta T$$

(a) Differentiating the first equation gives $\qquad d\rho = -\frac{m}{V^2}dV$

For very small changes in V and ρ, this can be written

$$\Delta\rho = -\frac{m}{V}\frac{\Delta V}{V} = -\left(\frac{m}{V}\right)\frac{\Delta V}{V}$$

Substituting both of our initial equations, we find that

$$\Delta\rho = -\rho\beta\Delta T \qquad \text{so} \qquad \Delta\rho/\rho = -\beta\Delta T \qquad\blacksquare$$

(b) The negative sign means that if β is positive, any increase in temperature causes the density to decrease and vice versa. ∎

(c) We apply the equation $\beta = -\dfrac{\Delta\rho}{\rho\Delta T}$ for the specific case of water:

$$\beta = -\frac{\left(1.000\ 0\ \text{g/cm}^3 - 0.999\ 7\ \text{g/cm}^3\right)}{\left(1.000\ 0\ \text{g/cm}^3\right)\left(4.00°\text{C} - 10.0°\text{C}\right)}$$

Calculating, we find that $\beta = 5.0 \times 10^{-5}\,°\text{C}^{-1}$ ∎

(d) Between 0 and 4°C, water contracts with increasing temperature, so we find a negative coefficient of volume expansion by the same method:

$$\beta = -\frac{\Delta\rho}{\rho\Delta T} = -\frac{1.000\ 0\ \text{g/cm}^3 - 0.999\ 9\ \text{g/cm}^3}{\left(0.999\ 9\ \text{g/cm}^3\right)\left(4.0°\text{C} - 0.0°\text{C}\right)} = -2.5 \times 10^{-5}\,°\text{C}^{-1}$$ ∎

Finalize: The volume expansion coefficients of water in this range are less than many other liquids.

---

49. The rectangular plate shown in Figure P19.49 has an area A_i equal to ℓw. If the temperature increases by ΔT, each dimension increases according to Equation 19.4, where α is the average coefficient of linear expansion. (a) Show that the increase in area is $\Delta A = 2\alpha A_i \Delta T$. (b) What approximation does this expression assume?

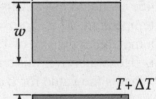

Figure P19.49

Solution

Conceptualize: We expect the area to increase in thermal expansion. It is neat that the coefficient of area expansion is just twice the coefficient of linear expansion.

Categorize: We will use the definitions of coefficients of linear and area expansion.

Analyze: From the diagram in Figure P19.49, we see that the **change** in area is

$$\Delta A = \ell\Delta w + w\Delta\ell + \Delta w\Delta\ell$$

Since $\Delta\ell$ and Δw are each small quantities, the product $\Delta w\Delta\ell$ will be very small compared to the original or final area.

Therefore, we assume $\quad \Delta w\Delta\ell \approx 0$ ∎

Since $\quad\quad\quad\quad\quad\quad\quad \Delta w = w\alpha\Delta T \quad$ and $\quad \Delta\ell = \ell\alpha\Delta T$

we then have $\qquad\qquad \Delta A = \ell w \alpha \Delta T + w \ell \alpha \Delta T$

(a) Finally, since $\qquad A = \ell w \qquad$ we have $\qquad \Delta A = 2\alpha A \Delta T$ ∎

(b) A clearer way to state the approximation we have made is that $\alpha \Delta T$ is much less than 1. ∎

Then the change in area is small compared to the original area and $\Delta w \Delta \ell$ is negligibly small.

Finalize: Increase in length and increase in width both contribute to the increase in area, through the narrow ribbons of extra area in the diagram. Their contributions accumulate to give the 2 in $\Delta A = 2\alpha A \Delta T$. The bit we ignored is the little rectangle off in the corner of the diagram. It is truly small if the temperature change is small.

---

52. A vertical cylinder of cross-sectional area A is fitted with a tight-fitting, frictionless piston of mass m (Fig. P19.52). The piston is not restricted in its motion in any way and is supported by the gas at pressure P below it. Atmospheric pressure is P_0. We wish to find the height h in Figure P19.52. (a) What analysis model is appropriate to describe the piston? (b) Write an appropriate force equation for the piston from this analysis model in terms of P, P_0, m, A, and g. (c) Suppose n moles of an ideal gas are in the cylinder at a temperature of T. Substitute for P in your answer to (b) to find the height h of the piston above the bottom of the cylinder.

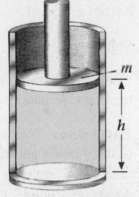

Figure P19.52

Solution

Conceptualize: The "spring of air," as Robert Boyle called it, does not follow the same equation as Hooke's law, but the weight of the piston compresses the gas just enough to raise its pressure enough to support the weight.

Categorize: We will use the definition of pressure and the ideal gas law.

Analyze:

(a) The air above the piston remains at atmospheric pressure, P_0. Model the piston as a particle in equilibrium. ∎

(b) $\Sigma F_y = ma_y$ yields $-P_0 A - mg + PA = 0$ ∎

where P is the pressure exerted by the gas contained. Noting that $V = Ah$, and that n, T, m, g, A, and P_0 are given,

(c) $PV = nRT \qquad$ becomes $\qquad P = \dfrac{nRT}{Ah}$

so we substitute to eliminate P

$$-P_0A - mg + \frac{nRT}{Ah}A = 0 \qquad \text{and} \qquad h = \frac{nRT}{P_0A + mg} \qquad \blacksquare$$

Finalize: It is reasonable that n is on top in the fraction in the derived equation, because more gas will occupy more volume. Expansion makes it reasonable that T is on top. A heavier piston would compress the gas, so it is reasonable that the piston mass is on the bottom.

---

59.   Starting with Equation 19.10, show that the total pressure P in a container filled with a mixture of several ideal gases is $P = P_1 + P_2 + P_3 + \cdots$, where $P_1, P_2, \cdots$, are the pressures that each gas would exert if it alone filled the container. (These individual pressures are called the *partial pressures* of the respective gases.) This result is known as *Dalton's law of partial pressures.*

Solution

Conceptualize: This problem emphasizes the reality of gas pressure. The gas is always tending to expand, so it or any part of it is always exerting pressure.

Categorize: We will use the equation of state of an ideal gas for each of the components of the mixture and for the whole sample.

Analyze: For each gas alone, $P_1 = \dfrac{N_1 k_B T}{V_1}$, $P_2 = \dfrac{N_2 k_B T}{V_2}$, $P_3 = \dfrac{N_3 k_B T}{V_3}$, $\cdots$

For the gases combined, the number of molecules is

$$N_1 + N_2 + N_3 + \cdots = N = \frac{PV}{k_B T}$$

Therefore,

$$\frac{P_1 V_1}{k_B T} + \frac{P_2 V_2}{k_B T} + \frac{P_3 V_3}{k_B T} + \cdots = \frac{PV}{k_B T}$$

But $V_1 = V_2 = V_3 = \cdots = V$ so $P_1 + P_2 + P_3 + \cdots = P$ $\qquad \blacksquare$

Finalize: We will later attribute free random motion to the molecules, in the kinetic theory of gases. This idea makes the law of partial pressures very natural.

---

65.   Review.   A steel guitar string with a diameter of 1.00 mm is stretched between supports 80.0 cm apart. The temperature is 0.0°C. (a) Find the mass per unit length of this string. (Use the value 7.86×10^3 kg/m³ for the density.) (b) The fundamental frequency of transverse oscillations of the string is 200 Hz. What is the tension in the string? Next, the temperature is raised to 30.0°C. Find the resulting values of (c) the tension and (d) the fundamental frequency. Assume both the Young's modulus of 20.0×10^{10} N/m^2 and the average coefficient of expansion $\alpha = 11.0 \times 10^{-6}$ (°C)$^{-1}$ have constant values between 0.0°C and 30.0°C.

Solution

Conceptualize: The tension will drop as the wire tends to expand, so the frequency will drop.

Categorize: The actual length does not change, but the unstretched length does as the temperature changes. So the strain changes.

Analyze:

(a) We find the linear density from the volume density as the mass-per-volume multiplied by the volume-per-length, which is the cross-sectional area.

$$\mu = \tfrac{1}{4}\rho(\pi d^2) = \tfrac{1}{4}\pi(1.00 \times 10^{-3}\text{ m})^2(7.86 \times 10^3\text{ kg/m}^3) = 6.17 \times 10^{-3}\text{ kg/m} \quad \blacksquare$$

(b) Since $f_1 = \dfrac{v}{2L}$, $v = \sqrt{T/\mu}$ and $f_1 = \dfrac{1}{2L}\sqrt{T/\mu}$

we have for the tension

$$F = T = \mu(2Lf_1)^2 = (6.17 \times 10^{-3}\text{ kg/m})[(2)(0.800\text{ m})(200\text{ s}^{-1})]^2 = 632\text{ N} \quad \blacksquare$$

(c) At 0°C, the length of the guitar string will be

$$L_{\text{actual}} = L_{0°C}\left(1 + \frac{F}{AY}\right) = 0.800\text{ m}$$

Where $L_{0°C}$ is the unstressed length at the low temperature. We know the string's cross-sectional area

$$A = \left(\frac{\pi}{4}\right)(1.00 \times 10^{-3}\text{ m})^2 = 7.85 \times 10^{-7}\text{ m}^2$$

and modulus $Y = 20.0 \times 10^{10}$ N/m^2

Therefore,

$$\frac{F}{AY} = \frac{632\text{ N}}{(7.85 \times 10^{-7}\text{ m}^2)(20.0 \times 10^{10}\text{ N/m}^2)} = 4.02 \times 10^{-3} \quad \text{and}$$

$$L_{0°C} = \frac{0.800\text{ m}}{1 + 4.02 \times 10^{-3}} = 0.796\ 8\text{ m}$$

Then at 30°C, the unstressed length is

$$L_{30°C} = (0.796\ 8\text{ m})\left(1 + (30.0°C)(11.0 \times 10^{-6}°C^{-1})\right) = 0.797\ 1\text{ m}$$

With the same clamping arrangement, $0.800 \text{ m} = (0.797 \text{ 1 m})\left[1 + \dfrac{F'}{A'Y}\right]$

where F' and A' are the new tension and the new (expanded) cross-sectional area.

Then $\quad \dfrac{F'}{A'Y} = \dfrac{0.800\ 0}{0.797\ 1} - 1 = 3.693 \times 10^{-3} \quad$ and $\quad F' = A'Y\left(3.693 \times 10^{-3}\right)$

$F' = \left(7.85 \times 10^{-7} \text{ m}^2\right)\left(20.0 \times 10^{10} \text{ N/m}^2\right)\left(3.693 \times 10^{-3}\right)(1 + \alpha \Delta T)^2$

$F' = (580 \text{ N})\left(1 + 3.30 \times 10^{-4}\right)^2 = 580 \text{ N}$ ∎

(d)   Also the new frequency f_1' is given by $\quad \dfrac{f_1'}{f_1} = \sqrt{\dfrac{F_*'}{F}}$

so $\quad f_1' = (200 \text{ Hz})\sqrt{\dfrac{580 \text{ N}}{632 \text{ N}}} = 192 \text{ Hz}$ ∎

Finalize: The string has gone flat by a bit less than a semitone. In the calculation we wrote down $A(1 + \alpha \Delta T)^2$ for A', the cross-sectional area at the high temperature, but the difference in area does not affect the answer to three digits.

20

The First Law of Thermodynamics

EQUATIONS AND CONCEPTS

The **specific heat** of a substance is the heat capacity per unit mass. Each substance requires a specific quantity of energy to change the temperature of 1 kg of the substance by 1.00°C. *Heat capacity, C, refers to a sample of material. Specific heat, c, is the heat capacity per unit mass, and is characteristic of a particular type of material.*

$$c \equiv \frac{Q}{m\Delta T} \tag{20.3}$$

The **energy Q that must be transferred** between a system of mass m and its surroundings to produce a temperature change ΔT varies with the substance.

$$Q = mc\Delta T \tag{20.4}$$

Calorimetry, a technique to measure specific heat, is based on the conservation of energy in an isolated system. *The negative sign in Equation 20.5 is required so that each side of the equation will be positive.*

$$Q_{cold} = -Q_{hot} \tag{20.5}$$

Q_{cold} *is a positive quantity*

Q_{hot} *is a negative quantity*

Latent heat, L, is a thermal property of a substance that determines the quantity of energy Q required to change the phase of a unit mass of that substance. *A phase change occurs at constant temperature and is accompanied by a change in internal energy.*

$$Q = L\Delta m \tag{20.7}$$

L is expressed in units of J/kg

Δm = change in mass of the higher phase substance

The **value of L for a given substance** depends on the nature of the phase change.

L_f = heat of fusion (phase change between solid and liquid)

L_v = heat of vaporization (phase change between liquid and vapor)

The work done on a gas when the volume changes from V_i to V_f depends on the path taken between the initial and final states. In order to evaluate the integral in Equation 20.9, the manner in which pressure varies with volume must be known.

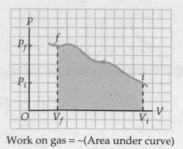

Work on gas = −(Area under curve)

$$W = -\int_{V_i}^{V_f} P\,dV \qquad (20.9)$$

The first law of thermodynamics is stated in mathematical form by Equation 20.10. The change in the internal energy of a system equals the sum of the energy transferred across the boundary by heat and the energy transferred by work. *The values of both Q and W depend on the path taken between the initial and final states; ΔE_{int} is path independent.*

$$\Delta E_{int} = Q + W \qquad (20.10)$$

Q = the energy transferred to the system by heat.

W = the work done on the system.

ΔE_{int} = change in internal energy.

Applications of the laws of thermodynamics will be described using some of the terms and equations described and shown below.

An **isolated system** does not interact with the surroundings. *The internal energy of an isolated system remains constant.*

Isolated system

$Q = W = 0$

$\Delta E_{int} = 0$

In a **cyclic process** a system is returned to the initial state and there is no change in the internal energy of the system.

Cyclic process

$Q = -W$

$\Delta E_{int} = 0$

PV **diagrams for basic thermal processes** are illustrated in the figure at right.

A = isovolumetric process
B = adiabatic process
C = isothermal process
D = isobaric process

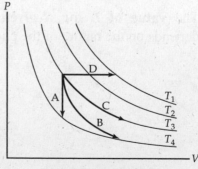

Basic thermal processes

A system can undergo an **adiabatic process** if it is thermally insulated from its surroundings.

Adiabatic process

$$E_{int} = W \qquad (20.11)$$

In an **isobaric process** energy is transferred into (out of) the gas as it expands (contracts).

Isobaric process

$$W = -P(V_f - V_i) \qquad (20.12)$$

W = work done on the gas

In an **isovolumetric process** zero work is done and the net energy added to the system by heat at constant volume increases the internal energy of the system.

Isovolumetric process

$$E_{int} = Q \qquad (20.13)$$

In an **isothermal process** no change occurs in the internal energy of a system; any energy that enters the system by heat is transferred out of the system by work, $(Q = -W)$.

Isothermal process

$$W_{env} = nRT \ln\left(\frac{V_f}{V_i}\right) \qquad (20.14)$$

Rate of energy transfer by conduction through a slab or rod of material is proportional to the temperature difference between the hot and cold faces and inversely proportional to the thickness of the slab or length of the rod (see figure). The thermal conductivity, k, is characteristic of a particular material; *good thermal conductors have large values of k.* The units of k are J/s · m · °C

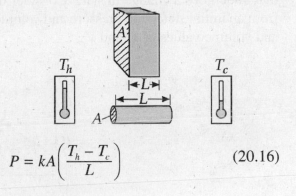

$$P = kA\left(\frac{T_h - T_c}{L}\right) \qquad (20.16)$$

Rate of energy conduction through a compound slab of area A is found by using Equation. 20.17. The temperatures of the outer surfaces are T_c and T_h. The thicknesses of the slabs are $L_1, L_2, L_3, \ldots$ and $k_1, k_2, k_3, \ldots$ are the respective thermal conductivities.

$$P = \frac{A(T_h - T_c)}{\sum_i (L_i / k_i)} \qquad (20.17)$$

The **R-value** of a material is the ratio of thickness to thermal conductivity. *Materials with large R-values are good insulators.* In the US, R-values are given in engineering units: ft² · °F · h/BTU.

$$R_i = \frac{L_i}{k_i}$$

Using the definition of R as stated above, Equation 20.17 can be written as shown in Equation 20.18.

$$P = \frac{A(T_h - T_c)}{\sum_i R_i} \qquad (20.18)$$

The **rate of energy transfer by radiation** (radiated power) from a surface of area A and temperature T is given by Stefan's law, Equation 20.19. The surface temperature T is measured in kelvins and the value of e (the emissivity of the radiating body) is determined by the nature of the surface.

$$P = \sigma A e T^4 \tag{20.19}$$

$$\sigma = 5.669\ 6 \times 10^{-8}\ \text{W/m}^2 \cdot \text{K}^4$$

$$e = \text{emissivity (values from 0 to 1)}$$

Net radiated power by an object at temperature T, in an environment at temperature T_0, is the difference between the rate of emission to the environment and rate of absorption. *At equilibrium $T = T_0$, the two rates are equal, and the temperature of the object remains constant.*

$$P_{net} = \sigma A e \left(T^4 - T_0^{\ 4}\right) \tag{20.20}$$

SUGGESTIONS, SKILLS, AND STRATEGIES

Many applications of the first law of thermodynamics deal with the work done on a system that undergoes a change in state. Consider the PV diagram showing the path taken by a gas from an initial state with pressure and volume values of P_i and V_i to a final state with pressure and volume values of P_f and V_f.

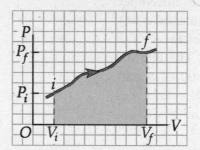

If the pressure is a known function of the volume, the work done on the gas can be calculated directly using Equation 20.9. If the path taken by the gas from its initial to its final state can be shown in a PV diagram, the work done on the gas during the expansion is equal to the negative of the area under the PV curve (the shaded region) shown in the figure. *It is important to recognize that the work depends on the path taken as the gas goes from i to f.*

PROBLEM SOLVING STRATEGY: CALORIMETRY PROBLEMS

When solving calorimetry problems, consider the following points:

*   Be sure your units are consistent throughout. For instance, if you are using specific heats in cal/g · °C, be sure that masses are in grams and temperatures are Celsius throughout.

*   Energy losses and gains are found by using $Q = mc\Delta T$ only for those intervals during which no phase changes occur. The equations $Q = \Delta m L_f$ and $Q = \Delta m L_v$ are to be used only when phase changes are taking place.

- Often sign errors occur in calorimetry problems. In Equation 20.5 $\left(Q_{cold} = -Q_{hot}\right)$, remember to include the negative sign in the equation.

REVIEW CHECKLIST

You should be able to:

- Write equations to describe a calorimetry experiment and calculate values of specific heat or change in temperature. (Section 20.2)

- Make calculations of energy exchange resulting in change of phase. (Section 20.3)

- Calculate the work done on (or by) a gas: (i) when the equation for $P = P(V)$ is given, (ii) for a constant temperature process, and (iii) by evaluating the area under a PV curve. (Sections 20.4 and 20.6)

- Calculate the change in the internal energy of a system for adiabatic, constant volume, and constant pressure processes. (Section 20.6)

- Sketch PV diagrams for cases when a gas is taken from an intial state to a final state via combinations of constant pressure and constant volume processes or by isothermal and adiabatic processes. (Sections 20.4 and 20.6)

- Make calculations of the rate of energy transfer due to conduction and radiation. (Section 20.7)

ANSWER TO AN OBJECTIVE QUESTION

12. If a gas is compressed isothermally, which of the following statements is true? (a) Energy is transferred into the gas by heat. (b) No work is done on the gas. (c) The temperature of the gas increases. (d) The internal energy of the gas remains constant. (e) None of those statements is true.

Answer (d) is true for a gas that is far from liquefaction. By definition, the temperature is constant in an isothermal process. For an ideal gas the internal energy is a function of temperature only and will also stay constant in this process. Choice (c) is untrue, because "isothermal" means constant-temperature. Choice (b) is untrue, because in any compression process the environment does positive work on the gas. To keep constant internal energy, the gas must then transfer energy out by heat, so choice (a) is untrue.

☐   ☐   ☐   ☐

ANSWERS TO SELECTED CONCEPTUAL QUESTIONS

3. What is wrong with the following statement: "Given any two bodies, the one with the higher temperature contains more heat."

Answer The statement shows a misunderstanding of the concept of heat. Heat is a process by which energy is transferred, not a form of energy that is held or contained. If you wish to speak of energy that is "contained," you speak of **internal energy**, not **heat**.

Further, even if the statement used the term "internal energy," it would still be incorrect, since the effects of specific heat and mass are both ignored. A 1-kg mass of water at 20°C has more internal energy than a 1-kg mass of air at 30°C. Similarly, the Earth has far more internal energy than a drop of molten titanium metal.

Correct statements would be: (1) "Given any two bodies in thermal contact, the one with the higher temperature will transfer energy to the other by heat." (2) "Given any two bodies of equal mass, the one with the higher product of absolute temperature and specific heat contains more internal energy."

□   □   □   □

5. Using the first law of thermodynamics, explain why the *total* energy of an isolated system is always constant.

Answer The first law of thermodynamics says that the net change in internal energy of a system is equal to the energy added by heat, plus the work done on the system.

$$\Delta E_{int} = Q + W$$

It applies to a system for which the internal energy is the only energy that is changing, and work and heat are the only energy-transfer processes that are occurring. On the other hand, an isolated system is defined as an object or set of objects for which there is no exchange of energy with its surroundings. In the case of the first law of thermodynamics, this means that $Q = W = 0$, so the change in internal energy of the system at all times must be zero.

As it stays constant in amount, an isolated system's energy may change from one form to another or move from one object to another within the system. For example, a "bomb calorimeter" is a closed system that consists of a sturdy steel container, a water bath, an item of food, and oxygen, all inside a jacket of thermal insulation. The food is burned in the presence of an excess of oxygen, and chemical energy is converted to energy of more rapid random molecular motion. In the process of oxidation, energy is transferred by heat to the water bath, raising its temperature. The change in temperature of the water bath is used to determine the caloric content of the food. The process works specifically **because** the total energy remains unchanged in a closed system.

□   □   □   □

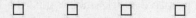

SOLUTIONS TO SELECTED END-OF-CHAPTER PROBLEMS

2. The highest waterfall in the world is the Salto Angel Falls in Venezuela. Its longest single falls has a height of 807 m. If water at the top of the falls is at 15.0°C, what is the maximum temperature of the water at the bottom of the falls? Assume all of the kinetic energy of the water as it reaches the bottom goes into raising its temperature.

Solution

Conceptualize: Water has a high specific heat, so the difference in water temperature between the top and bottom of the falls is probably on the order of 1°C or less. (Besides, if the difference was significantly large, people would warm their houses by building them under waterfalls.)

Categorize: The temperature change can be found from the gravitational potential energy that is converted to internal energy. The final temperature is this change added to the initial temperature of the water.

Analyze: The absolute value of the change in potential energy is $|\Delta U| = mgy$. It will turn entirely into kinetic energy of the water at the bottom of the falls, and then produce the same temperature change as the same amount of heat entering the water from a stove, as described by $Q = mc\Delta T$

Thus, $mgy = mc\Delta T$

Isolating ΔT, $\Delta T = \dfrac{gy}{c} = \dfrac{(9.80 \text{ m/s}^2)(807 \text{ m})}{4.186 \times 10^3 \text{ J/kg} \cdot {}^\circ\text{C}} = 1.89{}^\circ\text{C}$

The final temperature is then

$$T_f = T_i + \Delta T = 15.0{}^\circ\text{C} + 1.89{}^\circ\text{C} = 16.9{}^\circ\text{C} \qquad \blacksquare$$

Finalize: For this exceptionally high falls, the rise in water temperature is only on the order of 1°C, as expected. The final temperature might be less than we calculated because our solution does not account for cooling of the water due to evaporation as it falls. The change in temperature is independent of the amount of water; you can think of each unit of mass as going over the falls in an insulated unbreakable Styrofoam jug, and warming just itself as it sloshes around inside the jug. Except for the evaporation complication, the surrounding parcels of water undergoing the same temperature change form just such a perfect insulating jacket. On his honeymoon in Switzerland in 1847, James Joule tried without success to measure the characteristic temperature change of a fairly high waterfall. In a remarkable coincidence, on the path to the waterfall the amateur experimenter Joule met Lord Kelvin, the great academic physicist who became his scientific patron and collaborator.

---

6. The temperature of a silver bar rises by 10.0°C when it absorbs 1.23 kJ of energy by heat. The mass of the bar is 525 g. Determine the specific heat of silver from these data.

Solution

Conceptualize: Silver is far down in the periodic table, so we expect its specific heat to be on the order of one-tenth of that of water, or some hundreds of J/kg·°C.

Categorize: We find its specific heat from the definition, which is…

Analyze: ...contained in the equation $Q = mc_{silver}\Delta T$ for energy input by heat to produce a temperature change. Solving, we have

$$c_{silver} = \frac{Q}{m\Delta T}$$

$$c_{silver} = \frac{1.23 \times 10^3 \text{ J}}{(0.525 \text{ kg})(10.0°\text{C})} = 234 \text{ J/kg} \cdot °\text{C} \qquad \blacksquare$$

Finalize: We were right about the order of magnitude.

---

9. A 1.50-kg iron horseshoe initially at 600°C is dropped into a bucket containing 20.0 kg of water at 25.0°C. What is the final temperature of the water-horseshoe system? Ignore the heat capacity of the container and assume a negligible amount of water boils away.

Solution

Conceptualize: Even though the horseshoe is much hotter than the water, the mass of the water is significantly greater, so we might expect the water temperature to rise less than 10°C.

Categorize: The energy lost by the iron will be gained by the water. From complementary expressions for this single quantity of heat, the change in water temperature can be found.

Analyze: We assume that the water-horseshoe system is thermally isolated (insulated) from the environment for the short time required for the horseshoe to cool off and the water to warm up. Then the total energy input from the surroundings is zero, as expressed by $Q_{iron} + Q_{water} = 0$

or

$$(mc\Delta T)_{iron} + (mc\Delta T)_{water} = 0$$

or

$$m_{Fe}c_{Fe}(T - 600°\text{C}) + m_w c_w(T - 25.0°\text{C}) = 0$$

Note that the first term in this equation is a negative number of joules, representing energy lost by the originally hot subsystem, and the second term is a positive number with the same absolute value, representing energy gained by heat by the cold stuff. Solving for the final temperature gives

$$T = \frac{m_w c_w(25.0°\text{C}) + m_{Fe}c_{Fe}(600°\text{C})}{m_{Fe}c_{Fe} + m_w c_w}$$

$$T = \frac{(20.0 \text{ kg})(4\,186 \text{ J/kg} \cdot °\text{C})(25.0°\text{C}) + (1.50 \text{ kg})(448 \text{ J/kg} \cdot °\text{C})(600°\text{C})}{(1.50 \text{ kg})(448 \text{ J/kg} \cdot °\text{C}) + (20.0 \text{ kg})(4\,186 \text{ J/kg} \cdot °\text{C})}$$

$$= 29.6°\text{C} \qquad \blacksquare$$

Finalize: The temperature only rose about 5°C, so our answer seems reasonable. The specific heat of the water is about 10 times greater than the iron, so this effect also reduces the change in water temperature. In this problem, we assumed that a negligible amount of water boiled away. In reality, the final temperature of the water would be somewhat less than we calculated, since some of the energy would be used to vaporize a bit of water.

---

16. A 3.00-g lead bullet at 30.0°C is fired at a speed of 240 m/s into a large block of ice at 0°C, in which it becomes embedded. What quantity of ice melts?

Solution

Conceptualize: The amount of ice that melts is probably small, maybe only a few grams based on the size, speed, and initial temperature of the bullet.

Categorize: We will assume that all of the initial kinetic and excess internal energy of the bullet goes into internal energy to melt the ice, the mass of which can be found from the latent heat of fusion. Because the ice does not all melt, in the final state everything is at 0°C.

Analyze: At thermal equilibrium, the energy lost by the bullet equals the energy gained by the ice: $|\Delta K_b| + |Q_b| = Q_{ice}$ gives

$$\tfrac{1}{2}m_b v_b^2 + m_b c_{Pb}|\Delta T| = L_f \Delta m_{melted} \qquad \text{or} \qquad \Delta m_{melt} = m_b \left(\frac{\tfrac{1}{2}v_b^2 + c_{Pb}|\Delta T|}{L_f} \right)$$

$$\Delta m_{melt} = \left(3.00 \times 10^{-3}\ \text{kg}\right)\left(\frac{\tfrac{1}{2}(240\ \text{m/s})^2 + (128\ \text{J/kg} \cdot \text{°C})(30.0\text{°C})}{3.33 \times 10^5\ \text{J/kg}} \right)$$

$$\Delta m_{melt} = \frac{86.4\ \text{J} + 11.5\ \text{J}}{3.33 \times 10^5\ \text{J/kg}} = 2.94 \times 10^{-4}\ \text{kg} = 0.294\ \text{g} \qquad \blacksquare$$

Finalize: The amount of ice that melted is less than a gram, which agrees with our prediction. It appears that most of the energy used to melt the ice comes from the kinetic energy of the bullet (88%), while the excess internal energy of the bullet only contributes 12% to melt the ice. Small chips of ice probably fly off when the bullet makes impact. Some of the energy is transferred to their kinetic energy, so in reality, the amount of ice that would melt should be less than what we calculated. If the block of ice were colder than 0°C (as is often the case), then the melted ice would refreeze.

---

19. In an insulated vessel, 250 g of ice at 0°C is added to 600 g of water at 18.0°C. (a) What is the final temperature of the system? (b) How much ice remains when the system reaches equilibrium?

Solution

Conceptualize: It takes a lot of energy to melt ice at 0°C without changing its temperature. If any bit of the ice remains solid when the originally warmer stuff reaches 0°C, the final temperature will be 0°C.

Categorize: The liquid water is losing energy by heat according to $mc\Delta T$. The ice is gaining energy described by $L\Delta m$. We must compare quantities to decide whether all the ice melts.

Analyze:

(a) If all 250 g of ice is melted it must absorb energy

$$Q_f = L_f\Delta m = (0.250 \text{ kg})(3.33 \times 10^5 \text{ J/kg}) = 83.3 \text{ kJ}$$

The energy released when 600 g of water cools from 18.0°C to 0°C is in absolute value

$$|Q| = |mc\Delta T| = (0.600 \text{ kg})(4\ 186 \text{ J/kg} \cdot °\text{C})(18.0°\text{C}) = 45.2 \text{ kJ}$$

Since the energy required to melt 250 g of ice at 0°C **exceeds** the energy released by cooling 600 g of water from 18.0°C to 0°C, not all the ice melts and the final temperature of the system (water + ice) must be 0°C. ∎

(b) The originally warmer water will cool all the way to 0°C, so it puts 45.2 kJ into the ice. This energy lost by the water will melt a mass of ice Δm, where $Q = L_f\Delta m$.

Solving for the mass,

$$\Delta m = \frac{Q}{L_f} = \frac{45.2 \times 10^3 \text{J}}{3.33 \times 10^5 \text{ J/kg}} = 0.136 \text{ kg}$$

Therefore, the ice remaining is

$$m' = 0.250 \text{ kg} - 0.136 \text{ kg} = 0.114 \text{ kg}$$ ∎

Finalize: Step-by-step reasoning is essential for solving a problem like this. It would be easy enough to write the equation $Lm_{ice} + m_{ice}c(T - 0) + m_{water}c(T - 18) = 0$. With it we would be assuming that all the ice melts. We would get a nonzero answer for final temperature T with no direct indication that it is wrong, and we would have made no progress toward solving the problem. The answer to part (a) was not read from a calculator, but identified by logic.

═══════════════

21. An ideal gas is enclosed in a cylinder with a movable piston on top of it. The piston has a mass of 8 000 g and an area of 5.00 cm² and is free to slide up and down, keeping the pressure of the gas constant. How much work is done on the gas as the temperature of 0.200 mol of the gas is raised from 20.0°C to 300°C?

Solution

Conceptualize: The gas is expanding to put out a positive quantity of work in lifting the piston. The work *on* the gas is this number of joules but negative. The original internal energy of the gas is some multiple of $nRT_i = (0.2 \text{ mol})(8.314 \text{ J/mol} \cdot \text{K})(293 \text{ K}) \sim 10^3 \text{ J}$. The work the gas puts out does not come just from this fund of internal energy, but from some of the energy that must be put into the sample by heat to raise its temperature. But still we can estimate the work as on the order of –1 kJ.

Categorize: The integral of PdV is easy because the pressure is constant. Then the equation of state will let us evaluate what we need about the pressure and the change in volume.

Analyze: For constant pressure, $W = -\int_i^f PdV = -P\Delta V = -P(V_f - V_i)$. Rather than evaluating the pressure numerically from atmospheric pressure plus the pressure due to the weight of the piston, we can just use the ideal gas law to write in the volumes, obtaining

$$W = -P\left(\frac{nRT_h}{P} - \frac{nRT_c}{P}\right) = -nR(T_h - T_c)$$

Therefore, $W = -nR\Delta T = -(0.200 \text{ mol})(8.314 \text{ J/mol} \cdot \text{K})(280 \text{ K}) = -466 \text{ J}$ ∎

Finalize: We did not need to use the mass of the piston or its face area. The answer would be the same if the gas had to lift a heavier load over a smaller volume change or a lighter load through a larger volume change. The calculation turned out to be simple, but only because the pressure stayed constant during the expansion. Do not try to use $-P\Delta V$ for work in any case other than constant pressure, and never try to use $-V\Delta P$.

---

23. An ideal gas is taken through a quasi-static process described by $P = \alpha V^2$, with $\alpha = 5.00 \text{ atm/m}^6$, as shown in Figure P20.23. The gas is expanded to twice its original volume of 1.00 m^3. How much work is done on the expanding gas in this process?

Solution

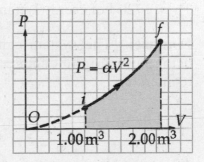

Figure P20.23

Conceptualize: The final pressure is $5 \cdot 4 = 20 \text{ atm} \approx 2 \text{ MPa}$. The average pressure during expansion we could estimate as 1 MPa, and then the work is like $-(1 \text{ MPa})(1 \text{ m}^3) = -1 \text{ MJ}$.

Categorize: The work done on the gas is the negative of the area under the curve $P = \alpha V^2$, from V_i to V_f. The work *on* the gas is negative, to mean that the expanding gas *does* positive work. We will find its amount by doing the integral...

Analyze: ... $W = -\int_{V_i}^{V_f} PdV$ with $V_f = 2V_i = 2(1.00 \text{ m}^3) = 2.00 \text{ m}^3$

$$W = -\int_{V_i}^{V_f} \alpha V^2 dV = -\tfrac{1}{3}\alpha(V_f^3 - V_i^3)$$

$$W = -\tfrac{1}{3}(5.00 \text{ atm/m}^6)(1.013 \times 10^5 \text{ Pa/atm})\left[\left(2.00 \text{ m}^3\right)^3 - \left(1.00 \text{ m}^3\right)^3\right]$$

$$= -1.18 \times 10^6 \text{ J}$$ ∎

Finalize: Our estimate was good. To do the integral of PdV, we start by identifying how P depends on V, just as to do the integral of ydx in mathematics you start by identifying how y depends on x.

---

29. A thermodynamic system undergoes a process in which its internal energy decreases by 500 J. Over the same time interval, 220 J of work is done on the system. Find the energy transferred from it by heat.

Solution

Conceptualize: Think of your personal wealth decreasing by $500 on a day when someone gives you salary $220. It is clear that you are paying out $720, analogous to this system transferring out 720 J by heat.

Categorize: It is the first law that keeps account of energy inputs and outputs by heat and work.

Analyze: We have $\Delta E_{int} = Q + W$, where W is positive, because work is done **on** the system. Then

$$Q = \Delta E_{int} - W = -500 \text{ J} - 220 \text{ J} = -720 \text{ J}$$

Thus, +720 J of energy is transferred **from** the system by heat. ∎

Finalize: The system can be visualized as a gas in a cylinder with the piston being pushed in to compress it while it is put into a cold environment and drops in temperature.

---

31. An ideal gas initially at 300 K undergoes an isobaric expansion at 2.50 kPa. If the volume increases from 1.00 m³ to 3.00 m³ and 12.5 kJ is transferred to the gas by heat, what are (a) the change in its internal energy and (b) its final temperature?

Solution

Conceptualize: The gas pressure is much less than one atmosphere. This could be managed by having the gas in a cylinder with a piston on the bottom end, supporting a constant load that hangs from the piston. The (large) cylinder is put into a warmer environment to make the gas expand as its temperature rises to 900 K and its internal energy rises by thousands of joules.

Categorize: The first law will tell us the change in internal energy. The ideal gas law will tell us the final temperature.

Analyze: We use the energy version of the nonisolated system model.

(a) $\Delta E_{int} = Q + W$

where $\quad W = -P\Delta V \quad$ for a constant-pressure process

so that $\quad \Delta E_{int} = Q - P\Delta V$

$\Delta E_{int} = 1.25 \times 10^4 \, \text{J} - (2.50 \times 10^3 \, \text{N/m}^2)(3.00 \, \text{m}^3 - 1.00 \, \text{m}^3) = 7\,500 \, \text{J}$ ■

(b) Since pressure and quantity of gas are constant, we have from the equation of state

$$\frac{V_1}{T_1} = \frac{V_2}{T_2} \quad \text{and} \quad T_2 = \left(\frac{V_2}{V_1}\right)T_1 = \left(\frac{3.00 \, \text{m}^3}{1.00 \, \text{m}^3}\right)(300 \, \text{K}) = 900 \, \text{K}$$ ■

Finalize: We could have found the quantity of gas n from the initial pressure, volume, and temperature, but our solution was shorter.

---

32. (a) How much work is done on the steam when 1.00 mol of water at 100°C boils and becomes 1.00 mol of steam at 100°C at 1.00 atm pressure? Assume the steam to behave as an ideal gas. (b) Determine the change in internal energy of the system of the water and steam as the water vaporizes.

Solution

Conceptualize: Imagine the liquid water vaporizing inside a cylinder with a piston. The material expands greatly to lift the piston and do positive work. Many hundreds of joules of negative work is done on the material. Even more positive energy is put into it by heat, to produce a net increase in its internal energy.

Categorize: We use the idea of the negative integral of PdV to find the work, identifying the volumes of the liquid and the gas as we do so. Then $L\Delta m$ will tell us the heat input and the first law of thermodynamics will tell us the change in internal energy.

Analyze:

(a) We choose as a system the H_2O molecules that all participate in the phase change. For a constant-pressure process, $\quad W = -P\Delta V = -P(V_s - V_w)$

where V_s is the volume of the steam and V_w is the volume of the liquid water.

We can find them respectively from

$$PV_s = nRT \quad \text{and} \quad V_w = m/\rho = nM/\rho$$

Calculating each work term,

$$PV_s = (1.00 \, \text{mol})\left(8.314\frac{\text{J}}{\text{K}\cdot\text{mol}}\right)(373 \, \text{K}) = 3\,101 \, \text{J}$$

$$PV_w = (1.00 \, \text{mol})(18.0 \, \text{g/mol})\left(\frac{1.013 \times 10^5 \, \text{N/m}^2}{1.00 \times 10^6 \, \text{g/m}^3}\right) = 1.82 \, \text{J}$$

Thus the work done is $\quad W = -3\,101 \, \text{J} + 1.82 \, \text{J} = -3.10 \, \text{kJ}$ ■

(b) The energy input by heat is $\quad Q = L_V\Delta m = (18.0 \, \text{g})(2.26 \times 10^6 \, \text{J/kg}) = 40.7 \, \text{kJ}$

so the change in internal energy is $\quad \Delta E_{int} = Q + W = 37.6 \, \text{kJ}$ ■

Finalize: Steam at 100°C is on the point of liquefaction, so it probably behaves differently from the ideal gas we were told to assume. Still, the volume increase by more than a thousand times means that a significant amount of the "heat of vaporization" is coming out of the sample as it boils, as work done on the environment in its expansion. For many students the hardest thing to remember is that we start with 18 cubic centimeters of liquid water. Visualize it!

---

33. A 2.00-mol sample of helium gas initially at 300 K and 0.400 atm is compressed isothermally to 1.20 atm. Noting that the helium behaves as an ideal gas, find (a) the final volume of the gas, (b) the work done on the gas, and (c) the energy transferred by heat.

Solution

Conceptualize: The final volume will be one-third of the original, several liters. The positive work will be hundreds of joules. For the temperature to stay constant, the sample must put out as many joules by heat as it takes in by work.

Categorize: The ideal gas law can tell us the original and final volumes. The negative integral of PdV will tell us the work input, and the first law will tell us the energy output by heat.

Analyze:

(a) Rearranging $PV = nRT$ we get $V_i = \dfrac{nRT}{P_i}$

The initial volume is

$$V_i = \frac{(2.00 \text{ mol})(8.314 \text{ J/mol} \cdot \text{K})(300 \text{ K})}{(0.400 \text{ atm})(1.013 \times 10^5 \text{ Pa/atm})}\left(\frac{1 \text{ Pa}}{\text{N/m}^2}\right) = 0.123 \text{ m}^3$$

For isothermal compression, PV is constant, so $P_iV_i = P_fV_f$ and the final volume is

$$V_f = V_i\left(\frac{P_i}{P_f}\right) = (0.123 \text{ m}^3)\left(\frac{0.400 \text{ atm}}{1.20 \text{ atm}}\right) = 0.041\ 0 \text{ m}^3 \quad ■$$

(b) $W = -\displaystyle\int PdV = -\int \frac{nRT}{V}dV = -nRT \ln\left(\frac{V_f}{V_i}\right) = -(4\ 988 \text{ J})\ln\left(\tfrac{1}{3}\right) = +5.48 \text{ kJ} \quad ■$

(c) The ideal gas keeps constant temperature so $\Delta E_{int} = 0 = Q + W$ and the heat is
$Q = -5.48 \text{ kJ}$ ■

Finalize: Visualize squashing a piston down on the sample in a cylinder surrounded by a constant-temperature bath. With all that compression, the gas would tend to rise in temperature, so energy must leave it by heat for the process to be isothermal.

---

49. A bar of gold (Au) is in thermal contact with a bar of silver (Ag) of the same length and area (Fig. P20.49). One end of the compound bar is maintained at 80.0°C, and the opposite end is at 30.0°C. When the energy transfer reaches steady state, what is the temperature at the junction?

Solution

Conceptualize: Silver is a better conductor than gold, so the junction temperature will be a bit less than the 55°C that is numerically halfway between the temperatures of the ends.

Categorize: We must think of the temperature difference across each bar as the driving force for an energy current of heat going through the bar by conduction.

Analyze: Call the gold bar Object 1 and the silver bar Object 2. Each is a nonisolated system in steady state. When energy transfer by heat reaches a steady state, the flow rate through each will be the same, so that the junction can stay at constant temperature thereafter, with as much heat coming in through the gold as goes out through the silver.

$$P_1 = P_2 \quad \text{or} \quad \frac{k_1 A_1 \Delta T_1}{L_1} = \frac{k_2 A_2 \Delta T_2}{L_2}$$

In this case, $L_1 = L_2$ and $A_1 = A_2$

so $k_1 \Delta T_1 = k_2 \Delta T_2$

Let T_3 be the temperature at the junction; then

$$k_1(80.0°C - T_3) = k_2(T_3 - 30.0°C)$$

Rearranging, we find

$$T_3 = \frac{(80.0°C)\,k_1 + (30.0°C)k_2}{k_1 + k_2}$$

$$T_3 = \frac{(80.0°C)(314 \text{ W/m} \cdot °C) + (30.0°C)(427 \text{ W/m} \cdot °C)}{(314 \text{ W/m} \cdot °C) + (427 \text{ W/m} \cdot °C)} = 51.2°C \quad \blacksquare$$

Finalize: We did not need to know the value of the length or area of either bar, or the value of the heat current, to identify the current as equal in the two materials in steady state. And that equality was enough to predict the junction temperature. Solving a problem like this about the voltage at the junction between electric resistors in a series circuit will be routine later in the course.

---

53. An aluminum rod 0.500 m in length and with a cross-sectional area of 2.50 cm^2 is inserted into a thermally insulated vessel containing liquid helium at 4.20 K. The rod is initially at 300 K. (a) If one half of the rod is inserted into the helium, how many liters of helium boil off by the time the inserted half cools to 4.20 K? Assume the upper half does not yet cool. (b) If the circular surface of the upper end of the rod is maintained at 300 K, what is the approximate boil-off rate of liquid helium after the lower half has reached 4.20 K? (Aluminum has thermal conductivity of 3 100 W/m · K at 4.20 K; ignore its temperature variation. The density of liquid helium is 125 kg/m^3.)

Solution

Conceptualize: Demonstrations with liquid nitrogen give us some indication of the phenomenon described. Since the rod is much hotter than the liquid helium and of significant size (almost 2 cm in diameter), a substantial volume (maybe as much as a liter) of helium will boil off before thermal equilibrium is reached. Likewise, since aluminum conducts rather well, a significant amount of helium will continue to boil off as long as the upper end of the rod is maintained at 300 K.

Categorize: Until thermal equilibrium is reached, the excess internal energy of the rod will go into vaporizing liquid helium, which is already at its boiling point (so there is no change in the temperature of the helium).

Analyze: As you solve this problem, be careful not to confuse L (the **conduction length** of the rod) with L_v (the **heat of vaporization** of the helium), or with L, the symbol for the unit liter.

(a) Before heat conduction has time to become important, we suppose the energy lost by heat by half the rod equals the energy gained by the helium. Therefore,

$$(L_v \Delta m)_{\text{He}} = |mc\Delta T|_{\text{Al}} \quad \text{or} \quad (\rho V L_v)_{\text{He}} = |\rho V c \Delta T|_{\text{Al}}$$

So
$$V_{\text{He}} = \frac{|\rho V c \Delta T|_{\text{Al}}}{(\rho L_v)_{\text{He}}} = \frac{(2.70 \text{ g/cm}^3)(62.5 \text{ cm}^3)(900 \text{ J/kg} \cdot °\text{C})(295.8 °\text{C})}{(0.125 \text{ g/cm}^3)(2.09 \times 10^4 \text{ J/kg})}$$

$V_{\text{He}} = 1.72 \times 10^4 \text{ cm}^3 = 17.2 \text{ liters}$ ∎

(b) Heat will be conducted along the rod at the rate

$$\frac{dQ}{dt} = P = \frac{kA\Delta T}{L} \quad\quad\quad [1]$$

and will boil off helium according to

$$Q = L_v \Delta m \quad \text{so} \quad \frac{dQ}{dt} = \left(\frac{dm}{dt}\right) L_v \quad\quad\quad [2]$$

Combining [1] and [2] gives us the "boil-off" rate: $\dfrac{dm}{dt} = \dfrac{kA\Delta T}{L \cdot L_v}$

Set the conduction length $L = 25$ cm, and use $k = 3\,100$ W/m · K

to find
$$\frac{dm}{dt} = \frac{(3\,100 \text{ W/100 cm} \cdot \text{K})(2.50 \text{ cm}^2)(295.8 \text{ K})}{(25.0 \text{ cm})(20.9 \text{ J/g})} = 43.9 \text{ g/s}$$

or
$$\frac{dV}{dt} = \frac{43.9 \text{ g/s}}{0.125 \text{ g/cm}^3} = 351 \text{ cm}^3/\text{s} = 0.351 \text{ liters/s}$$ ∎

Finalize: The volume of helium boiled off initially is much more than expected. If our calculations are correct, that sure is a lot of liquid helium that is wasted! Since liquid helium is much more expensive than liquid nitrogen, most low-temperature equipment is designed to avoid unnecessary loss of liquid helium by surrounding the liquid with a Dewar inside a container of liquid nitrogen.

61. Water in an electric teakettle is boiling. The power absorbed by the water is 1.00 kW. Assuming the pressure of vapor in the kettle equals atmospheric pressure, determine the speed of effusion of vapor from the kettle's spout if the spout has a cross-sectional area of 2.00 cm². Model the steam as an ideal gas.

Solution

Conceptualize: A story says that James Watt observed that the jet from his mother's teakettle could do work, and went on to invent the steam engine that started the Industrial Revolution. Not many grams will come out of the spout each second, but the density of the steam is low enough for us to suppose that the speed is a few meters per second.

Categorize: The temperature of the teakettle is staying constant at 100°C. We can call the water and steam inside a nonisolated system, in energy terms, in steady state, gaining energy from the electric resistor by heat and losing energy by matter transfer. The equality of the rates of energy input and output will give us the answer. We use *Power* as a symbol for the rate of energy transfer, and P_0 to represent atmospheric pressure.

Analyze: From $Q = L_V \Delta m$ the rate of boiling is described by

$$Power = \frac{Q}{\Delta t} = \frac{L_V \Delta m}{\Delta t}$$ so that the mass flow rate of steam from the kettle is

$$\frac{\Delta m}{\Delta t} = \frac{Power}{L_V}$$

The symbols Δm for mass vaporized and m for mass leaving the kettle have the same meaning, but recall that M represents the molar mass. Even though it is on the point of liquefaction, we model the water vapor as an ideal gas. The volume flow rate $V/\Delta t$ of the fluid is the cross-sectional area of the spout multiplied by the speed of flow, forming the product Av.

$$P_0 V = nRT = \left(\frac{m}{M}\right) RT$$

$$\frac{P_0 V}{\Delta t} = \frac{m}{\Delta t}\left(\frac{RT}{M}\right)$$

$$P_0 Av = \frac{Power}{L_V}\left(\frac{RT}{M}\right)$$

$$v = \frac{Power\, RT}{ML_V P_0 A} = \frac{1\,000\ \text{W}(8.314\ \text{J/mol}\cdot\text{K})(373\ \text{K})}{(0.018\,0\ \text{kg/mol})(2.26\times10^6\ \text{J/kg})(1.013\times10^5\ \text{N/m}^2)(2.00\times10^{-4}\ \text{m}^2)}$$

$$v = 3.76\ \text{m/s} \qquad\blacksquare$$

Finalize: The speed is on the order of 10 m/s, not just 1 m/s. In Chapter 22 we will treat in some detail the practical application of thermodynamics to heat engines like this one. Heat engines are devices that convert energy input by heat into mechanical energy output. They always have limited efficiency.

68. (a) The inside of a hollow cylinder is maintained at a temperature T_a and the outside is at a lower temperature T_b (Fig. P20.68). The wall of the cylinder has a thermal conductivity k. Ignoring end effects, show that the rate of energy conduction from the inner surface to the outer surface in the radial direction is

$$\frac{dQ}{dt} = 2\pi Lk \left[\frac{T_a - T_b}{\ln(b/a)} \right]$$

Suggestions: The temperature gradient is dT/dr. A radial energy current passes through a concentric cylinder of area $2\pi rL$. (b) The passenger section of a jet airliner is in the shape of a cylindrical tube with a length of 35.0 m and an inner radius of 2.50 m. Its walls are lined with an insulating material 6.00 cm in thickness and having a thermal conductivity of 4.00×10^{-5} cal/s·cm·°C. A heater must maintain the interior temperature at 25.0°C while the outside temperature is −35.0°C. What power must be supplied to the heater?

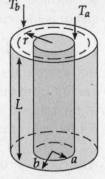

Figure P20.68

Solution

Conceptualize: In the equation for part (a), we identify the reasonable proportionalities of the energy current to the temperature difference and to the thermal conductivity. In part (b), Baby, it's cold outside! We might estimate hundreds of watts.

Categorize: In part (a), we apply the general law of thermal conduction to the annulus of insulating material. Part (b) will be a direct application of the equation derived in (a).

Analyze:

(a) For a thin cylindrical shell within the annulus, having radius r, height L, and thickness dr, the equation for thermal conduction,

$$\frac{dQ}{dt} = -kA\frac{dT}{dx} \quad \text{becomes} \quad \frac{dQ}{dt} = -k(2\pi rL)\frac{dT}{dr}$$

Under steady-state conditions, $\dfrac{dQ}{dt}$ is constant for all radii. Therefore, we can separate variables and integrate from a point on the inner wall to a point on the outer wall.

$$dT = -\frac{dQ}{dt}\left(\frac{1}{2\pi kL}\right)\left(\frac{dr}{r}\right) \quad \text{and} \quad \int_{T_a}^{T_b} dT = -\frac{dQ}{dt}\left(\frac{1}{2\pi kL}\right)\int_a^b \frac{dr}{r}$$

$$T_b - T_a = -\frac{dQ}{dt}\left(\frac{1}{2\pi kL}\right)\ln\left(\frac{b}{a}\right)$$

But $T_a > T_b$, so we can more naturally write the energy current as

$$\frac{dQ}{dt} = \frac{2\pi kL(T_a - T_b)}{\ln(b/a)} \qquad \blacksquare$$

(b) By substitution, here

$$\frac{dQ}{dt} = \frac{2\pi(3\,500\text{ cm})(4.00\times10^{-5}\text{ cal/s}\cdot\text{cm}\cdot{}^\circ\text{C})(60.0{}^\circ\text{C})}{\ln(2.56/2.50)}$$

and $\dfrac{dQ}{dt} = 2.23\times10^3\text{ cal/s} = 9.32\text{ kW}$

■

This is the rate of energy loss by heat from the plane, and consequently the rate at which energy must be supplied in order to maintain a steady-state temperature.

Finalize: Anticipating part (b) helps many students to make their way through part (a). In particular, think throughout of k, L, a, b, T_a, and T_b as known constants. The quantity dQ/dt is also a constant, but it is the unknown. The r and T refer to the unknown pattern of change of temperature with radius within the insulating jacket. Mathematically, r and T are variables of integration that disappear when the integral is done. Note that the constant temperature inside is not an equilibrium temperature. The air inside is not isolated, but constantly taking in energy by heat from the heater and putting out energy through the insulating cylinder into the cold environment.

21

The Kinetic Theory of Gases

EQUATIONS AND CONCEPTS

The **pressure of an ideal gas** is proportional to the number of molecules per unit volume and the average translational kinetic energy per molecule.

$$P = \frac{2}{3}\left(\frac{N}{V}\right)\left(\frac{1}{2}m_o\overline{v^2}\right) \tag{21.2}$$

m_o = molecular mass

The **average kinetic energy per molecule** in a gas is proportional to the absolute temperature. *This result follows from Equation 21.2 and the equation of state for an ideal gas, $PV = Nk_BT$.*

$$\frac{1}{2}m_o\overline{v^2} = \frac{3}{2}k_BT \tag{21.4}$$

$\overline{v^2}$ = root-mean-square speed

The **internal energy** of a sample of an ideal monatomic gas depends only on the absolute temperature. *In Equation 21.10, N is the number of molecules and n is the number of moles.*

$$E_{int} = K_{tot\ trans} = \frac{3}{2}Nk_BT = \frac{3}{2}nRT \tag{21.10}$$
(monatomic gas)

The **molecular specific heat of an ideal gas** is defined for two processes:

C_V (specific heat at constant volume)
C_P (specific heat at constant pressure)

At constant volume

$$Q = nC_V\Delta T \tag{21.8}$$

$$C_V = \frac{3}{2}R \quad \text{(monatomic)} \tag{21.14}$$

$$C_V = \frac{5}{2}R \quad \text{(diatomic)} \tag{21.21}$$

The energy transferred by heat required to take a gas between two given temperatures at constant pressure will be greater than that required to take the gas between the same two temperatures at constant volume; $C_P > C_V$. See *Suggestions, Skills, and Strategies* for an explanation of this result.

At constant pressure

$$Q = nC_P\Delta T \tag{21.9}$$

$$C_P = \frac{5}{2}R \quad \text{(monatomic)}$$

An adiabatic processes for an ideal gas involves changes in the variables P, V, and T. During the adiabatic process, no energy is transferred by heat between the gas and its environment.

$$PV^{\gamma} = \text{constant} \qquad (21.18)$$

$$TV^{\gamma-1} = \text{constant} \qquad (21.20)$$

The **Boltzmann distribution law** (which is valid for a system containing a large number of particles) states that the probability of finding the particles in a particular arrangement (distribution of energy values) varies exponentially as the negative of the energy divided by $k_B T$. In Equation 21.23, $n_V(E)$ is a function called the number density.

$$n_V(E) = n_0 e^{-E/k_B T} \qquad (21.23)$$

$n_V(E)dE$ = number of molecules per unit volume with energy within the range between E and $E + dE$.

The **Maxwell–Boltzmann speed distribution function** describes the most probable distribution of speeds of N gas molecules at temperature T (where k_B is the Boltzmann constant). *The probability that a molecule has a speed in the range between v and $v + dv$ is equal to dN/N.*

$$N_v = 4\pi N \left(\frac{m_o}{2\pi k_B T}\right)^{3/2} v^2 e^{-m_o v^2/2k_B T}$$
$$(21.24)$$

$dN = N_v dv$

dN = number of molecules with speeds between v and $v + dv$.

Root-mean-square speed, **average speed**, and **most probable speed** can be calculated from the Maxwell–Boltzmann distribution function. At a given temperature, lighter molecules have higher average speeds than heavier ones.

$$v_{\text{rms}} = 1.73 \sqrt{\frac{k_B T}{m_o}} \qquad (21.25)$$

$$v_{\text{avg}} = 1.60 \sqrt{\frac{k_B T}{m_o}} \qquad (21.26)$$

$$v_{\text{mp}} = 1.41 \sqrt{\frac{k_B T}{m_o}} \qquad (21.27)$$

SUGGESTIONS, SKILLS, AND STRATEGIES

CALCULATING SPECIFIC HEATS

It is important to recognize that, depending on the type of molecule (and hence the number of degrees of freedom), the ratio of specific heats for a gas (γ) might be 1.67, 1.40, or some other value. However, in any ideal gas, two equations are always valid, and can be used to determine the specific heats C_P and C_V from a known value of γ.

$$\gamma = \frac{C_P}{C_V} \qquad (21.17) \qquad \text{and} \qquad C_P - C_V = R \qquad (21.16)$$

Combining the equations above, we find

$$C_V = \frac{R}{\gamma - 1} \qquad \text{and} \qquad C_P = \frac{\gamma R}{\gamma - 1}$$

These equations may be used for all ideal gases.

C_P **is greater than** C_V **for the following reason.** The temperature of a sample of gas can be increased by a given amount ΔT by either a constant pressure process or by a constant volume process. The resulting increase in internal energy of the gas, ΔE_{int} will be the same in both cases. When energy is transferred to a gas by heat at constant pressure, the volume increases and therefore a portion of the added energy leaves the system by work. Only the remaining portion of the energy transferred by heat contributes to an increase in the internal energy of the gas. If the temperature increase is accomplished by a constant volume process, no work is done; and all of the energy transferred by heat contributes to an increase in the internal energy of the gas. Therefore the added energy per mole required to effect the same increase in internal energy, and hence the same temperature increase, must be greater for the constant pressure process.

REVIEW CHECKLIST

You should be able to:

- State and understand the assumptions made in developing the molecular model of an ideal gas. (Section 21.1)

- Calculate pressure, average translational kinetic energy per molecule and rms speed per molecule and quantity (moles) of a confined gas. (Section 21.1)

- Calculate Q and ΔE_{int} when the temperature of a given number of moles of a gas is increased by an amount ΔT through a constant volume or a constant pressure process. (Section 21.2)

- Make calculations involving an adiabatic process for an ideal gas. (Section 21.3)

- Find values for average speed, root-mean-square speed, and most probable speed when given the individual values of speed for a number of particles. (Section 21.5)

Some important points to remember:

- The temperature of an ideal gas is proportional to the average molecular kinetic energy. In a gas sample containing molecules with different values of mass, the average speed of the more massive molecules will be smaller.

- According to the theorem of equipartition of energy, each degree of freedom of a molecule contributes equally to the total internal energy of a gas.

ANSWER TO AN OBJECTIVE QUESTION

4. A helium-filled latex balloon initially at room temperature is placed in a freezer. The latex remains flexible. (i) Does the balloon's volume (a) increase, (b) decrease, or (c) remain the same? (ii) Does the pressure of the helium gas (a) increase significantly, (b) decrease significantly, or (c) remain approximately the same?

Answer The helium is far from liquefaction. Therefore, we model it as an ideal gas, described by $PV = nRT$. In the process described, the pressure stays nearly constant. A gauge reveals that the interior pressure is only slightly larger than 1 atm in a balloon as it is ordinarily inflated. When the balloon wall goes slack and moves inward with no acceleration, the interior pressure is precisely 1 atm. Since the temperature may decrease by 10% (from about 293 K to 263 K), the volume should also decrease by 10%. This process is called "isobaric cooling," or "isobaric contraction." The answers are (i) b and (ii) c.

□   □   □   □

ANSWERS TO SELECTED CONCEPTUAL QUESTIONS

3. When alcohol is rubbed on your body, it lowers your skin temperature. Explain this effect.

Answer As the alcohol evaporates, high-speed molecules leave the liquid. This reduces the average speed of the remaining molecules. Since the average speed is lowered, the temperature of the alcohol is reduced. This process helps to carry energy away from your skin, resulting in cooling of the skin. The alcohol plays the same role of evaporative cooling as does perspiration, but alcohol evaporates much more quickly than perspiration.

□   □   □   □

4. What happens to a helium-filled latex balloon released into the air? Does it expand or contract? Does it stop rising at some height?

Answer As the balloon rises it encounters air at lower and lower pressure, because the bottom layers of the atmosphere support the weight of all the air above. The flexible wall of the balloon moves and then stops moving to make the pressures equal on both sides of the wall. Therefore, as the balloon rises, the pressure inside it will decrease. At the same time, the temperature decreases slightly, but not enough to overcome the effects of the pressure.

By the ideal gas law, $PV = nRT$. Since the decrease in temperature has little effect, the decreasing pressure results in an increasing volume; thus the balloon expands quite dramatically. By the time the balloon reaches an altitude of 10 miles, its volume will be 90 times larger.

Long before that happens, one of two events will occur. The first possibility, of course, is that the balloon will break, and the pieces will fall to the earth. The other possibility is that the balloon will come to a spot where the average density of the balloon and its payload is equal to the average density of the air. At that point, buoyancy will cause the balloon to "float" at that altitude, until it loses helium and descends.

People who remember releasing balloons with pen-pal notes will perhaps also remember that replies most often came back when the balloons were low on helium, and were barely able to take off. Such balloons found a fairly low flotation altitude, and were the most likely to be found intact at the end of their journey.

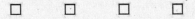

SOLUTIONS TO SELECTED END-OF-CHAPTER PROBLEMS

1. Calculate the mass of an atom of (a) helium, (b) iron, and (c) lead. Give your answers in kilograms. The atomic masses of these atoms are 4.00 u, 55.9 u, and 207 u, respectively.

Solution

Conceptualize: The mass of an atom of any element is essentially the mass of the protons and neutrons that make up its nucleus, because the mass of the electrons is negligible (less than a 0.05% contribution). Since most atoms have about the same number of neutrons as protons, the atomic mass is approximately double the atomic number (the number of protons). We should expect that the mass of a single atom is a very small fraction of a gram ($\sim 10^{-23}$ g) since one mole (6.02×10^{23}) of atoms has a mass on the order of several grams.

Categorize: An atomic mass unit is defined as 1/12 of the mass of a carbon-12 atom. The mass in kilograms can be found by multiplying the atomic mass by the mass of one atomic mass unit (u): 1 u = 1.66×10^{-27} kg.

Analyze:

For He, $m_o = 4.00\ u = (4.00\ u)(1.66 \times 10^{-27}\ kg/u) = 6.64 \times 10^{-27}\ kg$ ∎

For Fe, $m_o = 55.9\ u = (55.9\ u)(1.66 \times 10^{-27}\ kg/u) = 9.28 \times 10^{-26}\ kg$ ∎

For Pb, $m_o = 207\ u = (207\ u)(1.66 \times 10^{-27}\ kg/u) = 3.44 \times 10^{-25}\ kg$ ∎

Finalize: As expected, the mass of the atoms is larger for higher atomic numbers. If we did not know the conversion factor for atomic mass units, we could use the mass of a proton as a close approximation: $1\ u \approx 1.67 \times 10^{-27}\ kg$. The conversion factor between atomic mass units and grams is exactly the reciprocal of Avogadro's number. This is true because Avogadro's number ($N_A = 6.02 \times 10^{23}$) of carbon-12 atoms together have a mass of 12 g exactly. So one atom, with mass 12 u exactly, has mass $12\ g/N_A$. Dividing both sides by 12 gives $1\ u = 1\ g/N_A$

---

2. A cylinder contains a mixture of helium and argon gas in equilibrium at 150°C. (a) What is the average kinetic energy for each type of gas molecule? (b) What is the rms speed of each type of molecule?

Solution

Conceptualize: Molecules are so small that their energies are tiny fractions of a joule. The speeds will be many hundreds of meters per second.

Categorize: The kinetic-theory identification of the meaning of temperature, $\frac{1}{2}m_o\overline{v^2} = \frac{3}{2}k_BT$, will give us all the answers.

Analyze:

(a) Both kinds of molecules have the same average kinetic energy. It is

$$\frac{1}{2}m_o\overline{v^2} = \frac{3}{2}k_BT = \frac{3}{2}\left(1.38 \times 10^{-23}\ J/K\right)(273 + 150)\ K$$

$$\overline{K} = 8.76 \times 10^{-21}\ J$$ ∎

(b) The root-mean square velocity can be calculated from the kinetic energy:

$$v_{rms} = \sqrt{\overline{v^2}} = \sqrt{\frac{2\overline{K}}{m_o}}$$

These two gases are noble, and therefore monatomic. The masses of the molecules are:

$$m_{He} = \frac{(4.00\ g/mol)\left(10^{-3}\ kg/g\right)}{6.02 \times 10^{23}\ atoms/mol} = 6.64 \times 10^{-27}\ kg$$

$$m_{\text{Ar}} = \frac{(39.9 \text{ g/mol})(10^{-3} \text{ kg/g})}{6.02 \times 10^{23} \text{ atoms/mol}} = 6.63 \times 10^{-26} \text{ kg}$$

Substituting these values,

$$v_{\text{rms, He}} = \sqrt{\frac{2(8.76 \times 10^{-21} \text{ J})}{6.64 \times 10^{-27} \text{ kg}}} = 1.62 \times 10^{3} \text{ m/s} \qquad \blacksquare$$

and $\qquad v_{\text{rms, Ar}} = \sqrt{\frac{2(8.76 \times 10^{-21} \text{ J})}{6.63 \times 10^{-26} \text{ kg}}} = 514 \text{ m/s} \qquad \blacksquare$

Finalize: Observe that argon has molecules ten times massive than helium and that their rms speed is smaller by the square root of ten times. A dust particle or a floating paper airplane is bombarded with molecules and participates in the random motion with the same average kinetic energy but with vastly smaller rms speed.

---

4. In an ultrahigh vacuum system (with typical pressures lower than 10^{-7} pascal), the pressure is measured to be 1.00×10^{-10} torr (where 1 torr = 133 Pa). Assuming the temperature is 300 K, find the number of molecules in a volume of 1.00 m³.

Solution

Conceptualize: Since high vacuum means low pressure as a result of a low molecular density, we should expect a relatively low number of molecules. At 273 K and atmospheric pressure one mole occupies 22.4 L, so the density of molecules is

$$6.02 \times 10^{23} \text{ molecules}/0.022\,4 \text{ m}^3 \sim 10^{25} \text{ molecules/m}^3$$

By comparison, in the ultrahigh vacuum the absolute pressure is 13 orders of magnitude lower than atmospheric pressure, so we might expect a value of $N \sim 10^{12}$ molecules/m³.

Categorize: The equation of state for an ideal gas can be used with the given information to find the number of molecules in a specific volume.

Analyze: $\quad PV = \left(\dfrac{N}{N_A}\right) RT \qquad$ means $\qquad N = \dfrac{PVN_A}{RT}$, so that

$$N = \frac{(1.00 \times 10^{-10})(133)(1.00)(6.02 \times 10^{23})}{(8.314)(300)} = 3.21 \times 10^{12} \text{ molecules} \qquad \blacksquare$$

Finalize: The calculated result has the expected order of magnitude. This ultrahigh vacuum is an environment very different from the floor of the ocean (of air) to which we are accustomed, as bottom-feeders living here below the Earth's atmosphere. The vacuum conditions can provide a "clean" environment for a variety of experiments and manufacturing processes that would otherwise be impossible.

---

5. A spherical balloon of volume 4.00×10^3 cm^3 contains helium at a pressure of 1.20×10^5 Pa. How many moles of helium are in the balloon if the average kinetic energy of each helium atom is 3.60×10^{-22} J?

Solution

Conceptualize: The balloon has a volume of 4.00 L and a diameter of about 20 cm, which seems like a reasonable size for a large toy balloon. The pressure of the balloon is only slightly more than 1 atm, and if the temperature is anywhere close to room temperature, we can use the estimate of 22 L/mol for an ideal gas at STP conditions. If this is valid, the balloon should contain about 0.2 moles of helium.

Categorize: The average kinetic energy can be used to find the temperature of the gas, which can be used with $PV = nRT$ to find the number of moles.

Analyze: The gas temperature must be that implied by $\frac{1}{2}m_o\overline{v^2} = \frac{3}{2}k_B T$ for a monatomic gas like helium.

$$T = \frac{2}{3}\left(\frac{\frac{1}{2}m_o\overline{v^2}}{k_B}\right) = \frac{2}{3}\left(\frac{3.60 \times 10^{-22} \text{ J}}{1.38 \times 10^{-23} \text{ J/K}}\right) = 17.4 \text{ K}$$

Now $PV = nRT$ gives

$$n = \frac{PV}{RT} = \frac{\left(1.20 \times 10^5 \text{ N/m}^2\right)\left(4.00 \times 10^{-3} \text{ m}^3\right)}{(8.314 \text{ J/mol} \cdot \text{K})(17.4 \text{ K})} = 3.32 \text{ mol}$$ ∎

Finalize: This result is more than ten times the number of moles we predicted, primarily because the temperature of the helium is much colder than room temperature. In fact, T is only slightly above the temperature at which the helium would liquefy (4.2 K at 1 atm). We should hope this balloon is not being held by a child; not only would the balloon sink in the air, it is cold enough to cause frostbite!

---

9. (a) How many atoms of helium gas fill a spherical balloon of diameter 30.0 cm at 20.0°C and 1.00 atm? (b) What is the average kinetic energy of the helium atoms? (c) What is the rms speed of the helium atoms?

Solution

Conceptualize: The balloon will have a volume of a few liters, so the number of atoms is of the same order of magnitude as Avogadro's number $\sim 10^{24}$. The speed will be some kilometers per second, faster than any reasonable airplane.

Categorize: The equation of state will tell us the quantity of gas as a number of moles, so multiplying by Avogadro's number gives us the number of molecules. $(1/2)m_o(v^2)_{avg} = (3/2)k_B T$ will tell us both the average kinetic energy and the rms speed.

Analyze:

(a) The volume is $V = \dfrac{4}{3}\pi r^3 = \dfrac{4}{3}\pi(0.150 \text{ m})^3 = 1.41 \times 10^{-2} \text{ m}^3$

Now $PV = nRT$ gives for the quantity of gas

$$n = \frac{PV}{RT} = \frac{\left(1.013 \times 10^5 \text{ N/m}^2\right)\left(1.41 \times 10^{-2} \text{ m}^3\right)}{(8.314 \text{ N} \cdot \text{m/mol} \cdot \text{K})(293 \text{ K})} = 0.588 \text{ mol}$$

The number of molecules is

$$N = nN_A = (0.588 \text{ mol})\left(6.02 \times 10^{23} \text{ molecules/mol}\right)$$

$$N = 3.54 \times 10^{23} \text{ helium atoms} \qquad\blacksquare$$

(b) The kinetic energy is $\bar{K} = \dfrac{1}{2}m_o\overline{v^2} = \dfrac{3}{2}k_BT$

$$\bar{K} = \frac{3}{2}\left(1.38 \times 10^{-23} \text{ J/K}\right)(293 \text{ K}) = 6.07 \times 10^{-21} \text{ J} \qquad\blacksquare$$

(c) An atom of He has mass

$$m_o = \frac{M}{N_A} = \frac{4.002\,6 \text{ g/mol}}{6.02 \times 10^{23} \text{ molecules/mol}} = 6.65 \times 10^{-24} \text{ g} = 6.65 \times 10^{-27} \text{ kg}$$

So the root-mean-square speed is given by

$$v_{\text{rms}} = \sqrt{\overline{v^2}} = \sqrt{\frac{2\bar{K}}{m_o}} = \sqrt{\frac{2 \times 6.07 \times 10^{-21} \text{ J}}{6.65 \times 10^{-27} \text{ kg}}} = 1.35 \text{ km/s} \qquad\blacksquare$$

Finalize: The quantity of internal energy in the sample is much more than two grams of material usually possesses as mechanical energy. The fact that helium is chemically inert and can be said to have no chemical energy makes the contrast sharper. To tap into the internal energy to use it for human purposes would mean to cool off the helium. If the helium is warmer than its environment, a heat engine can systematically extract some but not all of its internal energy, as we will study in Chapter 22.

---

15. A 1.00-mol sample of hydrogen gas is warmed at constant pressure from 300 K to 420 K. Calculate (a) the energy transferred to the gas by heat, (b) the increase in its internal energy, and (c) the work done on the gas.

Solution

Conceptualize: The gas expands in this process, so it puts out positive work. The work done *on* it is negative. In absolute value, we expect some hundreds of joules for all three answers. The heat input will be the largest of the three, enough to supply the other two together.

Categorize: The specific heats of hydrogen at constant pressure and at constant volume will give us the first two answers. Their difference will be the energy escaping as work.

Analyze:

(a) Since this is a constant-pressure process, $Q = nC_p\Delta T$

The temperature rises by $\Delta T = 420 \text{ K} - 300 \text{ K} = 120 \text{ K}$

$$Q = (1.00 \text{ mol})(28.8 \text{ J/mol} \cdot \text{K})(120 \text{ K}) = 3.46 \text{ kJ}$$ ∎

(b) For any gas $\Delta E_{int} = nC_V\Delta T$

so $\Delta E_{int} = (1.00 \text{ mol})(20.4 \text{ J/mol} \cdot \text{K})(120 \text{ K}) = 2.45 \text{ kJ}$ ∎

(c) The first law says $\Delta E_{int} = Q + W$

so $W = \Delta E_{int} - Q = 2.45 \text{ kJ} - 3.46 \text{ kJ} = -1.01 \text{ kJ}$ ∎

Finalize: The gas puts out work by expanding. Another way to find the work in this special constant-pressure case is to calculate

$$W = -\int_i^f P\,dV = -P\int_i^f dV = -P\left(V_f - V_i\right) = -nRT_f + nRT_i = -nR\Delta T$$

$$= -(1 \text{ mol})(8.314 \text{ J/mol} \cdot \text{K})(120 \text{ K}) = -0.998 \text{ kJ}$$

in good agreement with our first-law result.

---

20. A 2.00-mol sample of a diatomic ideal gas expands slowly and adiabatically from a pressure of 5.00 atm and a volume of 12.0 L to a final volume of 30.0 L. (a) What is the final pressure of the gas? (b) What are the initial and final temperatures? Find (c) Q, (d) ΔE_{int}, and (e) W for the gas during this process.

Solution

Conceptualize: The volume gets 2.5 times larger in the expansion. If the gas were at constant temperature, the pressure would become 2.5 times smaller (and ΔE_{int} would be zero, and Q would be positive). In the adiabatic process Q is zero. No heat comes in to supply the joules of work output, so the temperature will drop and the final pressure will be significantly smaller than 2 atm.

Categorize: We use the equation $PV^\gamma = $ constant for an adiabatic path. Then we use the ideal gas law to find the temperatures. From the temperatures we can find ΔE_{int}. With the known $Q = 0$, the first law will give us W.

Analyze:

(a) In an adiabatic process $P_iV_i^\gamma = P_fV_f^\gamma$

$$P_f = P_i\left(\frac{V_i}{V_f}\right)^\gamma = (5.00 \text{ atm})\left(\frac{12.0 \text{ L}}{30.0 \text{ L}}\right)^{1.40} = 1.39 \text{ atm}$$ ∎

(b) The initial temperature is

$$T_i = \frac{P_i V_i}{nR} = \frac{(5.00 \text{ atm})(1.013 \times 10^5 \text{ Pa/atm})(12.0 \times 10^{-3} \text{ m}^3)}{(2.00 \text{ mol})(8.314 \text{ N} \cdot \text{m/mol K})} = 366 \text{ K} \quad \blacksquare$$

and similarly the final temperature is $\quad T_f = \frac{P_f V_f}{nR} = 253 \text{ K} \quad \blacksquare$

(c) This is an adiabatic process, so by the definition $Q = 0$ $\qquad\qquad\qquad\qquad$ $\blacksquare$

(d) For any process, $\qquad\qquad\qquad\qquad\qquad \Delta E_{int} = nC_V \Delta T$

and for this diatomic ideal gas, $\qquad C_V = \frac{R}{\gamma - 1} = \tfrac{5}{2}R$

Thus, $\Delta E_{int} = \frac{5}{2}(2.00 \text{ mol})(8.314 \text{ J/mol} \cdot \text{K})(253 \text{ K} - 366 \text{ K}) = -4\,660 \text{ J} \quad \blacksquare$

(e) Now, $W = \Delta E_{int} - Q = -4\,660 \text{ J} - 0 = -4\,660 \text{ J} \qquad\qquad$ $\blacksquare$

Finalize: The work done *on* the gas is negative, so positive work is done *by* the gas on its environment as the gas expands.

---

21. Air in a thundercloud expands as it rises. If its initial temperature is 300 K and no energy is lost by thermal conduction on expansion, what is its temperature when the initial volume has doubled?

Solution

Conceptualize: The air should cool as it expands, so we should expect $T_f < 300$ K.

Categorize: The air expands adiabatically, losing no heat but dropping in temperature as it does work on the air around it, so we have $PV^\gamma = $ constant, where $\gamma = 1.40$ for an ideal diatomic gas. Even as the triatomic water molecules condense in a thundercloud, the overwhelming majority of molecules are oxygen and nitrogen far from liquefaction, so a diatomic ideal gas is a good model for the air.

Analyze: Combining $PV^\gamma = $ constant with the ideal gas law gives one of the textbook equations describing adiabatic processes, $T_1 V_1^{\gamma-1} = T_2 V_2^{\gamma-1}$

$$T_2 = T_1 \left(\frac{V_1}{V_2}\right)^{\gamma-1} = 300 \text{ K} \left(\frac{1}{2}\right)^{(1.40-1)} = 227 \text{ K} \quad \blacksquare$$

Finalize: The air does cool, but the temperature is not inversely proportional to the volume. The temperature drops 24% while the volume doubles. The pressure drops also in the expansion.

---

31. The relationship between the heat capacity of a sample and the specific heat of the sample material is discussed in Section 20.2. Consider a sample containing 2.00 mol of an ideal diatomic gas. Assuming the molecules rotate but do not vibrate, find (a) the total heat capacity of the sample at constant volume and (b) the total heat capacity at constant pressure. (c) **What If?** Repeat parts (a) and (b), assuming the molecules both rotate and vibrate.

Solution

Conceptualize: The molar heat capacity of an ideal gas at constant volume with single-atom molecules is $(3/2)R$. With two moles the total heat capacity will be twice as large. Having to supply energy to drive the diatomic molecules' rotation will raise the heat capacity farther. The heat capacity will be still larger if the molecules vibrate. And the heat capacity at constant pressure is always larger than that for a constant-volume process, because some of the energy put in as heat leaks out of the expanding gas as work while the gas is warming up in a constant-pressure process.

Categorize: We use the kinetic theory structural model of an ideal gas.

Analyze:

(a) Count degrees of freedom. A diatomic molecule oriented along the y axis can possess energy by moving in x, y, and z directions and by rotating around x and z axes. Rotation around the y axis does not represent an energy contribution because the moment of inertia of the molecule about this axis is essentially zero. The molecule will have an average energy $\frac{1}{2}k_BT$ for each of these five degrees of freedom.

The gas with N molecules will have internal energy

$$E_{int} = N\left(\frac{5}{2}\right)k_BT = nN_A\left(\frac{5}{2}\right)\left(\frac{R}{N_A}\right)T = \left(\frac{5}{2}\right)nRT$$

so the constant-volume heat capacity of the whole sample is

$$\frac{\Delta E_{int}}{\Delta T} = \frac{5}{2}nR = \frac{5}{2}(2.00 \text{ mol})(8.314 \text{ J/mol} \cdot \text{K}) = 41.6 \text{ J/K} \qquad \blacksquare$$

(b) For one mole, $C_P = C_V + R$ For the sample, $nC_P = nC_V + nR$

With P constant,

$$nC_P = 41.6 \text{ J/K} + (2.00 \text{ mol})(8.314 \text{ J/mol} \cdot \text{K}) = 58.2 \text{ J/K} \qquad \blacksquare$$

(c) Vibration adds a degree of freedom for kinetic energy and a degree of freedom for elastic energy.

Now the molecule's average energy is $\frac{7}{2}k_BT$

the sample's internal energy is $N\frac{7}{2}k_BT = \frac{7}{2}nRT$

and the sample's constant-volume heat capacity is

$$\frac{7}{2}nR = \frac{7}{2}(2.00 \text{ mol})(8.314 \text{ J/mol} \cdot \text{K}) = 58.2 \text{ J/K}$$ ∎

At constant pressure, its heat capacity is

$$nC_V + nR = 58.2 \text{ J/K} + (2.00 \text{ mol})(8.314 \text{ J/mol} \cdot \text{K}) = 74.8 \text{ J/K}$$ ∎

Finalize: It is the randomness of the collisions among molecules that makes the average energy equal for each different kind of motion, for each degree of freedom. One can say that the diatomic molecule does not possess any energy of rotation about an axis along the line joining the atoms because any outside force has no lever arm to set the massive atomic nuclei into rotation about this axis. When we study energy quantization at the end of the book we will be able to give a better account of the reason.

---

32. Fifteen identical particles have various speeds: one has a speed of 2.00 m/s, two have speeds of 3.00 m/s, three have speeds of 5.00 m/s, four have speeds of 7.00 m/s, three have speeds of 9.00 m/s, and two have speeds of 12.0 m/s. Find (a) the average speed, (b) the rms speed, and (c) the most probable speed of these particles.

Solution

Conceptualize: All of the measures of 'central tendency' will be around 5 m/s, but all will be different because of the scatter in speeds.

Categorize: We find each kind of average from its definition.

Analyze:

(a) The average is

$$\bar{v} = \frac{\sum n_i v_i}{\sum n_i} = \frac{1(2.00) + 2(3.00) + 3(5.00) + 4(7.00) + 3(9.00) + 2(12.0)}{1 + 2 + 3 + 4 + 3 + 2} \text{ m/s}$$

$$\bar{v} = 6.80 \text{ m/s}$$ ∎

(b) To find the average squared speed we work out

$$\overline{v^2} = \frac{\sum n_i v_i^2}{\sum n_i}$$

$$\overline{v^2} = \frac{1\left(2.00^2\right) + 2\left(3.00^2\right) + 3\left(5.00^2\right) + 4\left(7.00^2\right) + 3\left(9.00^2\right) + 2\left(12.0^2\right)}{15} \text{ m}^2/\text{s}^2$$

$$\overline{v^2} = 54.9 \text{ m}^2/\text{s}^2$$

Then the rms speed is $v_{\text{rms}} = \sqrt{\overline{v^2}} = \sqrt{54.9 \text{ m}^2/\text{s}^2} = 7.41 \text{ m/s}$ ∎

(c)   More particles have $v_{mp} = 7.00$ m/s than any other speed.   ∎

Finalize: The average in (a) is called the mean. The most probable value in (c) is called the mode. We could also find the median, as the speed of the seventh of the fifteen molecules ranked by speed. It is 7 m/s.

---

38.   From the Maxwell–Boltzmann speed distribution, show that the most probable speed of a gas molecule is given by Equation 21.27. *Note:* The most probable speed corresponds to the point at which the slope of the speed distribution curve dN_v/dv is zero.

Solution

Conceptualize: We are determining the coordinate of one point on the Maxwell–Boltzmann graph of the distribution of the number of molecules with different speeds.

Categorize: From the Maxwell speed distribution function,

$$N_v = 4\pi N \left(\frac{m_o}{2\pi k_B T}\right)^{3/2} v^2 e^{-m_o v^2/2k_B T}$$

we locate the peak in the graph of N_v versus v by evaluating dN_v/dv and setting it equal to zero, to solve for the most probable speed.

Analyze: The derivative is

$$\frac{dN_v}{dv} = 4\pi N \left(\frac{m_o}{2\pi k_B T}\right)^{3/2}\left(\frac{-2m_o v}{2k_B T}\right)v^2 e^{-m_o v^2/2k_B T} + 4\pi N \left(\frac{m_o}{2\pi k_B T}\right)^{3/2} 2v e^{-m_o v^2/2k_B T} = 0$$

This equation has solutions $v = 0$ and $v \to \infty$, but those correspond to minimum-probability speeds. Then we divide by v and by the exponential function, obtaining

$$v_{mp}\left(-\frac{m_o\left(2v_{mp}\right)}{2k_B T}\right) + 2 = 0 \qquad \text{or} \qquad v_{mp} = \sqrt{\frac{2k_B T}{m_o}}$$

This is Equation 21.27.   ∎

Finalize: It might help you to keep a table of symbols. k_B, T, N, and m_o are known quantities. N_v is the variable function of the variable v. As soon as we set its variable derivative dN_v/dv equal to zero, we start looking for the unknown v that is the most probable speed.

---

48.   In a cylinder, a sample of an ideal gas with number of moles n undergoes an adiabatic process. (a) Starting with the expression $W = -\int P\,dV$ and using the condition $PV^\gamma =$ constant, show that the work done on the gas is

$$W = \left(\frac{1}{\gamma - 1}\right)\left(P_f V_f - P_i V_i\right)$$

(b) Starting with the first law of thermodynamics, show that the work done on the gas is equal to $nC_v(T_f - T_i)$. (c) Are these two results consistent with each other? Explain.

Solution

Conceptualize: The gas does positive work if it expands. Its environment does positive work on it if the gas contracts.

Categorize: We will start as instructed with the integral expression for work in part (a) and with $\Delta E_{int} = Q + W$ in part (b). In each case we specialize to an adiabatic process and evaluate.

Analyze:

(a) For the adiabatic process $PV^\gamma = k$, a constant.

The work is

$$W = -\int_i^f P\,dV = -k\int_{V_i}^{V_f} \frac{dV}{V^\gamma} = \frac{-kV^{1-\gamma}}{1-\gamma}\Bigg|_{V_i}^{V_f}$$

For k we can substitute $P_i V_i^\gamma$ and also $P_f V_f^\gamma$ to have

$$W = -\frac{P_f V_f^\gamma V_f^{1-\gamma} - P_i V_i^\gamma V_i^{1-\gamma}}{1-\gamma} = \frac{P_f V_f - P_i V_i}{\gamma - 1} \qquad \blacksquare$$

(b) For an adiabatic process $\Delta E_{int} = Q + W$ and $Q = 0$

Therefore, $W = \Delta E_{int} = nC_V\Delta T = nC_V(T_f - T_i)$ $\qquad \blacksquare$

(c) They are consistent. To prove consistency between these two equations,

consider that $\qquad \gamma = \dfrac{C_P}{C_V} \qquad$ and $\qquad C_P - C_V = R$

so that $\qquad \dfrac{1}{\gamma - 1} = \dfrac{C_V}{R}$

Thus part (a) becomes $\qquad W = (P_f V_f - P_i V_i)\dfrac{C_V}{R}$

Then, $\quad \dfrac{PV}{R} = nT \quad$ so that $\quad W = nC_V(T_f - T_i)$ $\qquad \blacksquare$

Finalize: In an adiabatic compression the temperature rises, and the final value of PV is greater than the initial value. Thus both expressions give positive results as required.

52. The compressibility κ of a substance is defined as the fractional change in volume of that substance for a given change in pressure:

$$\kappa = -\frac{1}{V}\frac{dV}{dP}$$

(a) Explain why the negative sign in this expression ensures κ is always positive. (b) Show that if an ideal gas is compressed isothermally, its compressibility is given by $\kappa_1 = 1/P$. (c) **What If?** Show that if an ideal gas is compressed adiabatically, its compressibility is given by $\kappa_2 = 1/\gamma P$. Determine values for (d) κ_1 and (e) κ_2 for a monatomic ideal gas at a pressure of 2.00 atm.

Solution

Conceptualize: The equation of state contains all information about how an ideal gas changes in any process, so we can use it to evaluate any property of the gas.

Categorize: We will evaluate the required derivatives from the ideal gas law.

Analyze:

(a) The pressure increases as volume decreases (and vice versa), so dV/dP is always negative.

In equation form, $\dfrac{dV}{dP} < 0$ and $-\left(\dfrac{1}{V}\right)\left(\dfrac{dV}{dP}\right) > 0$ ∎

(b) For an ideal gas, $V = \dfrac{nRT}{P}$ and $\kappa_1 = -\dfrac{1}{V}\dfrac{d}{dP}\left(\dfrac{nRT}{P}\right)$

For isothermal compression, T is constant and the derivative gives us

$$\kappa_1 = -\frac{nRT}{V}\left(\frac{-1}{P^2}\right) = \frac{1}{P} \qquad ∎$$

(c) For an adiabatic compression, $PV^\gamma = C$ (where C is a constant) and we evaluate dV/dP like this:

$$\kappa_2 = -\left(\frac{1}{V}\right)\frac{d}{dP}\left(\frac{C}{P}\right)^{1/\gamma} = \left(\frac{1}{V\gamma}\right)\frac{C^{1/\gamma}}{\left(P^{1/\gamma+1}\right)} = \frac{V}{V\gamma P} = \frac{1}{\gamma P}$$

For a monatomic ideal gas, $\gamma = C_p/C_V = 5/3$:

(d) $\kappa_1 = \dfrac{1}{P} = \dfrac{1}{2.00\text{ atm}} = 0.500\text{ atm}^{-1}$ ∎

(e) $\kappa_2 = \dfrac{1}{\gamma P} = \dfrac{1}{(5/3)(2.00\text{ atm})} = 0.300\text{ atm}^{-1}$ ∎

Finalize: Parts (d) and (e) help you to see that we have a definite result for comparison to experiment. The compressibility is different for a sample of gas in different states—in particular, at different pressures. Measuring the adiabatic compressibility is a good way to find the value of gamma for a gas. Newton did not understand the extra factor of gamma in the adiabatic compressibility and had to gloss over it in comparing his theoretical speed of sound with experiment. It might be better to call the quantity 'incompressibility' because a higher value describes a sample that is harder to compress.

---

59. For a Maxwellian gas, use a computer or programmable calculator to find the numerical value of the ratio $N_v(v)/N_v(v_{mp})$ for the following values of v: (a) $v = (v_{mp}/50.0)$, (b) $(v_{mp}/10.0)$, (c) $(v_{mp}/2.00)$, (d) v_{mp}, (e) $2.00v_{mp}$, (f) $10.0v_{mp}$, and (g) $50.0v_{mp}$. Give your results to three significant figures.

Solution

Conceptualize: We are finding coordinates of points on the graph of the Maxwell–Boltzmann speed distribution function. The values will be small for speed values well below and well above the most probable speed, with, by definition, the maximum value at v_{mp}.

Categorize: In each case we substitute and evaluate. Getting the numerical value in each case emphasizes that the graph has a definite shape.

Analyze: We first substitute the most probable speed

$$v_{mp} = \left(\frac{2k_BT}{m_o}\right)^{1/2}$$

into the distribution function

$$N_v(v) = 4\pi N \left(\frac{m_o}{2\pi k_BT}\right)^{3/2} v^2 e^{-m_o v^2/2k_BT}$$

and evaluate

$$N_v(v_{mp}) = 4\pi N \left(\frac{m_o}{2\pi k_BT}\right)^{3/2} \frac{2k_BT}{m_o} e^{-1}$$

simplifying,

$$N_v\left(v_{mp}\right) = 4\pi^{-1/2} e^{-1} N \left(\frac{m_o}{2k_BT}\right)^{1/2} = 4\pi^{-1/2} e^{-1} N v_{mp}^{-1}$$

Then the fraction we are assigned to think about has the simpler expression

$$\frac{N_v(v)}{N_v\left(v_{mp}\right)} = \frac{4\pi^{-1/2} N v_{mp}^{-3} v^2 e^{-(v/v_{mp})^2}}{4\pi^{-1/2} N v_{mp}^{-1} e^{-1}} = \left(\frac{v^2}{v_{mp}^2}\right) e^{1-v^2/v_{mp}^2}$$

(a) For $v = \dfrac{v_{mp}}{50}$, $\dfrac{N_v(v)}{N_v\left(v_{mp}\right)} = \left(\dfrac{1}{50}\right)^2 e^{1-(1/50)^2} = 1.09 \times 10^{-3}$ ∎

(b) For $v = \dfrac{v_{mp}}{10}$, $\dfrac{N_v(v)}{N_v\left(v_{mp}\right)} = \left(\dfrac{1}{10}\right)^2 e^{1-1/100} = 2.69 \times 10^{-2}$ ∎

(c) For $v = \dfrac{v_{mp}}{2}$, $\dfrac{N_v(v)}{N_v(v_{mp})} = \left(\dfrac{1}{2}\right)^2 e^{1-1/4} = 0.529$ ∎

(d) For $v = v_{mp}$, $\dfrac{N_v(v)}{N_v(v_{mp})} = (1)^2 e^{1-1} = 1.00$ ∎

Similarly we fill in the rest of the table:

v/v_{mp}	$N_v(v)/N_v(v_{mp})$
1/50	1.09×10^{-3}
1/10	2.69×10^{-2}
1/2	0.529
1	1.00
(e) 2	0.199
(f) 10	1.01×10^{-41}
(g) 50	$1.25 \times 10^{-1\,082}$

(e), (f), (g) marked ∎

To find the last we calculate

$$\left(50^2\right)e^{1-2\,500} = \left(50^2\right)e^{-2\,499} = 10^{\log 2\,500}\, e^{(\ln 10)(-2\,499/\ln 10)} = 10^{(\log 2\,500 - 2\,499/\ln 10)}$$

$$= 10^{-1\,081.904}$$

Thus, $\left(50^2\right)e^{1-2\,500} = 10^{-1\,081.904} = 10^{0.096\,0}10^{-1\,082} = 1.25 \times 10^{-1\,082}$ ∎

Finalize: The distribution function gets very small as it goes to zero for speeds approaching zero. And the distribution function gets most remarkably small for speeds several times larger than the most probable speed. Think about the distribution function for wealth among the human population. If it were like that for speeds among gas molecules that exchange energy in random collisions, no one in the world would have more than a few times the average amount of money.

22

Heat Engines, Entropy, and the Second Law of Thermodynamics

EQUATIONS AND CONCEPTS

The **net work done by a heat engine** during one cycle equals the net energy, Q_{net}, transferred to the engine. In Equation 22.1 $|Q_h|$ is the quantity of energy absorbed from a hot reservoir and $|Q_c|$ is the amount of energy expelled to the cold reservoir.

$$W_{eng} = |Q_h| - |Q_c| \qquad (22.1)$$

The **thermal efficiency**, e, of a heat engine is defined as the ratio of the net work done by the engine to the energy input at the higher temperature during one cycle. *For any real engine, $e < 1$.*

$$e \equiv \frac{W_{eng}}{|Q_h|} = \frac{|Q_h| - |Q_c|}{|Q_h|} = 1 - \frac{|Q_c|}{|Q_h|} \qquad (22.2)$$

The **effectiveness of a heat pump** is described in terms of the coefficient of performance, COP. W = the work done on the heat pump.

$$\text{COP (cooling mode)} = \frac{|Q_c|}{W} \qquad (22.3)$$

$|Q_c|$ = energy transferred at low temperature

$$\text{COP (heating mode)} = \frac{|Q_h|}{W} \qquad (22.4)$$

$|Q_h|$ = energy transferred at high temperature

The **Carnot efficiency** is the maximum theoretical efficiency of a heat engine operating reversibly in a cycle between T_c and T_h. *No real heat engine operating irreversibly between two reservoirs $(T_c$ and $T_h)$ can be more efficient than a Carnot engine operating between the same two reservoirs.*

$$e_C = 1 - \frac{T_c}{T_h} \qquad (22.6)$$

The **efficiency of the Otto cycle** (gasoline engine) depends on the compression ratio (V_1/V_2) and the ratio of molar specific heats(γ).

$$e = 1 - \frac{1}{(V_1/V_2)^{\gamma-1}} \quad \text{(Otto cycle)} \qquad (22.7)$$

Entropy, S, is a thermodynamic variable which characterizes the degree of disorder in a system.

The **change in entropy** of a system depends on the process which takes the system between the initial and final states. When energy is absorbed by the system, dQ_r is positive and the entropy of the system increases; when energy is expelled by the system, dQ_r is negative and the entropy of the system decreases.

When a system moves along a reversible path between infinitesimally separated equilibrium states, it is assumed that T remains constant.

Reversible process

$$dS = \frac{dQ_r}{T} \qquad (22.8)$$

When a system moves between initial and final states via a finite process, T may not be constant. However, the change in entropy is the *same for all paths* connecting the two states.

Finite process

$$\Delta S = \int_i^f dS = \int_i^f \frac{dQ_r}{T} \qquad (22.9)$$

When a system is carried around a reversible cycle the change in entropy is identically zero. *The integration must be carried out around a closed path.*

Reversible cycle

$$\oint \frac{dQ}{T} = 0 \qquad (22.10)$$

The adiabatic free expansion of a gas is an example of an irreversible process. The free expansion results in an increase in the entropy (and disorder) of the gas.

Irreversible process
(Example: Free expansion of a gas)

$$\Delta S = nR \ln\left(\frac{V_f}{V_i}\right) \qquad (22.11)$$

The microscopic definition of entropy is based on the number of *microstates* corresponding to a given *macrostate*. A greater number of microstates corresponds to increased level of entropy of the macrostate.

Entropy of a macrostate

$$S \equiv k_B \ln W \qquad (22.14)$$

W = number of microstates

REVIEW CHECKLIST

You should be able to:

- State the Kelvin-Planck and Clausius forms of the second law of thermodynamics, and discuss the difference between reversible and irreversible processes. (Sections 22.1, 22.2, and 22.3)

- Determine the efficiency of a heat engine operating between two reservoirs of known temperatures and calculate the net work output per cycle. (Section 22.1)

- Calculate the COP for a refrigerator and for a heat pump operating in the heating cycle. (Sections 22.2)

- Determine the efficiency of a Carnot engine operating between two reservoirs of given temperatures. (Section 22.4)

- Describe the sequence of processes which comprise the Carnot and Otto cycles; and sketch a *PV* diagram for each cycle. (Sections 22.4 and 22.5)

- Determine the efficiency of an engine operating in the Otto cycle using an ideal gas as the working substance. (Section 22.5)

- Calculate entropy changes for reversible processes (e.g., melting) and for irreversible processes (e.g., free expansion). (Sections 22.6 and 22.7)

ANSWER TO AN OBJECTIVE QUESTION

5.   Consider cyclic processes completely characterized by each of the following net energy inputs and outputs. In each case, the energy transfers listed are the *only* ones occurring. Classify each process as (a) possible, (b) impossible according to the first law of thermodynamics, (c) impossible according to the second law of thermodynamics, or (d) impossible according to both the first and second laws. **(i)** Input is 5 J of work, and output is 4 J of work. **(ii)** Input is 5 J of work, and output is 5 J of energy transferred by heat. **(iii)** Input is 5 J of energy transferred by electrical transmission, and output is 6 J of work. **(iv)** Input is 5 J of energy transferred by heat, and output is 5 J of energy transferred by heat. **(v)** Input is 5 J of energy transferred by heat, and output is 5 J of work. **(vi)** Input is 5 J of energy transferred by heat, and output is 3 J of work plus 2 J of energy transferred by heat.

Answer   Each process is cyclic, returning the sample to its starting point. The internal energy of the working substance undergoes no net change. The first law then says that the total energy input must be equal in amount to the total energy output. The first law does not refer to forms of energy. The second law, we can say, implies that if the input energy is transferred by heat, at least some of the output energy must be transferred by heat. Process **(i)** is (b) impossible according to the first law because 5 J is different from 4 J. Process **(ii)** is (a) possible: for example, give a box a push to set it sliding across the floor, wait for it to stop, and wait some more for the rubbing surfaces to cool off to reach the environmental temperature again. Process **(iii)** is (b) an example of imagined energy creation, with 6 J different from 5 J. Process **(iv)** is (a) possible: warm up some soup on the stove and let it cool off again in the bowl of a child who won't eat it. Process **(v)** is (c) an impossible heat engine with 100% efficiency. Process **(vi)** is (a) possible, representing a 60%-efficient heat engine. We could imagine a process violating both laws: let a sample take in 5 J by heat and put out 6 J carried by mechanical waves as it returns to its initial state.

□　　□　　□　　□

ANSWERS TO SELECTED CONCEPTUAL QUESTIONS

1. What are some factors that affect the efficiency of automobile engines?

Answer A gasoline engine does not exactly fit the definition of a heat engine. It takes in energy by mass transfer. Nevertheless, it converts this chemical energy into internal energy, so it can be modeled as taking in energy by heat. Therefore, Carnot's limit on the efficiency of a heat engine applies to a gasoline engine. The fundamental limit on its efficiency is $1 - T_c/T_h$, set by the maximum temperature the engine block can stand and the temperature of the surroundings into which exhaust heat must be dumped.

The Carnot efficiency can only be attained by a reversible engine. To run with any speed, a gasoline engine must carry out irreversible processes and have an efficiency below the Carnot limit. In order for the compression and power strokes to be adiabatic, we would like to minimize irreversibly losing heat to the engine block. We would like to have the processes happen very quickly, so that there is negligible time for energy to be conducted away. But in a standard piston engine the compression and power strokes must take one-half of the total cycle time, and the angular speed of the crankshaft is limited by the time intervals required to open and close the valves to get the fuel into each cylinder and the exhaust out.

Other limits on efficiency are imposed by irreversible processes like friction, both inside and outside the engine block. If the ignition timing is off, then the effective compressed volume will not be as small as it could be. If the burning of the gasoline is not complete, then some of the chemical energy will never enter the process.

The assumption used in obtaining $1 - T_c/T_h$, that the intake and exhaust gases have the same value of γ, hides other possibilities. Combustion breaks up the gasoline and oxygen molecules into a larger number of simpler water and CO_2 molecules, raising γ. By increasing γ and N, combustion slightly improves the efficiency. For this reason a gasoline engine can be more efficient than a methane engine, and also more efficient after the gasoline is refined to remove double carbon bonds (which would raise γ for the fuel).

□　　□　　□　　□

2. A steam-driven turbine is one major component of an electric power plant. Why is it advantageous to have the temperature of the steam as high as possible?

Answer   The most optimistic limit on efficiency in a steam engine is the ideal (Carnot) efficiency. This can be calculated from the high and low temperatures of the steam, as it expands:

$$e_C = \frac{T_h - T_c}{T_h} = 1 - \frac{T_c}{T_h}$$

The engine will be most efficient when the low temperature is extremely low, and the high temperature is extremely high. However, since the electric power plant is typically placed on the surface of the Earth, the temperature and pressure of the surroundings limit how much the steam can expand, and how low the lower temperature can be. The only way to further increase the efficiency, then, is to raise the temperature of the hot steam, as high as possible.

□   □   □   □

7.   The device shown in Figure CQ22.7, called a thermoelectric converter, uses a series of semiconductor cells to transform internal energy to electric potential energy, which we will study in Chapter 25. In the picture on the left, both legs of the device are at the same temperature and no electric potential energy is produced. When one leg is at a higher temperature than the other as shown in the picture on the right, however, electric potential energy is produced as the device extracts energy from the hot reservoir and drives a small electric motor. (a) Why is the difference in temperature necessary to produce electric potential energy in this demonstration? (b) In what sense does this intriguing experiment demonstrate the second law of thermodynamics?

Figure CQ22.7

Answer

(a)   The semiconductor converter operates essentially like a thermocouple, which is a pair of wires of different metals, with a junction at each end. When the junctions are at different temperatures, a small voltage appears around the loop, so that the device can be used to measure temperature or (here) to drive a small motor.

(b)   The second law states that an engine operating in a cycle cannot absorb energy by heat from one reservoir and expel it entirely by work. This exactly describes the first situation, where both legs are in contact with a single reservoir, and the thermocouple fails to produce electrical work. To expel energy by work, the device must transfer energy by heat from a hot reservoir to a cold reservoir, as in the second situation. The device must have heat exhaust.

□   □   □   □

9. Discuss the change in entropy of a gas that expands (a) at constant temperature and (b) adiabatically.

Answer

(a) The expanding gas is doing work. If it is ideal, its constant temperature implies constant internal energy, and it must be taking in energy by heat equal in amount to the work it is doing. As energy enters the gas by heat its entropy increases. The change in entropy is in fact $\Delta S = nR \ln(V_f/V_i)$.

(b) In a reversible adiabatic expansion there is no entropy change. We can say this is because the heat input is zero, or we can say it is because the temperature drops to compensate for the volume increase. In an irreversible adiabatic expansion the entropy increases. In a free expansion the change in entropy is again $\Delta S = nR \ln(V_f/V_i)$.

□   □   □   □

SOLUTIONS TO SELECTED END-OF-CHAPTER PROBLEMS

5. A particular heat engine has a mechanical power output of 5.00 kW and an efficiency of 25.0%. The engine expels 8.00×10^3 J of exhaust energy in each cycle. Find (a) the energy taken in during each cycle and (b) the time interval for each cycle.

Solution

Conceptualize: Visualize the input energy as dividing up into work output and wasted output. The exhaust by heat must be the other three-quarters of the energy input, so the input will be a bit more than 10 000 J in each cycle. If each cycle took one second, the mechanical output would be around 3 000 J each second. A cycle must take less than a second for the useful output to be 5 000 J each second.

Categorize: We use the first law and the definition of efficiency to solve part (a). We use the definition of power for part (b).

Analyze: In $Q_h = W_{eng} + |Q_c|$ we are given that $|Q_c| = 8\ 000$ J

(a) We have $\quad e = \dfrac{W_{eng}}{|Q_h|} = \dfrac{|Q_h| - |Q_c|}{|Q_h|} = 1 - \dfrac{|Q_c|}{|Q_h|} = 0.250$

Isolating $|Q_h|$, we have $\quad |Q_h| = \dfrac{|Q_c|}{1 - e} = \dfrac{8\ 000\ \text{J}}{1 - 0.250} = 10.7$ kJ ∎

(b) The work per cycle is $\quad W_{eng} = |Q_h| - |Q_c| = 2\ 667$ J

From the definition of output power $\quad P = \dfrac{W_{eng}}{\Delta t}$,

we have the time for one cycle $\quad \Delta t = \dfrac{W_{eng}}{P} = \dfrac{2\ 667\ \text{J}}{5\ 000\ \text{J/s}} = 0.533$ s ∎

Finalize: It would be correct to write efficiency as useful output power divided by total input power. This would amount to thinking about energy transfer per second instead of energy transfer per cycle.

---

13. One of the most efficient heat engines ever built is a coal-fired steam turbine in the Ohio River valley, operating between 1 870°C and 430°C. (a) What is its maximum theoretical efficiency? (b) The actual efficiency of the engine is 42.0%. How much mechanical power does the engine deliver if it absorbs 1.40×10^5 J of energy each second from its hot reservoir?

Solution

Conceptualize: The brakes on your car can have 100% efficiency in converting kinetic energy into internal energy. We do not get so much as 50% efficiency from this heat engine. Its useful output power will be far less than 10^5 joules each second.

Categorize: We use the Carnot expression for maximum possible efficiency, and the definition of efficiency to find the useful output.

Analyze: The engine is a steam turbine in an electric generating station with

$$T_c = 430°C = 703 \text{ K} \qquad \text{and} \qquad T_h = 1\,870°C = 2\,143 \text{ K}$$

(a) $\quad e_C = \dfrac{\Delta T}{T_h} = \dfrac{1\,440 \text{ K}}{2\,143 \text{ K}} = 0.672 \qquad$ or $\qquad 67.2\%$ ∎

(b) $\quad e = W_{\text{eng}}/|Q_h| = 0.420 \qquad$ and $\qquad |Q_h| = 1.40 \times 10^5$ J

for one second of operation, so $\qquad W_{\text{eng}} = 0.420\,|Q_h| = 5.88 \times 10^4$ J

and the power is $\qquad P = \dfrac{W_{\text{eng}}}{\Delta t} = \dfrac{5.88 \times 10^4 \text{ J}}{1 \text{ s}} = 58.8 \text{ kW}$ ∎

Finalize: Your most likely mistake is forgetting to convert the temperatures to kelvin. Consumers buying electricity in the Midwest pay for the fuel and the operation of the turbine and the generator it runs. The wasted heat output makes a very small contribution to global warming, and the carbon dioxide output makes a larger contribution. Special technology may be required to get coal to burn so hot. The plant may put out sulfur and nitrogen oxides that contribute to acid rain in the northeastern states and the Maritime Provinces.

---

20. An ideal refrigerator or ideal heat pump is equivalent to a Carnot engine running in reverse. That is, energy $|Q_c|$ is taken in from a cold reservoir and energy $|Q_h|$ is rejected to a hot reservoir. (a) Show that the work that must be supplied to run the refrigerator or heat pump is

$$W = \frac{T_h - T_c}{T_c} |Q_c|$$

(b) Show that the coefficient of performance (COP) of the ideal refrigerator is

$$COP = \frac{T_c}{T_h - T_c}$$

Solution

Conceptualize: The Carnot cycle makes the highest-performance refrigerator as well as the highest-efficiency heat engine.

Categorize: We solve the problem by using $|Q_h|/|Q_c| = T_h/T_c$, the first-law equation accounting for total energy input equal to energy output, and the definition of refrigerator COP.

Analyze:

(a)   For a complete cycle $\Delta E_{int} = 0$, and

$$W = |Q_h| - |Q_c| = |Q_c|\left(\frac{|Q_h|}{|Q_c|} - 1\right)$$

For a Carnot cycle (and only for a Carnot cycle), $|Q_h|/|Q_c| = T_h/T_c$.

Then, $W = |Q_c|(T_h - T_c)/T_c$ ∎

(b)   The coefficient of performance of a refrigerator is defined as $COP = \dfrac{|Q_c|}{W}$,

so its best possible value is $COP = \dfrac{T_c}{T_h - T_c}$ ∎

Finalize: The result of part (b) says that it can be easy to pump heat across a small temperature difference, but the refrigerator's effectiveness goes to zero as the low temperature it produces approaches absolute zero. Does thermodynamics seem to pull a rabbit out of a hat? From Joule and Kelvin meeting by chance at a Swiss waterfall, through Planck and Einstein at the beginning of the twentieth century, to you, physicists have been impressed with how an axiom-based chain of logic can have such general things to say about order and disorder, energy, and practical devices.

---

22. How much work does an ideal Carnot refrigerator require to remove 1.00 J of energy from liquid helium at 4.00 K and expel this energy to a room-temperature (293 K) environment?

Solution

Conceptualize: The refrigerator must lift the joule of heat across a large temperature difference, so many joules of work must be put in to make it run.

Categorize: We use the definition of a refrigerator's coefficient of performance and the identification of its maximum possible value.

Analyze: As the chapter text notes and as problem 20 proves,

$$(COP)_{\text{Carnot refrig}} = \frac{T_c}{\Delta T} = \frac{4.00 \text{ K}}{293 \text{ K} - 4.00 \text{ K}} = 0.013\ 8 = \frac{|Q_c|}{W}$$

so $\quad W = \dfrac{|Q_c|}{COP} = \dfrac{1.00 \text{ J}}{0.013\ 8} = 72.2 \text{ J}$ ∎

Finalize: Liquid nitrogen at 77 K sells for about the same price as beer, but a cryogenic refrigerator at liquid-helium temperatures is an expensive thing to run.

---

25. An ideal gas is taken through a Carnot cycle. The isothermal expansion occurs at 250°C, and the isothermal compression takes place at 50.0°C. The gas takes in 1.20×10^3 J of energy from the hot reservoir during the isothermal expansion. Find (a) the energy expelled to the cold reservoir in each cycle and (b) the net work done by the gas in each cycle.

Solution

Conceptualize: The two answers will add up to 1 200 J, and may be approximately equal to each other.

Categorize: We will use just the Carnot expression for the maximum efficiency. The temperatures quoted are the maximum and minimum temperatures.

Analyze:

(a) For a Carnot cycle, $\qquad\qquad\qquad e_C = 1 - \dfrac{T_c}{T_h}$

For any engine, $\qquad\qquad\qquad e = \dfrac{W_{\text{eng}}}{|Q_h|} = 1 - \dfrac{|Q_c|}{|Q_h|}$

Therefore, for a Carnot engine, $\quad 1 - \dfrac{T_c}{T_h} = 1 - \dfrac{|Q_c|}{|Q_h|}$

Then we have $\qquad |Q_c| = |Q_h|\left(\dfrac{T_c}{T_h}\right) = (1\ 200 \text{ J})\left(\dfrac{323 \text{ K}}{523 \text{ K}}\right) = 741 \text{ J}$ ∎

(b) The work we can calculate as $\quad W_{\text{eng}} = |Q_h| - |Q_c| = 1\ 200 \text{ J} - 741 \text{ J} = 459 \text{ J}$ ∎

Finalize: We could equally well compute the efficiency numerically as our first step. We could equally well use $W_{\text{eng}} = eQ_h$ in the last step. Seeing that 459 J is considerably less than 741 J, we see that the efficiency of this Carnot cycle is considerably less than 50%.

---

33. In a cylinder of an automobile engine, immediately after combustion, the gas is confined to a volume of 50.0 cm^3 and has an initial pressure of 3.00×10^6 Pa. The piston moves outward to a final volume of 300 cm^3, and the gas expands without energy transfer by heat. (a) If $\gamma = 1.40$ for the gas, what is the final pressure? (b) How much work is done by the gas in expanding?

Solution

Conceptualize: The pressure will decrease as the volume increases. If the gas were to expand at constant temperature, the final pressure would be $\frac{50}{300}(3 \times 10^6 \text{ Pa}) = 5 \times 10^5$ Pa

But the temperature must drop a great deal as the gas spends internal energy in doing work. Thus the final pressure must be lower than 500 kPa. The order of magnitude of the work can be estimated from the average pressure and change in volume:

$$W \sim (10^6 \text{ N/m}^2)(100 \text{ cm}^3)(10^{-6} \text{ m}^3/\text{cm}^3) = 100 \text{ J}$$

Categorize: The gas expands adiabatically (there is not enough time for significant heat transfer),

so we will use $\qquad P_i V_i^\gamma = \text{constant}$

to find the final pressure. With $Q - 0$, the amount of work can be found from the change in internal energy.

Analyze:

(a) For adiabatic expansion, $P_i V_i^\gamma = P_f V_f^\gamma$

Therefore, $P_f = P_i \left(\dfrac{V_i}{V_f}\right)^\gamma = (3.00 \times 10^6 \text{ Pa})\left(\dfrac{50.0 \text{ cm}^3}{300 \text{ cm}^3}\right)^{1.40} = 2.44 \times 10^5$ Pa ∎

(b) Since $Q = 0$,

we have $\qquad W_{\text{eng}} = Q - \Delta E = -\Delta E = -nC_V \Delta T = -nC_V(T_f - T_i)$

From $\qquad \gamma = \dfrac{C_P}{C_V} = \dfrac{C_V + R}{C_V} \qquad$ we get $(\gamma - 1)C_V = R$

So that $\qquad C_V = \dfrac{R}{1.40 - 1} = 2.50 R$

$$W_{\text{eng}} = n(2.50\, R)(T_i - T_f) = 2.50\, P_i V_i - 2.50\, P_f V_f$$

$$W_{\text{eng}} = 2.50[(3.00 \times 10^6 \text{ Pa})(50.0 \times 10^{-6} \text{ m}^3) - (2.44 \times 10^5 \text{ Pa})(300 \times 10^{-6} \text{ m}^3)]$$

$$W_{\text{eng}} - 192 \text{ J} \qquad\qquad ∎$$

Finalize: The final pressure is about half of the 500 kPa, in agreement with our qualitative prediction. The order of magnitude of the work is as we predicted.

From the work done by the gas in part (b), the average power of the engine could be calculated if the time for one cycle was known. Adiabatic expansion is the power stroke of our industrial civilization.

---

35. An idealized diesel engine operates in a cycle known as the *air-standard diesel* cycle shown in Figure P22.35. Fuel is sprayed into the cylinder at the point of maximum compression, *B*. Combustion occurs during the expansion *B→C*, which is modeled as an isobaric process. Show that the efficiency of an engine operating in this idealized diesel cycle is

$$e = 1 - \frac{1}{\gamma}\left(\frac{T_D - T_A}{T_C - T_B}\right)$$

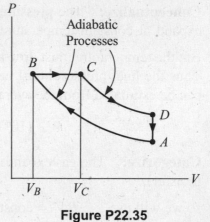

Figure P22.35

Solution

Conceptualize: One reasonable feature of the result to be derived is that it is less than 1.

Categorize: We will do an analysis of heat input and output for each process, then add up heat input and work output for the whole cycle, and use the definition of efficiency.

Analyze: The energy transfers by heat over the paths *CD* and *BA* are zero since they are adiabats.

Over path *BC*: $Q_{BC} = nC_P(T_C - T_B) > 0$

Over path *DA*: $Q_{DA} = nC_V(T_A - T_D) < 0$

Therefore, $|Q_c| = |Q_{DA}|$ and $|Q_h| = Q_{BC}$

Hence, the efficiency is

$$e = 1 - \frac{|Q_c|}{Q_h} = 1 - \left(\frac{T_D - T_A}{T_C - T_B}\right)\frac{C_V}{C_P}$$

$$e = 1 - \frac{1}{\gamma}\left(\frac{T_D - T_A}{T_C - T_B}\right) \qquad\blacksquare$$

Finalize: Easier than you expected? Shorter, at least? The high temperature is T_C. Since it is in the bottom of the fraction in the negative term, raising that temperature will raise the efficiency, just as in a Carnot cycle.

---

37. A Styrofoam cup holding 125 g of hot water at 100°C cools to room temperature, 20.0°C. What is the change in entropy of the room? Neglect the specific heat of the cup and any change in temperature of the room.

Solution

Conceptualize: The answer will be several joules per kelvin, a subtle measure of the increased disorder associated with the outward diffusion of the excess internal energy of the originally hot water.

Categorize: We proceed from the definition of entropy change as reversible-energy-input-by-heat divided by absolute temperature.

Analyze: The hot water has negative energy input by heat, given by $Q = mc\Delta T$. The surrounding room has positive energy input of this same number of joules, which we can write as $Q_{room} = (mc|\Delta T|)_{water}$. Imagine the room absorbing this energy reversibly by heat, from a stove at 20.001°C. Then its entropy increase is $\dot{Q}_{room}/T$. We compute

$$\Delta S = \frac{Q_{room}}{T} = \frac{(mc|\Delta T|)_{water}}{T} = \frac{0.125 \text{ kg}(4\,186 \text{ J/kg} \cdot °C)(100 - 20)°C}{293 \text{ K}} = 143 \text{ J/K} \quad \blacksquare$$

Finalize: The water in the cup loses entropy as it cools, but its negative entropy change is smaller in magnitude than this 143 J/K, because the water's temperature is a bigger number than 293 K. The absolute temperature at which the energy is transferred goes into the bottom of the fraction in $\int dQ/T$. Then the entropy change of the universe (here, the water and the room) is a net positive quantity, describing the irreversibility of allowing energy to flow by heat from a higher-temperature object to one at lower temperature.

---

40. A 1.00-mol sample of H_2 gas is contained in the left side of the container shown in Figure P22.40, which has equal volumes left and right. The right side is evacuated. When the valve is opened, the gas streams into the right side. (a) What is the entropy change of the gas? (b) Does the temperature of the gas change? Assume the container is so large that the hydrogen behaves as an ideal gas.

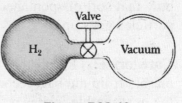

Figure P22.40

Solution

Conceptualize: One mole is only a couple of grams, so we expect an entropy increase of only a few joules per kelvin.

Categorize: The process is irreversible, so it must create entropy. We must think of a reversible process that carries the gas to the same final state. Then the definition of entropy change applied to that reversible process will give us the answer. We need not think about the environment outside the container, because it is unchanged.

Analyze:

(a) This is an example of free expansion; from the chapter text we have $\Delta S = nR \ln\left(\dfrac{V_f}{V_i}\right)$

$$\Delta S = (1.00 \text{ mole})(8.314 \text{ J/mole} \cdot \text{K}) \ln\left(\frac{2}{1}\right)$$

evaluating, $\quad \Delta S = 5.76 \text{ J/K} \quad$ or $\quad \Delta S = 1.38 \text{ cal/K}$ ∎

(b) The gas is expanding into an evacuated region. Therefore, $W = 0$. It expands so fast that energy has no time to flow by heat: $Q = 0$. But $\Delta E_{int} = Q + W$, so in this case $\Delta E_{int} = 0$. For an ideal gas, the internal energy is a function of the temperature and no other variables, so with $\Delta E_{int} = 0$, the temperature remains constant.

Finalize: When the gas was confined in one half of the container, we could have let it blow through a turbine as it expanded to double in volume. The amount of its entropy increase in the process considered here can be thought of as a measure of the loss of opportunity to extract work from the original system.

---

47. Prepare a table like Table 22.1 by using the same procedure (a) for the case in which you draw three marbles from your bag rather than four and (b) for the case in which you draw five marbles rather than four.

Solution

Conceptualize: At one roulette wheel in Monte Carlo in the 1920s, the red won twenty-something times in succession. Identifying possibilities in this problem will show how that sort of spontaneous order becomes less and less probable when the number of particles goes up.

Categorize: Recall that the table describes the results of an experiment of drawing a marble from the bag, recording its color, and then returning the marble to the bag before the next marble is drawn. Therefore the probability of the result of one draw is constant. There is only one way for the marbles to be all red, but we will list several ways for more even breaks to show up.

Analyze:

(a)

Result	Possible combinations	Total
All Red	RRR	1
2R, 1G	RRG, RGR, GRR	3
1R, 2G	RGG, GRG, GGR	3
All Green	GGG	1 ∎

(b)

Result	Possible combinations	Total
All Red	RRRRR	1
4R, 1G	RRRRG, RRRGR, RRGRR, RGRRR, GRRRR	5
3R, 2G	RRRGG, RRGRG, RGRRG, GRRRG, RRGGR, RGRGR, GRRGR, RGGRR, GRGRR, GGRRR	10
2R, 3G	GGGRR, GGRGR, GRGGR, RGGGR, GGRRG, GRGRG, RGGRG, GRRGG, RGRGG, RRGGG	10
1R, 4G	GGGGR, GGGRG, GGRGG, GRGGG, RGGGG	5
All Green	GGGGG	1

Finalize: The 'possible combinations' in the middle column are microstates and the 'results' in the first column are macrostates. The table in (b) shows that it is a good bet that drawing five marbles will yield three of one color and two of the other.

---

53. Energy transfers by heat through the exterior walls and roof of a house at a rate of 5.00×10^3 J/s = 5.00 kW when the interior temperature is 22.0°C and the outside temperature is −5.00°C. (a) Calculate the electric power required to maintain the interior temperature at 22.0°C if the power is used in electric resistance heaters that convert all the energy transferred in by electrical transmission into internal energy. (b) **What If?** Calculate the electric power required to maintain the interior temperature at 22.0°C if the power is used to drive an electric motor that operates the compressor of a heat pump that has a coefficient of performance equal to 60.0% of the Carnot-cycle value.

Solution

Conceptualize: To stay at constant temperature, the house must take in 5 000 J by heat every second, to replace the 5 kW that the house is losing to the cold environment. The electric heater should be 100% efficient, so $P = 5$ kW in part (a). It sounds as if the heat pump is only 60% efficient, so we might expect $P = 9$ kW in (b).

Categorize: Power is the rate of energy transfer per unit of time, so we can find the power in each case by examining the energy input as heat required for the house as a non-isolated system in steady state.

Analyze:

(a) We know that $P_{electric} = T_{ET}/\Delta t$, where T_{ET} stands for energy transmitted electrically. All of the energy transferred into the heater by electrical transmission becomes internal energy, so

$$P_{electric} = \frac{T_{ET}}{\Delta t} = 5.00 \text{ kW}$$

(b)   Now let T stand for absolute temperature.

For a heat pump, $(\text{COP})_{\text{Carnot}} = \dfrac{T_h}{\Delta T} = \dfrac{295\text{ K}}{27.0\text{ K}} = 10.93$

$$\text{Actual COP} = (0.600)(10.93) = 6.56 = \frac{|Q_h|}{W} = \frac{|Q_h|/\Delta t}{W/\Delta t}$$

Therefore, to bring 5 000 W of heat into the house only requires input power

$$P_{\text{heat pump}} = \frac{W}{\Delta t} = \frac{|Q_h|/\Delta t}{\text{COP}} = \frac{5\ 000\text{ W}}{6.56} = 763\text{ W} \qquad \blacksquare$$

Finalize: The result for the electric heater's power is consistent with our prediction, but the heat pump actually requires **less** power than we expected. Since both types of heaters use electricity to operate, we can now see why it is more cost effective to use a heat pump even though it is less than 100% efficient!

---

57.   In 1816, Robert Stirling, a Scottish clergyman, patented the *Stirling engine,* which has found a wide variety of applications ever since, including the solar power application illustrated on the cover of the textbook. Fuel is burned externally to warm one of the engine's two cylinders. A fixed quantity of inert gas moves cyclically between the cylinders, expanding in the hot one and contracting in the cold one. Figure P22.57 represents a model for its thermodynamic cycle. Consider n moles of an ideal monatomic gas being taken once through the cycle, consisting of two isothermal processes at temperatures $3T_i$ and T_i and two constant-volume processes. Let us find the efficiency of this engine. (a) Find the energy transferred by heat into the gas during the isovolumetric process AB. (b) Find the energy transferred by heat into the gas during the isothermal process BC. (c) Find the energy transferred by heat into the gas during the isovolumetric process CD. (d) Find the energy transferred by heat into the gas during the isothermal process DA. (e) Identify which of the results from parts (a) through (d) are positive and evaluate the energy input to the engine by heat. (f) From the first law of thermodynamics, find the work done by the engine. (g) From the results of parts (e) and (f), evaluate the efficiency of the engine. A Stirling engine is easier to manufacture than an internal combustion engine or a turbine. It can run on burning garbage. It can run on the energy transferred by sunlight and produce no material exhaust. Stirling engines are not presently used in automobiles due to long startup times and poor acceleration response.

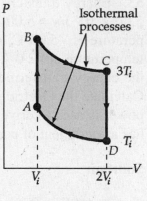

Figure P22.57

Solution

Conceptualize: A Carnot engine operating between these temperatures would have efficiency $2/3 = 67\%$. The actual engine must have lower efficiency, maybe 30%.

Categorize: We must think about energy inputs and outputs by heat and work for each process, add up heat input and work output for the whole cycle, and use the definition of efficiency.

Analyze: The internal energy of a monatomic ideal gas is $E_{int} = \frac{3}{2}nRT$

(a) In the constant-volume processes the work is zero, so

$$Q_{AB} = \Delta E_{int,AB} = nC_V \Delta T = \frac{3}{2} nR(3T_i - T_i) = 3nRT_i \qquad \blacksquare$$

(b) For an isothermal process $\Delta E_{int} = 0$ and

$$Q = -W = nRT \ln \left(\frac{V_f}{V_i} \right)$$

Therefore, $Q_{BC} = nR(3T_i) \ln 2$ ∎

(c) Similarly, $Q_{CD} = \Delta E_{int,CD} = \frac{3}{2} nR(T_i - 3T_i) = -3nRT_i$ ∎

(d) and $Q_{DA} = nRT_i \ln \left(\frac{1}{2} \right) = -nRT_i \ln 2$ ∎

(e) Q_h is the sum of the positive contributions to Q_{net}. In processes AB and BC energy is taken in from the hot source, amounting to

$$Q_h = Q_{AB} + Q_{BC} = 3nRT_i + 3nRT_i \ln 2 = 3nRT_i(1 + \ln 2) \qquad \blacksquare$$

(f) Since the change in temperature for the complete cycle is zero,

$$\Delta E_{int} = 0 \qquad \text{and} \qquad W_{eng} = Q_{net} = Q_{AB} + Q_{BC} + Q_{CD} + Q_{DA}$$

so $\quad W_{eng} = 2nRT_i \ln 2$ ∎

(g) Therefore the efficiency is $e = \dfrac{W_{eng}}{Q_h} = \dfrac{2 \ln 2}{3(1 + \ln 2)} = 0.273 = 27.3\%$ ∎

Finalize: Our estimate was good. The Sterling engine need have no material exhaust, but it has energy exhaust.

---

62. A 1.00-mol sample of a monatomic ideal gas is taken through the cycle shown in Figure P22.62. At point A, the pressure, volume, and temperature are P_i, V_i, and T_i, respectively. In terms of R and T_i find (a) the total energy entering the system by heat per cycle, (b) the total energy leaving the system by heat per cycle, and (c) the efficiency of an engine operating in this cycle. (d) Explain how the efficiency compares with that of an engine operating in a Carnot cycle between the same temperature extremes.

Solution

Conceptualize: This real engine will have lower efficiency than a comparable Carnot engine.

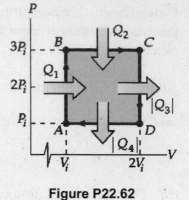

Figure P22.62

Categorize: The *PV* diagram shows the pressures and volumes at the beginning and end of each process. We will find the temperature at each corner. Then we will find the heat, input or output, for each process. That will tell us enough to identify the work output and heat input for the cycle, so that we can find the efficiency from its definition.

Analyze: At point *A*, $P_i V_i = nRT_i$ with $n = 1.00$ mol

At point *B*, $3P_i V_i = nRT_B$ so $T_B = 3T_i$

At point *C*, $(3P_i)(2V_i) = nRT_C$ and $T_C = 6T_i$

At point *D*, $P_i(2V_i) = nRT_D$ and $T_D = 2T_i$

We find the energy transfer by heat for each step in the cycle using

$$C_V = \tfrac{3}{2}R \quad \text{and} \quad C_P = \tfrac{5}{2}R$$

$$Q_1 = Q_{AB} = C_V(3T_i - T_i) = 3\,RT_i$$

$$Q_2 = Q_{BC} = C_P(6T_i - 3T_i) = 7.5\,RT_i$$

$$Q_3 = Q_{CD} = C_V(2T_i - 6T_i) = -6\,RT_i$$

$$Q_4 = Q_{DA} = C_P(T_i - 2T_i) = -2.5\,RT_i$$

(a) Therefore, $Q_{in} = |Q_h| = Q_{AB} + Q_{BC} = 10.5\,RT_i$ ∎

(b) $Q_{out} = |Q_c| = |Q_{CD} + Q_{DA}| = 8.5\,RT_i$ ∎

(c) $e = \dfrac{|Q_h| - |Q_c|}{|Q_h|} = 0.190 = 19.0\%$ ∎

(d) The Carnot efficiency is $e_c = 1 - \dfrac{T_c}{T_h} = 1 - \dfrac{T_i}{6T_i} = 0.833 = 83.3\%$ ∎

Finalize: The net work output is only $10.5\,RT_i - 8.5\,RT_i = 2\,RT_i$

The actual efficiency is a lot less than the Carnot efficiency. Irreversible processes go on as the gas absorbs and gives up heat to hot and cold reservoirs across large temperature differences.

---

68. A system consisting of *n* moles of an ideal gas with molar specific heat at constant pressure C_P undergoes two reversible processes. It starts with pressure P_i and volume V_i, expands isothermally, and then contracts adiabatically to reach a final state with pressure P_i and volume

$3V_i$. (a) Find its change in entropy in the isothermal process. (The entropy does not change in the adiabatic process.) (b) **What If?** Explain why the answer to part (a) must be the same as the answer to problem 65. (You do not need to solve Problem 65 to answer this question.)

Solution

Conceptualize: The final state has larger volume than the initial state. The molecules' motion has more randomness, so the entropy will be higher.

Categorize: We will use the equations describing the shape of the isothermal and adiabatic curves to find the volume after the isothermal expansion. Then $nR \ln(V_f/V_i)$ will tell us the change of entropy in the isothermal process.

Analyze: The diagram shows the isobaric process considered in problem 65 as AB. The processes considered in this problem are AC and CB.

(a)   For the isotherm (AC), $P_A V_A = P_C V_C$

For the adiabat (CB) $P_C V_C^{\gamma} = P_B V_B^{\gamma}$

Combining these equations by substitution gives

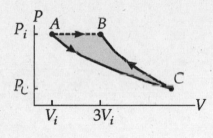

$$V_C = \left(\frac{P_B V_B^{\gamma}}{P_A V_A} \right)^{1/(\gamma-1)} = \left[\left(\frac{P_i}{P_i} \right) \frac{(3V_i)^{\gamma}}{V_i} \right]^{1/(\gamma-1)} = \left(3^{\gamma/(\gamma-1)} \right) V_i$$

Therefore,

$$\Delta S_{AC} = nR \ln\left(\frac{V_C}{V_A} \right) = nR \ln\left[3^{\gamma/(\gamma-1)} \right] = \frac{nR\gamma \ln 3}{\gamma - 1} \qquad \blacksquare$$

(b)   Since the change in entropy is path independent, $\Delta S_{AB} = \Delta S_{AC} + \Delta S_{CB}$

But because (CB) is adiabatic, $\Delta S_{CB} = 0$

Then $\Delta S_{AB} = \Delta S_{AC}$

The answer to problem 65 was stated as $\Delta S_{AB} = nC_P \ln 3$

Because $\gamma = C_P/C_V$ we have $C_V = C_P/\gamma$

along with $C_P - C_V = R$ this gives $C_P - C_P/\gamma = R$

So $\gamma C_P - C_P = \gamma R$ and $C_P = \gamma R/(\gamma - 1)$

Thus, the answers to problems 65 and 68 are in fact equal. $\blacksquare$

Finalize: For one case this has been an explicit demonstration of the general principle that entropy change depends only on the initial and final states, not on the path taken between them. Entropy is a *function of state*. Corollaries: The total change in entropy for a system undergoing a cycle must be zero. The change in a system's entropy for a process $A \rightarrow B$ is the negative of its change in entropy for the process $B \rightarrow A$.